Nuclear Emergency Planning and Response

Dr Keith Pearce

Published 29/4/2021 by Katwab Limited
ISBN 978-1-9164658-2-4

This is an updated and greatly extended version of "Nuclear Emergency Response for Local Authorities: An Introduction" (2018) by Keith Pearce

Main cover picture Torness NPP
Taras Young, CC BY-SA 4.0 <https://creativecommons.org/licenses/by-sa/4.0>, via Wikimedia Commons

About the Author

Keith Pearce originally trained as a nuclear physicist exploring the mysteries of trition induced charge exchange reactions on 2S-1D shell nuclei to gain a PhD but his first real job, at the Royal Naval College in Greenwich, London, was to teach nuclear submariners and nuclear dockyard officers about radiation; about what it is, how to measure it, how it interacts with matter, how it can harm human beings and how to stay safe in its presence.

He then moved to Nuclear Electric and joined their Health Physics Research team and spent several years developing and applying models of the movement of radioactivity in the atmosphere, seas and foodchains and the consequential dose to members of the public. This work supported the safety cases for gas cooled reactors, the Magnox (now shutdown) and Advanced Gas Cooled Reactors (AGR) classes as used in the UK.

In this role he helped to develop the tools used to estimate the dose to the public in an emergency to help formulate protective action advice. He joined the site emergency scheme as an on-site radiation monitor (an uncomfortable job involving the wearing of coveralls and respirator and walking around in a radiation plume that sensible people have run away from), an Assistant Health Physicist (a more comfortable job in an air-conditioned room and near to the coffee machine) and then as an Executive Officer (a job where other people keep your coffee cup charged).

He also joined the Company's central emergency response organisation, taking on dose assessment, health physics and command roles over time.

As Head of Emergency Planning, responsible for the emergency preparedness of ten Magnox nuclear reactor sites in, he worked with regulators, government officials, emergency services, local authorities and other expert groups to revise and maintain national standards and to keep the ten separate emergency plans and the corporate response up to standard. He played a significant role in the training of site and corporate emergency responders.

He also witnessed many Level 2 exercises from Strategic Co-ordination Centres (SCCs) and participated as Company Media Technical Briefer.

For several years he has worked as an independent consultant and author. It this time he has helped to write, review and maintain emergency schemes for new build companies, nuclear dockyards, operating companies, enrichment plant, local authorities and a farm.

His first book "How to survive a nuclear emergency" contains advice aimed at the people living near a nuclear site about how to prepare for and respond to an off-site nuclear emergency. His latest book "The Physics of the Chernobyl Accident" explains why reactor 4 at Chernobyl exploded.

This book is a significant revision, update and extension of *Nuclear Response for Local Authorities* published in 2018. In addition to the updated content of the earlier book it now discusses a wider range of nuclear sites, the application of REPPIR-19 and the planning process.

Contents

"We all recognise that, as an inescapable fact of life, we are surrounded by hazards – all with a potential to give rise to unwanted consequences" and, furthermore "consciously or unconsciously, we all view hazards and evaluate their risks to determine which ones we choose to notice, ignore or perhaps do something about". Tolerability of Risk (Ref. 1).

"Remember, if you fail to prepare, you are preparing to fail." Attributed to Pastor H.K. Williams among others

1. Introduction and Scope

This Section provides an introduction and route map for the work.

Imagine that you are a local authority Chief Executive or a member of the authority's emergency planning and response team and that you hear that an accident at a local power station has sent a cloud of radioactive material floating towards a village or town, or that a vehicle transporting radioactive sources to a hospital has left the road and overturned just outside the town. What do you do? How dangerous is this situation to the local population? How can you protect them? Who can help you make decisions and who can help you make the right things happen? These days we have mature plans to manage such situations but, at the time of the Windscale Accident there were none. We have since learned from a series of accidents and natural events that preparedness is valuable and we have learned that by working together the local authorities, emergency services, civil service, operators and a whole host of others can make important differences in crises situations. This document presents some of the wisdom collected over the years since the Windscale accident.

You cannot see, hear or smell radiation but just about everybody knows that it can be dangerous and should be treated with care. In high doses it will kill. In low doses it may enhance the likelihood of cancer induction. Already this work has talked about "high doses" and "cancer induction" the hinterland of a vast vocabulary of specialist terms that make nuclear science and radiological protection seem impenetrable to many. The two words that stand out in this paragraph, are widely understood and likely to be remembered are "kill" and "cancer".

The nuclear industry's reputation is somewhat tarnished after several high-profile accidents leading to loss of life, injury, damage to plant and extensive environmental harm but it works hard, and generally successfully, to maintain safety. The safety standards used in the design, construction, operation and maintenance of nuclear installations reduce to a very low level the risk of accidents, which could have a consequence for the general public. Nonetheless, prudence requires the preparation of plans for dealing with such events. In the UK the independent regulator, the Office for Nuclear Regulation (ONR) (Ref. 2) will not grant a license of the operation of a nuclear site until it is satisfied that the risks of operation have been carefully analysed and minimised and that adequate emergency arrangements are in place. Similar arrangements exist in the rest of the world with international expectations defined by the International Atomic Energy Agency (IAEA).

For the nuclear industry emergency preparedness is about putting in place mechanisms to recognise when situations on site are running out of control and being able to focus appropriate resources to make the site safe. "Appropriate resources" includes qualified and trained people, equipment, facilities and practiced processes. Where appropriate there are arrangements to ensure that personnel are available at short notice.

If the event may release significant amounts of radioactivity from the site or elevate the external dose rates in areas accessible to the public then the protection and reassurance of the public becomes a high priority. One way to do this is to fix the underlying problem and reduce the leakage of radioactive material and/or the external dose rates. But it may also be necessary to communicate to the public; to ask them to move away from the immediate vicinity of the site, for those nearby to shelter in their homes or offices while the radiation plume passes by, to take stable iodine tablets to protect their thyroid glands or to refrain from eating locally produced food and locally sourced drinking water for a period of time. These interventions require the help of the local authority to inform and to support the affected local community; the police service to manage roads and transport links; a consistent message to be promulgated to the public via all appropriate communications technologies including traditional and social media; they need health services and radiation specialists to reassure, advise and possibly treat affected members of the public; they need environmental experts to advise on short term and long term measures to reduce any harm done to the environment and all of these responders need to understand what they are doing, understand the risks of radiation and contamination, particularly if they are "on the front line" and they all need to be co-ordinated. Meanwhile the operators may need support from the emergency services to help rescue casualties, put out fires or cope with dangerous spills on the site.

Importantly, many faults give no warning time before the release begins. Protective action advice must get to the public without undue delay, and the public must be ready to act decisively to maximise its effectiveness. Hence all the organisations with a role to play in public protection should be contactable and able to start their emergency provisions quickly, no matter the time of day, day of week or holiday. These capabilities do not just fall into place naturally; they take a lot of hard work from several organisations, and it takes stamina to retain a high level of availability and capability year after year, particularly when the plans are not initiated for real.

In an emergency the major goals of responders are to protect the public, protect the people on site (personnel and responders) and to protect the environment. Protective actions (formerly called countermeasures) to protect the public include asking people in badly affected areas to take stable iodine tablets (if it is a reactor fault) and to shelter or, in extreme cases, evacuate. Within a few hours controls may be required to prevent the consumption of contaminated food or water. In the longer term there may be restrictions on the use of contaminated areas, including relocating people who live or work in the area.

Protective action - an action or actions taken in order to prevent or reduce the exposure of emergency workers, members of the public, the environment or the contamination of property from ionising radiation in the event of a radiation emergency and includes the provision of appropriate information to the public.
REPPIR - 2019 (Ref. 3)

Radiation emergency - a non-routine situation or event arising from work with ionising radiation that necessitates prompt action to mitigate the serious consequences—
(a) of a hazard resulting from that situation or event;
(b) of a perceived risk arising from such a hazard; or
(c) to any one or more of—
(i) human life;
(ii) health and safety;
(iii) quality of life;
(iv) property;
(v) the environment.
REPPIR - 2019 Reg. 2 (Ref. 3)

Although many of the faults envisaged will stop releasing radioactivity in a matter of hours, some more serious ones could last several days. Even with short releases the authorities may wish the public to remain in shelter until environmental monitoring can give a sufficiently accurate picture of the deposition for long term advice to be formulated, although this is debatable. Care and support of the public may require responders to enter and to work in areas subject to protective action advice and may expose them to elevated levels of radioactivity.

If you recognise that an industry has benefits to society but entails risks, how do you balance the benefits and the risks? That was a debate in the early years of nuclear power and remains a valid discussion point today. It was soon realised that the radioactive materials produced as a by-product of nuclear reactors would be dangerous and it was feared that the awesome power of atom would not be easy to tame. Indeed, some thought that there was a danger that once a nuclear reactor went critical it would be impossible to control and catastrophe was inevitable.

The public were sidestepped in the early debate by the secrecy surrounding the industry, only partly justified by the relationship with defence programs but also a side effect of the unhealthy competition between the Russians, Americans, British, Canadians and French to be seen as technology leaders and their willingness to take calculated, some would say reckless, risks to achieve that aim.

It is reported (Ref. 4) that in 1940s America the initial reactor siting criteria were based on the "worst conceivable accident". This was when reactors were generally small, the scientific and engineering knowledge of reactors was in its infancy and there was a fear of a runaway nuclear chain reaction. In 1950 the WASH-3[1] report (as discussed in reference 5) suggested an exclusion area around a nuclear reactor given by the following rule of thumb:

$$R = 0.016 \sqrt{P}$$

Where:

R is the radius of exclusion area (no residents permitted) in kilometres;

P is the reactor power (thermal) in kW.

For Sizewell B (P_{th} = 3479 MW), 2.5 km from Leiston which has a population 5,500, this gives an exclusion zone of about 30 km. How the situation evolved to give the current situation of Detailed Emergency Planning Zones (DEPZs) of a few kilometres is described in Section 3.

The current process for the scoping of nuclear emergency plans in the UK is discussed in Section 11.4. This process defines an area, the Detailed Emergency Planning Zone (DEPZ), where the off-site planners are expected to focus the bulk of their preparations and encompasses the ranges out to which the principle protective actions of shelter, stable iodine, evacuation and public information might be needed for accidents near the upper range of credible. In these "off-site plans" the local authority and the local Police Service will bear the brunt of the responsibility for managing the situation and supporting the populations affected. An Outline Planning Zone (OPZ), with lesser levels of preparedness expected, extends beyond the DEPZ. There are different prior information and outline planning requirements for this area.

During the time that shelter advice is in force tasks such as refuse collection and grass cutting will be stopped in the affected areas. However, those in shelter may include vulnerable people who require help from home visits throughout the shelter period. This includes the infirm, the bedridden, the sensory or mobility impaired and the ill (See Ref. 6 for UK government thoughts about what makes a person vulnerable). There may also be the "normal" emergencies such as the onset of labour, health crises, fires, gas or water leaks and so on that may require urgent attendance from an emergency service or a utility engineer.

Understandably there may be a reluctance to send people into areas affected by protective action advice and there may be a reluctance on the part of the person asked to carry out such a mission. Most people, including most potential responders, lack experience and training in the field of radiation protection. They often overestimate the harmful effects of low levels of additional radiation (which can be very easy to detect even at very low levels) and will go to great lengths to avoid it. This has been known to include refusing to work in affected areas with minimal associated radiation risks. To operate effectively and safely in a nuclear emergency requires that, for each task, the risks resulting from the presence of radiation are recognised, quantified, and sensibly compared to the potential gains resulting from the task.

There is a need to balance the real and imagined needs of the vulnerable person or group against the real and imagined risks to the worker. Knowledge is required in order to allow correct decisions about when to send people in to assist and when not to.

On-site there will be the requirement to work out what has happened or is happening. This could be a quite different experience on an AGR compared to the new reactors being built at Hinkley and planned elsewhere. On a gas cooled reactor, the worst reasonably foreseeable faults involve a breach of the cooling circuit. These result in a rapid release of hot carbon dioxide which is a direct threat to life and has the potential to cause additional local damage. Casualty rescue teams, if needed, must be dispatched without any delay. Damage assessment and then damage repair teams may be sent into the incident zone to isolate the leak. All entry teams will be in full protective equipment, including self-contained breathing apparatus as protection against carbon dioxide and radioactivity. These have a limited duration irrespective of the heat and radiation situation and hamper vision, movement and precise manipulation of tools and values. The Access Control Points, where these teams are briefed, equipped and dispatched and then returned, undressed, monitored, decontaminated (if required) and debriefed, are busy and stressed.

---

[1] "WASH" is short for Washington. The United States Atomic Energy Commission (AEC) used WASH as a prefix for its documents.

On a modern PWR or BWR (Boiling Water Reactor – not yet seen in UK), with multiple core cooling and shutdown systems, double containment and core catchers, most operator actions envisaged take place in the air-conditioned reactor control room. "On-plant" actions usually only require a small team or individual to walk along a clean and well-lit corridor to an instrument panel, control panel or valve. Significant off-site releases are only expected for very unlikely events and are expected to have at least a number of hours warning. This expectation of a warning time reduces some of the demands on the response targets for emergency plans.

On-site there will be teams of people co-ordinating and protecting the front-line responders. Many of these will be in the Access Control Point (ACP) as described above. There will be a more strategically focussed team in a facility such as an Emergency Control Centre (ECC) monitoring the situation, collating information, determining the likely course of the event and devising plans to bring the site back to a safe position. The ECC will have access to technical advice which might come from on-site, might come from the Operator's technical headquarters or might come from the reactor designer's facilities.

The on-site emergency plan has the role of allowing these diverse teams of people to work together, bringing a wide range of expertise to bear, to save and preserve life, protect the environment and make safe.

In part 1 this book provides an overview of the UK nuclear industry, concentrating on the risk assessments that help define the functional specification for emergency plans and provides sufficient health physics knowledge for people to understand the risks of elevated dose rates and how to minimise those risks while performing essential tasks.

Part 2 looks at the emergency preparedness community in the UK, discussing the key regulation and national and international advice and guidance and the general form of nuclear licensed site off-site plans within the UK. This section seeks to provide an overview of regulation and guidance. Read the source documents carefully or seek expert assistance if you are attempting to comply with the law.

Part 3 looks at responding to a nuclear emergency. How the emergency is detected, and the alert promulgated, the decision-making process, including the decision to work in an affected area or defer work, the roles of the key players and the information that might be available within the strategic co-ordination centre and how to interpret it. Transition and recovery are also discussed.

Appendices provide a review of a number of past accidents looking at what could be learned from them. Although we should avoid being guilty of "preparing for the last emergency", it is good practice to collectively consider how our current plans would cope in the circumstances resulting from past events. It is also worth those with response roles thinking about these events; considering how they would have responded and what equipment, information and other team members they would have wanted available to them. In this way they help get themselves ready to respond.

Security issues are not discussed in any depth. If an event has, or may have, a malicious cause the response at the scene and the various control centres will be complicated by the security needs and response but these should not greatly affect the off-site efforts to support the public.

Within this work I have included short notes and definitions as well as sections from regulations and international advice where I think it helpful. These are in boxes for clarity. I have also highlighted some key points with background shading for emphasis.

This is a quite long and complex document although it could be read cover to cover it may be better to select those bits that you need to know about or want to know about. Each section starts with a brief description of the section and its aims to help you judge if it is the right section for you.

Note on the use of the internet: There is a vast amount of information on the internet. Some sources are more trustworthy than others but, unfortunately it is not always possible to know how correct and complete internet papers are. It is impossible to read all the pages and attachments available. In this work I have made use of the internet with some care. I have tried to stick to internationally recognised organisations' works or papers published in peer review journals where possible. I have tried to always quote my source. These resources were there when I completed my research, but internet pages and attachments are prone to reorganisation or removal and may not be available at a later date.

Part 1 - Background

2. Background to the nuclear industry

"Transmutation of the elements,–unlimited power, ability to investigate the working of living cells by tracer atoms, the secret of photosynthesis about to be uncovered,–these and a host of other results all in 15 short years. ***It is not too much to expect that our children will enjoy in their homes electrical energy too cheap to meter,*** *–will know of great periodic regional famines in the world only as matters of history,–will travel effortlessly over the seas and under them and through the air with a minimum of danger and at great speeds,–and will experience a lifespan far longer than ours, as disease yields and man comes to understand what causes him to age. This is the forecast for an age of peace."*

Lewis L Struass, 1954 (Ref. 7)

The nuclear industry in the UK consists of:

- The generation of electricity using AGRs (Advanced Gas Cooled Reactors) and currently one Pressurized Water Reactor (PWR);
- The production of nuclear fuels (enrichment and fabrication plant);
- The storage and reprocessing of nuclear fuel;
- The decommissioning of nuclear reactors and other redundant facilities;
- The treatment and/or repackaging of existing radioactive waste materials;
- New nuclear build;
- Defence facilities including nuclear powered submarines, nuclear weapons and support facilities;
- Sites involved in the manufacture of radiopharmaceutical products and storage of waste arising;
- Sites recycling, processing or storing legacy wastes.

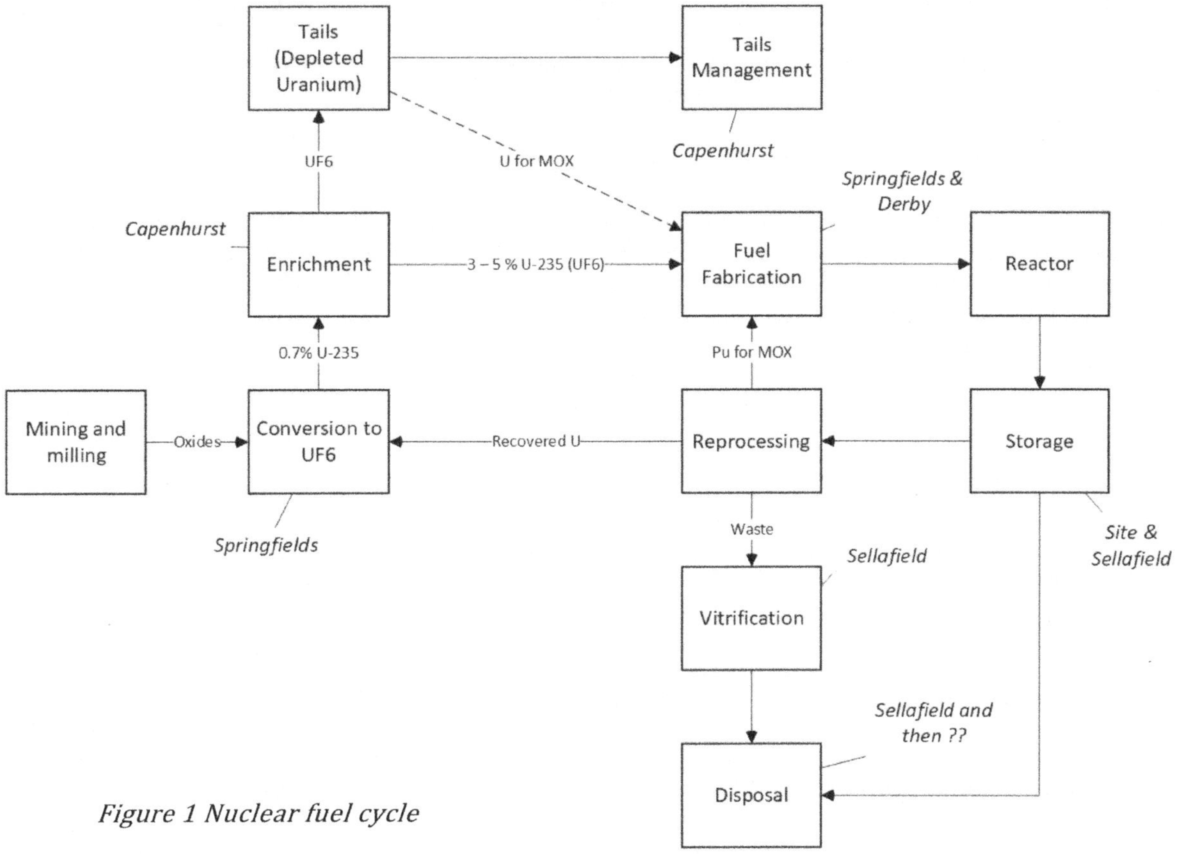

Figure 1 Nuclear fuel cycle

This section will explain the nuclear fuel cycle in the UK (Figure 1). It will cover the production of pure UF_6 at Springfields, the enrichment of UF_6 at Capenhurst, the production of fuel elements at Springfields and Derby, the operation of AGR and PWR reactors, the storage of used fuel and the reprocessing of fuel at Sellafield. It will also mention Dounreay and the facilities associated with the UK's nuclear deterrent. It will concentrate on the off-site risks of the processes, looking at the old and new REPPIR submissions, where available, for information to extract the "planning assumptions" valid for each site.

2.1. Nuclear fuel preparation

The important component of nuclear fuel is uranium. In 2016, Kazakhstan was the world's largest producer, reporting a production of 24 689 tonnes (40 % of the world production), more than the combined production of Canada and Australia, the second and third largest uranium producers (Ref. 8).

An early use of uranium ores was the extraction of radium to paint dials so they could be seen in poor light. (See reference 9 for an account of the US Radium Corporation and the "radium girls[2]").

Uranium consists of a number of isotopes (whose atoms have the same number of protons (92 in the case of uranium) but differing numbers of neutrons). Natural uranium is 99.27% U-238 and 0.72% U-235 with small amounts of U-233, U-234 and U-236. The isotope that is useful in nuclear reactors, because it can be made to fission if hit with thermal neutrons, is U-235. Other than CANDU and Magnox reactors all commercial reactors require the proportion of U-235 to be increased above the natural level through the process of fuel enrichment.

Enriched or not, the uranium has to be purified and manufactured into a form that can survive the harsh conditions of high temperatures and high radiation fields within reactors. Common forms of fuel are metal plates or rods, or ceramic rods (uranium dioxide, OU_2, is the most common but not the only ceramic used). Some reactors use molten metals, molten salts or solutions instead of solid forms. Other designs include "pebble-bed" reactors in which the fuel is in the form of pellets, each with a hard protective shell. Since the Fukushima accident there has been increased interest in "accident tolerant fuel", which can survive higher temperatures for longer (see, for example, Ref. 10). Even with decades of development, nuclear fuel has to be removed from the reactor long before all of the U-235 has been used up because of the increasing damage caused to the fuel matrix and some esoteric nuclear physics considerations.

From the early days of the industry reprocessing, in which used nuclear fuel is treated to recover unused fissile material, was seen to be necessary both for the uranium economy and for military purposes. However, time and politics moves on, reprocessing is coming to an end in the UK with the completion of Magnox reprocessing and the closure of THORP. The UK is still looking for a long-term solution to the fate of used nuclear fuel (Ref. 11).

Early fears about limits on the supply of uranium led to the fast breeder reactor programmes around the world. These could use nuclear physics tricks to make more fuel than they burned thus offering secure energy supplies. However, the technical challenges, both in the reactor design and in the fuel cycle, were significant. While there are still some countries working on the fast reactor (Ref. 12) the UK does not currently have a fast reactor programme. Dounreay was the UK lead centre for the fast reactors.

2.1.1. Mining and milling

Nuclear reactors utilise the nuclear chain reaction to generate heat. The main fuel used is uranium-235 which comprises a small fraction of natural uranium, which, in turn, is mined as uranium oxide ore containing about 1.5% uranium. The ore is crushed and ground to a fine slurry and uranium removed chemically and precipitated as purer uranium oxide called "yellowcake" in a process known as "milling". This is usually performed near the mine to reduce the weight and volume of material that has to be transported. Modern yellowcake typically contains 70% to 90%

---

[2] The "Radium Girls" were women who painted watch and clock dials with radium paint (a product called "undark") from about 1917. They suffered many health problems and in 1927 the company closed amid court cases bought by the victims, many of who died of the effects of radiation.

triuranium octoxide (U_3O_8). This passes through a number of chemical processes to produce uranium trioxide (UO_3).

2.1.2. Conversion and fuel fabrication (Springfields)

All the UK's Magnox and AGR fuel was manufactured at Springfields in Lancashire.

At Springfields the feed material of uranium trioxide (UO_3) undergoes a number of chemical processes in two sets of rotary kilns and a fluidised bed reactor (chemical not nuclear) yielding uranium hexafluoride (UF_6) (Ref. 13). Hydrogen fluoride (HF) and fluorine (F) are used as feed gases in these processes introducing chemically hazardous materials into the process. If required, the uranium hexafluoride is sent to Capenhurst for enrichment (See 2.1.3) before returning to be incorporated into nuclear fuel.

Fuel fabrication is described in reference 14. AGR fuel is produced by converting UF_6 into ceramic grade uranium dioxide powder by heating it to a vapour and mixing it first with steam to produce solid uranyl fluoride (UO_2F_2), and then in hydrogen and steam which removes the fluorine and chemically reduces the uranium to a pure microcrystalline UO_2 product.

The UO_2 powder is then pressed in dies into cylindrical pellet form using a load of several hundred MPa. The pellets are then sintered by heating in a furnace at about 1750 °C under a precisely controlled reducing atmosphere (usually argon-hydrogen) in order to consolidate them. This also has the effect of decreasing their volume. The pellets are then machined to the exact dimensions required.

AGR fuel assemblies consist of a circular array of 36 stainless steel-clad fuel pins each containing 20 enriched UO_2 fuel pellets, and the assembly weighs about 43 kilograms. Enrichment levels vary up to about 3.5%. The assembly is enclosed in a graphite sleeve.

Figure 2 AGR Fuel

A Sizewell B fuel assembly is made up of 264 zirconium alloy tubes, each containing about 300 pellets.

The UF_6 is mildly radioactive from spontaneous fissions of the U-235 and the decay of fission products. On release to the environment, UF_6 reacts to form Uranyl Fluoride (UO_2F_2) and gaseous Hydrogen Fluoride (HF). The chemotoxic hazards are therefore primarily from gaseous HF and the heavy metal toxicity of uranium.

In a REPPIR-01 HIRE assessed by ONR in 2017 Springfields Fuel Limited (managed by Westinghouse) identified the reference accident for the site as the accidental release of uranium hexafluoride due to a vehicle impacting a storage cylinder located on a storage area outside the Oxide Fuel Complex. The ONR agreed with the conclusion that the area in which a member of the public might potentially receive a radiation dose of up to 5 mSv as a result of a reasonably foreseeable radiation emergency is bounded by a distance of 185 metres from the site boundary (Ref. 15). They then defined the DEPZ as: "*A land area that varies from approximately 300 metres to 1.4 km from the boundary of the Springfields Fuels Ltd. site, which is defined by roads, footpaths, field boundaries and waterways. The area includes the villages Lea Town and Clifton*" (Ref. 16).

The REPPIR-19 Consequence Report (Ref. 17) recommended the retention of the REPPIR-01 DEPZ and the implementation of urgent sheltering in this area. Neither this document nor the Council's subsequent DEPZ determination, which introduced a few minor changes (Ref. 18), shed any further light on the risks of living nearby.

Springfields is also an upper tier COMAH site (Ref. 19).

2.1.3. Fuel Enrichment Plant (Capenhurst)

Natural uranium, as mined, contains about 0.7% of the fissile U-235 with the rest being mainly U-238 which is not fissile. It is hard to maintain a fission chain reaction with natural uranium and most modern reactors used "enriched fuel" in which the fraction of U-235 has been increased artificially.

The enrichment process involves either a gaseous diffusion process or a gaseous centrifuge process using UF_6 as the feed gas. These processes both use the small difference in mass or size between UF_6 with U-235 and U-238 to separate out the two forms. This separates the UF_6 "feed" into a "tails" stream of depleted uranium and a "product" stream of enriched uranium. The tails and product streams can have a range of final ratios of U-235/U-238. At Capenhurst they use the centrifuge process to produce a range of enrichments for the nuclear industry. A good description of the site and the perceived hazards can be found in the ENSREG "stress test" for the site (Ref. 20).

The sites where reprocessing takes place tend to have large stocks of HF_6 as feed and as tails stored on the site (product is valuable and is usually shipped to customers rather than stored for any length of time). These stocks are stored in large cylinders on storage rafts. The type 48Y cylinders used for transport, for example, hold about 12.5 tonnes of UF_6.

Capenhurst also has a large facility, the Tails Management Facility (Ref. 21) for the conversion of tails from UF_6 to the more stable U_3O_8 for safer long-term storage. The by-product, gaseous HF is cooled and liquefied as hydrofluoric acid, which is then sold. This facility is an upper tier COMAH site.

The report on the REPPIR-19 DEPZ determination by the local authority can be found as Reference 22. This is based on the Operator's Consequence report (Ref. 23).

The bounding reasonably foreseeable radiation emergency identified in REPPIR documents as the basis of the detailed emergency plan is based on a UF_6 release from a damaged cylinder. Small punctures to cylinders might result in a slow release of chemotoxic chemicals but UF_6 systems have a self-sealing / flow restriction capacity by formation of solid plugs of uranyl fluoride on exposure of UF_6 to atmosphere. The faults used to scope emergency plans are based on more violent events that produce large breaches in cylinders (drop from height, vehicle collision, and seismic event).

2.1.4. Emergency arrangements at fuel preparation sites

The major risk on sites enriching uranium and manufacturing fuel elements are the chemo-toxic hazards of the chemicals handled and stored. The reference accidents for these sites tend to be leakage of the UF_6 cylinders producing a radioactive and chemotoxic cloud which can disperse in the atmosphere. Leakage can be caused by a cylinder being punctured by accident or by corrosion leading to a slow release or by a more violent event such as a cylinder being dropped or having something heavy dropped on it leading to a large release at the moment of the event and a continuing release as the HF_6 exposed to the atmosphere reacts and the reaction products escape.

The process, at each site, has been assessed against internal (to the plant) and external (to the plant) hazards, including seismic events and severe weather. The principal controls and safeguards to prevent or mitigate release of radioactivity from confinement include robust design and integrity of cylinders and pipework systems commensurate with the level of potential hazard, UF_6 processes in systems below atmospheric pressure wherever reasonably practicable, secondary containment (e.g., autoclaves) wherever UF_6 is in the liquid state and for above atmospheric pressure UF_6 systems wherever reasonably practicable and strict procedural controls and operating instructions.

Gaseous uranium hexafluoride reacts rapidly with atmospheric moisture and hydrolyses to form a fine aerosol of uranyl fluoride particles with simultaneous liberation of gaseous hydrogen fluoride. Both uranyl fluoride and hydrogen fluoride are chemically toxic. Uranyl fluoride is also slightly radioactive. In the short term the irritant and toxic effects of the fluorides are the dominant consideration.

Inhalation of HF would cause irritation of the nose, throat and lungs with injury in the event of severe exposure. The toxicity is similar to that of chlorine. However, you can see and smell HF at concentrations well below those concentrations which can cause harm.

The emergency response will be to try to stop people inhaling the gases and dusts released. Workers near the scene will evacuate to a safe place. Suitably protected teams will start measuring down-wind for radiation and for the expected toxic gases. Response teams in the appropriate personal protective equipment and respiratory protective equipment will attend the scene to rescue casualties and provide first aid, to assess the situation and, if possible, to reduce the leakage and recover the spilt materials.

Protection of the public relies on preventing such releases, stopping the release if one occurs, and advising the public downwind to shelter from the plume. The health impact of the chemo-toxicity is of greater concern than the radioactivity as it can damage health and kill on the day of exposure, whereas the radioactivity, at these levels, could take decades before inducing cancers in a small minority of the people exposed.

The emergency responders will want to know how much HF_6 is "at risk", that is not yet released but might be if attempts to stop it are unsuccessful and how long the release may continue for. They will also want to know what is being done to slow or stop the release, when the attempt might be made and the chances of success.

In prior information members of the public are told (From the External Emergency Plan Urenco UK Ltd (previously obtained from internet but currently not available)).

"If you hear the sirens or if you smell unusually strong, irritating fumes you should:

- *Remain calm and stay indoors;*
- *Stay indoors, close all doors and windows. If possible go upstairs;*
- *Put out, or switch off, all heating and ventilation fans and air conditioning systems to avoid drawing in outside air;*
- *Remain indoors until you are told by an official source that the danger has passed or you hear the all clear siren (which is a continuous tone);*
- *Please try not to use telephones to keep lines clear for use by emergency services;*
- *Tune to your local radio station and listen for announcements telling you what to do;*
- *IF YOU CAN GIVE SHELTER TO ANYONE CAUGHT IN THE OPEN, PLEASE DO SO".*

2.2. Nuclear reactors

2.2.1. Introduction

This section aims to give the reader sufficient background knowledge about nuclear reactors, particularly those in use in the UK, to be able to follow some of the discussions in the SCC about the on-site situation and how it may develop.

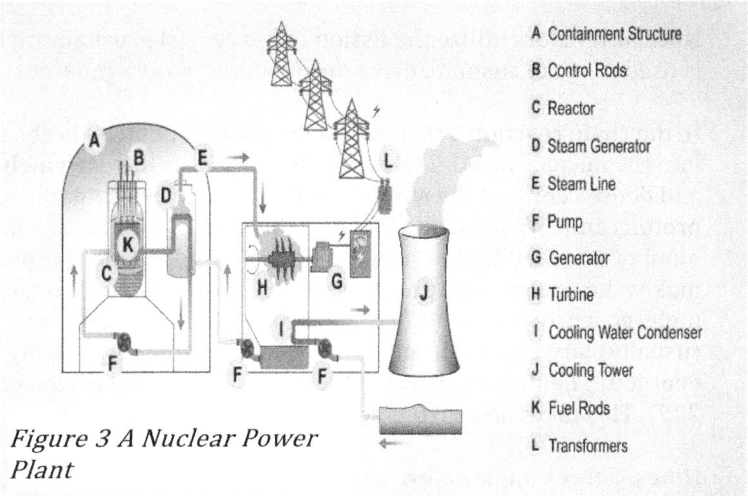

Figure 3 A Nuclear Power Plant

Figure 3 shows a schematic of a Pressurized Water Reactor (PWR). The reactor core (C) is composed of fuel elements (K), control rods, in-core monitoring devices and a structure to hold it all in place. The fission process takes place in the fuel rods (K). The heat is removed from the core by pressurised water pumped around the primary circuit (the pump is component F at the bottom of the containment building). The pressure prevents the water from boiling. Heat is removed from the primary circuit by heat exchangers ("steam generators" at D). Here the heat is used to turn water in the secondary circuit into steam. This steam circulates in the secondary circuit steam line (E) and drives turbines (H) which are connected to a generator (G) to produce electricity. Condensed water (I) from the turbines is returned to the steam generators by feed pumps (F). The reactor system is enclosed in a containment structure (A) which is designed to prevent the leakage of radioactive material in the case of reactor damage.

The terminology used here can cause confusion. Often a secondary system is a back-up to a primary system i.e. comes in and does the job when the primary is not functioning. That is not the case here. The primary circuit removes the heat from the reactor and the secondary circuit removes the heat from the primary circuit. They work in series. Ultimately the energy produced in the core is removed from the secondary circuit and exported as electricity, as slightly warmed sea water or as dumped steam.

It is important to understand that without cooling the heat generated in the core of the reactor will continue to raise the core temperature beyond the point at which damage occurs. The primary loop has to be working to some extent to remove the heat being generated in the core from the reactor and the secondary system has to be working to some extent to take the heat out of the primary. Without both loops in action the core may not be adequately cooled and damage is likely if the situation is not rectified.

For AGRs, which are massive and operate at a low power density, the grace time available to restore cooling is unusually long for a nuclear reactor design.

Commercial reactors have multiple systems to provide core cooling.

A great deal of work is undertaken to assess the safety of nuclear reactors and auxiliary systems such as the used fuel route and storage facilities. An idea of the extent of this work can be gleaned from the UK Regulator's Generic Design Assessment website (Ref. 24) and looking at the number and thickness of documents contained therein.

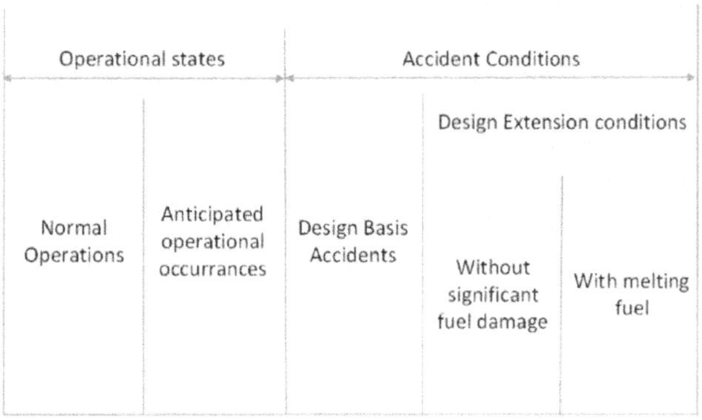

Figure 4 Plant states

Safety cases work with the concept of a reactor operating normally and safely and then systematically considering what could go wrong; which pumps might fail to pump, which valves might fail to open or close on demand, which pipes might leak or burst, which buttons the operators might press in error or not press when they should and how these might affect the plant, how the plant and operators would respond and how this response could go wrong. The intention is that any deviation from normal operation is detected and corrected promptly. There is a hierarchy of states that indicate how far from safe operation the reactor has strayed each increases the urgency of response. (Figure 4).

2.2.2. The chain reaction

Nuclear reactors utilize the fission of heavy nuclei, usually an isotope of uranium, namely U-235, to generate heat which is used to make steam to drive turbines and thus to generate electricity.

In the chain reaction a thermal (slow moving) neutron is absorbed into the nucleus of a U-235 atom. The nucleus is the extremely small and dense centre of the atom. In U-235 the nucleus contains 92 protons and 143 neutrons (giving a total of 235 nucleons, hence the number attached to the chemical symbol). Adding another neutron makes the nucleus into unstable U-236. Most of the U-236 nuclei undergo a process called "fission". That is, they split into two (usually) large fission fragments and a number of fast (highly energetic) neutrons (average about 2.45 for thermal fission of U-235). This takes about 10^{-14} seconds and liberates energy.

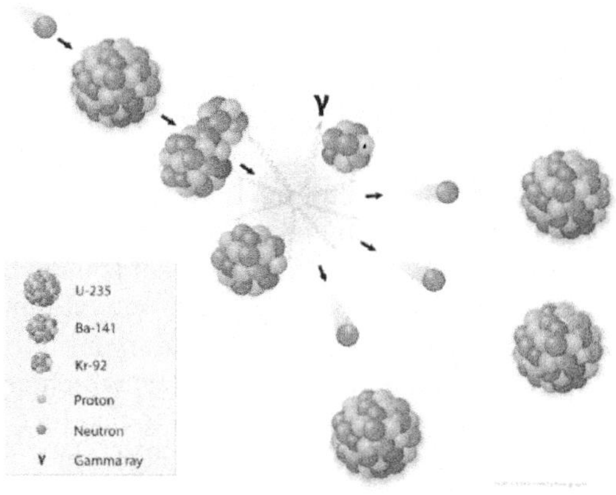

Figure 5 The chain reaction

Notes. This example shows barium-141 and krypton-92 being formed. This is only one of a large number of possible sets of fission products. The graphic suggests that the three neutrons may each hit another U-235 and cause three further fissions. This situation would be a nuclear bomb with a very rapid increase in the power level. In nuclear reactors at steady power the rate of fission is not changing with time, each fission leads to one more on average while the other neutrons are lost by other processes such as leaking from the system or being absorbed in a control rod, a distributed poison (something put in to absorb neutrons without fissioning) or part of the reactor structure.

Figure 5 suggests that the neutrons produced in fission can go on to cause the next fission immediately. This is true to an extent, fast reactors work in that manner but most do not. The neutrons produced in fission have a lot of kinetic energy, they are moving fast. At these energies they are far less likely to be absorbed by U-235 than slow moving neutrons. In thermal reactors (AGR, PWR, BWR, CANDU for example) the neutrons are slowed down, "moderated" to increase the chances of a further fission. Water and graphite are good moderators, hence their use in common reactor designs.

About 85% of the energy produced in fission is the kinetic energy of the fission products which is converted to heat as they interact with the fuel and stop within microscopic distances. More energy is found in gamma rays emitted at the moment of fission and in the kinetic energy of the neutrons. Additional energy and neutrons are liberated by the radioactive decay of fission products.

Kinetic Energy - Energy which a body possesses by virtue of being in motion. Compare with potential energy. Concise Oxford English Dictionary

Different fates can await the neutrons produced in fission; they can fly out of the reactor, they can be absorbed in the reactor internals, they can be slowed down (moderated or thermalized) by interaction with the materials they pass through, or they can go on to induce another fission (more likely after slowing down than before). If on average one neutron from each fission goes on to induce another then the rate of reaction will be constant, and we say that the reactor is *critical*. If the average is less than one then the power is reducing and the reactor is sub-critical.

For more on nuclear fission and its control see Reference 25.

2.2.3. Decay heat and radioactive products

The fission products are almost always unstable and decay sometime later giving off both radiation and heat. There are a large number of fission products with half-lives ranging from a fraction of a second to millions of years. These build up as the reactor operates. The short-lived ones reach an equilibrium level where their rate of decay equals their rate of production. The levels of them in a reactor are therefore dependent on the power at which the reactor is running. The very long-lived decay products do not reach equilibrium and continue to accumulate while the reactor is operating. Thus, at the time of any accident the concentrations of short-lived fission products depends on the recent power history of the reactor while the concentrations of long lived fission products depends on the power history since the fuel was loaded.

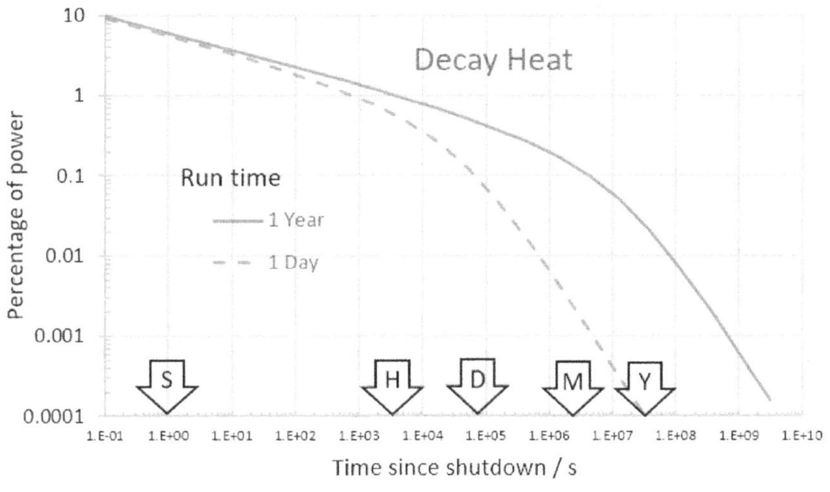

Figure 6 Decay heat post shutdown

When you shut down a reactor by inserting control rods or by injecting boron into the cooling circuit you stop the chain reaction, but you do not stop the decay of existing fission products. These continue to produce heat as they decay. Figure 6 shows how decay heat decreases post shutdown. At shutdown it is about 10% of full power and drops to 1% at about an hour. This is still sufficient to melt fuel and damage reactor internals, particularly in high power density reactors such as boiling water reactors and pressurised water reactors. Reactor designers take steps to maintain cooling, and in modern reactors take steps to cope with melting and damaged reactor cores.

The other problem with fission products is the radioactivity they emit. It is harmful to health so used nuclear fuel must be stored or reprocessed in such a manner that the fission products are kept safely away from the environment. It is reported, in a paper considering the potential for terrorists to misuse used nuclear fuel (Ref. 26), that even for light water reactor fuel that has cooled for 15 years, a lethal dose, taken to be 4.5 Sv, would be received at 1 m from the centre of the fuel assembly after several minutes.

The radiation and heat generation both cause problems for reactor designers and operators. The radioactive fission products must be isolated from the environment pretty much indefinitely to protect us from radiation and the heat, called "decay heat" in this context, is sufficient to damage the fuel unless adequate cooling is provided for an extended period after the reactor has shut down.

2.2.4. The Advanced Gas Cooled Reactor (AGR)

The Advanced Gas Cooled Reactor (AGR) is a UK design of reactor. It is the second generation of UK gas-cooled reactors after the Magnox fleet which have now all reached the end of their generating lives. There are 14 AGR reactors in service in the UK (as of September 2018)[3]. These are due to cease generation by 2030 (Ref. 27).

The core (Figure 7) is a massive structure of interlocking graphite blocks (3). These provide structure and moderate the neutrons. Channels leading top to bottom through the core (1) contain fuel, control rod and instrumentation channels and allow the coolant gas (carbon dioxide) to flow over the hot fuel (4) pushed by the gas circulators (pumps) (6). Control rods (2) enter at the top of the reactor. Water is pumped in at the bottom of the boilers (7) (8), takes heat from the carbon dioxide coolant (9) and exits as steam (10) which is used to drive turbines and generate electricity.

The reactor core, boilers and gas circulators are housed in a single pre-stressed concrete cavity known as the pressure vessel. The reactor core is a sixteen-sided stack of interconnected graphite bricks maintained in position by a steel restraint structure surrounding the core, and is supported by a system of steel plates known as the diagrid. The complete assembly of the core comprises of 332 fuel channels.

The fuel is slightly enriched U-235 in uranium dioxide ceramic pellets contained in stainless steel "pins". It is in these pellets that the fission process takes place generating heat. The fuel matrix and the clad are designed to keep hold of the fission products. The Heysham AGR reactors, for example, each contain approximately 100 tonnes of fuel when fully charged.

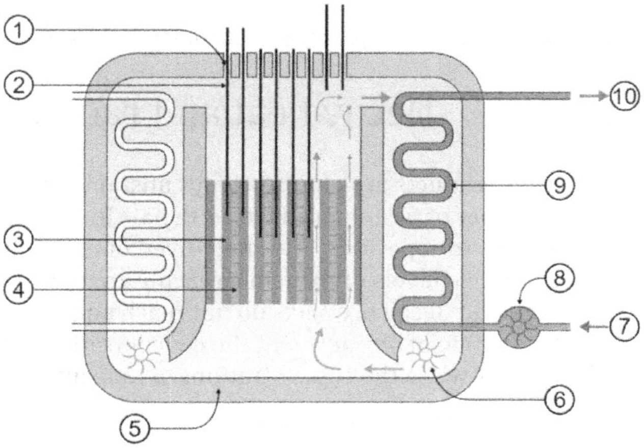

Figure 7 AGR Schematic

The coolant in an AGR is carbon dioxide (CO_2) pumped around the primary circuit by the "circulators". This cools the nuclear fuel and the reactor internals and transfers the heat to the "boilers" which cool the carbon dioxide while turning water into steam in the secondary circuit. The steam drives turbines to generate electricity and the cooled and condensed water is then fed back into the boilers.

The main control of reactivity is by the insertion of "control rods" made of materials that readily absorb neutrons -in the jargon of nuclear physics they have a high neutron capture cross-section.

The turbine condensers for all AGRs are cooled by seawater, which is the primary ultimate heat sink. A short but useful description of an AGR can be found in Reference 28, a more detailed report can be found in Reference 29 and the post-Fukushima "stress test" for Torness (Torness, in the east of Scotland hosts an AGR reactor site), for example, is reported in Reference 30. Other stress test results can also be found on the internet.

Spent fuel is moved to a cooled CO_2 environment and then to cooling ponds on site. The pond water, which has carefully controlled chemistry, provides both shielding and cooling for the fuel during the storage period[4]. After at least 90 days

[3] Two each at Dungeness B, Torness, Hunterston B, Heysham I, Heysham II, Hartlepool and Hinkley Point B.

[4] One of the problems at Fukushima was the overheating of fuel storage ponds when the electrical cooling failed leading to the pond water evaporating exposing the fuel and causing multiple fires.

cooling the spent fuel is transported to Sellafield.

Stress tests. In the wake of events at Fukushima, on 24-25 March 2011 the European Council (EC) requested a review of safety at all European nuclear power plants (NPP) and the European Nuclear Safety Regulators Group (ENSREG) produced a specification and a plan for this review based on preliminary work carried out by the Western European Nuclear Regulators Association (WENRA). The reviews are commonly referred to as the 'stress tests'. These reports, which are generally available on the internet, provide some useful information for emergency planners and responders.

Potential faults

The operators of AGRs are provided with, and trained in the use of, a series of operating procedures that detail how to respond to alarms and unforeseen events. These are organised in a progression:

1. Operating Rules (ORs), for normal operation;
2. Emergency Operating Procedures (EOPs), for responding to alarms;
3. Symptom Based Emergency Response Guidelines (SBERGs), for events beyond design basis; these concentrate on shutting the reactor down and adequately cooling it to prevent large releases of radioactivity; and
4. Severe Accident Guidance (SAGs); for severe accident conditions. These provide non-prescriptive advice and concentrate on ways in which safety functions could be re-established.

The intent of this hierarchy is to catch deviations from normal operation as soon as possible and return the plant to a safe condition with actions commensurate with the severity of the situation.

The AGR safety cases recognise a number of fault types some of which are described in the following sections.

Coolant leak

The primary coolant is carbon dioxide at high pressure. The containment is a steel lined concrete structure. If there were to be a leak in the primary coolant circuit the carbon dioxide gas would leak out with the rate depending on the size of the breach and the pressure difference across it. There is a limit to the size of breach in this circuit considered to be credible. AGRs are designed to be able to survive the maximum credible breach with some engineering margin.

A significant leak would lead to the reactor being shutdown (either automatically or manually depending on the severity of the event). This would reduce the heat generation and the need for cooling. The operators would then either make up the coolant volume from storage tanks while isolating the leak or *blow down* the reactor (through filters to remove particulate and iodine activity) to a lower pressure so that a repair could be completed. Importantly, there will only be an unplanned discharge of radioactivity above the trivial amounts normally found in the coolant circuit if there were to be coincident fuel damage or if fuel were to be damaged by the rapid depressurisation.

An “uncontrolled depressurisation” (the polite term for a leak) with some failed fuel provides the source term (list of nuclides released, the amounts released and (sometimes) the variation with time of release) originally used to determine the scale of the emergency plan for AGR sites.

Containment: Methods or physical structures designed to prevent or control the release and the dispersion of radioactive substances. IAEA Glossary
Blow down: The act of deliberately reducing the pressure in a gas cooled reactor by releasing some of the coolant gas, either through filters or not, is called a *blow down*.
Maximum credible leak: Safety cases consider a long list of things that can go wrong with the plant under consideration. To ensure that the results err on the side of over-estimating consequences, but not ridiculously so, values for parameters that can vary depending on the circumstances tend to be set at the highest or lowest value that is considered “credible” (or possible) to maximise the predicted harm. The maximum credible leak in a complex circuit would tend to be the diameter of the largest pipe that could be sheared, or otherwise fail, in an accident.

Failure to post-trip cool

In normal operation, when a reactor is tripped there is a sequence of events that takes place to ensure that the right amount of energy is removed from the core at all times so that it is neither over-cooled or under-cooled as the reactor shuts down and decay heat reduces. This involves the rate of forced circulation of coolant gas and the rate of water being pumped through the boilers both slowing significantly.

> **Reactor Trip:** In nuclear engineering terms, the automatic shutdown of a nuclear reactor is called a *reactor trip* or scram. A reactor trip causes all the control rods to insert into the reactor core and shut down the plant in a very short time. US NRC

If post-trip cooling is not established for any reason then the temperature in the fuel, and in the core in general, will begin to rise. Because the AGRs are large and massive they have a high heat capacity, and the rate of temperature rise will be gradual. However, if cooling is not re-established the temperature rise will continue.

At this stage, the operators have a number of options for cooling the reactor and surrounds and it is considered that it would require a very unlikely combination of faults, mishaps, errors or deliberate actions for this to reach the stage of reactor damage.

A pressurised AGR will remain adequately cooled provided that water can be pumped into the boilers and either cooled in the main condenser or released to atmosphere as steam. There are diverse methods to supply water to the boilers.

A depressurised AGR does require the primary coolant to be circulated by force.

As the heat is mainly generated in the fuel it is the fuel assemblies that will heat up the quickest, so it is likely that the first component to be damaged by the rising temperatures is the fuel cladding which has a lower melting point than the fuel. This will start in those pins with the worst combination of high internal heat generation and poor external cooling. This will increase the levels of radioactivity in the coolant circuit and, if that is leaking, increase the rate of radioactivity discharge.

Over time more and more fuel rods will reach the melting point of the cladding so this phase could last for a significant period. During this time, the release of radionuclides may be very variable, spiking as each fuel pin fails.

Components at the top of the reactor may be the next to fail. This would provide a shorter path to atmosphere for radioactivity from the damaged fuel, increasing the release rate of radioactivity. It may also further compromise reactor cooling by causing a reduction in pressure.

Further rises in temperature could cause the fuel itself to soften which both increases the rate of radioactivity leakage and further compromises cooling. Once again, this phase could be protracted as the large reactor heats up unevenly.

The next act in this very unlikely chain of events is for molten fuel to damage the bottom of the reactor, leading to further breaches and further releases.

This fault sequence would take several days to play through. The operators and their technical support will be doing everything they can to cool the reactor to reduce the consequences of this sequence of events and lengthen the time between stages. The success or failure of their actions does introduce another element of unpredictability into the process, so the operating company may not be able to accurately predict the timing and magnitude of future releases, much to the frustration of those trying to provide advice to protect the public.

Reactor trip with failure to shut-down

This sequence would involve an unlikely series of failures in control systems making the cooling circuit behave as if there was a reactor trip. This would dramatically reduce cooling whilst the reactor was still at power. The resulting rapid rise in core temperature would cause widespread core damage and result in a significant release of radioactivity. The release would occur in a short period of time and without warning.

Water ingress fault

If a boiler were to be damaged in such a way that the barrier between the primary gas circuit and the secondary water circuit were to be breached, then water would leak into the primary circuit and raise the pressure in the reactor rapidly. This in turn would lead to pressure relief valves opening to vent reactor coolant and prevent reactor damage.

The consequences, at this stage are limited to the release of some of the activity that was in the reactor circuit at the time of the ingress fault, but it is recognised that the fault could develop if the safety relief valves fail to reseat as designed leading to reactor pressure dropping. If this failure continues, and reactor cooling is compromised, the fault could be serious and would last as long as the operators failed to reseat the valve or otherwise stem the leak.

Untripped gag fault

On an AGR, devices known as "Gags" are used to control the flow of coolant gas into the individual fuel channels to influence the temperature distribution across the reactor.

If one of these were to fail or be wrongly set the result could be a channel without sufficient cooling which could heat up and eventually melt the fuel clad (the stainless-steel layer on the outside of a fuel element) within that channel.

If systems and operators failed to notice and rectify this situation then a significant release of radioactivity could occur over a small period.

Summary of AGR faults and risk assessment

There is a range of potential faults in an AGR reactor, all of which are unlikely due to the design and operation of the plant.

Some of these faults will give little or no warning before a release to atmosphere of radioactive material begins. Other faults will develop over many hours before a release occurs, in which time the operators will probably be able to prevent it but the authorities should consider putting in place precautionary protective actions in case the repair does not succeed.

For many fault types the operator should be able to give a reasonably accurate assessment of the duration of the release. This is likely to be in the order of a few hours as it depends on the time taken to depressurise the reactor and seal any breach.

For the "no-notice" faults leading to a release of up to a few hours or so, shelter and, for reactor faults only, taking of pre-distributed stable iodine are the most practical protective actions for most members of the public.

The operator will know if a fault is likely to last for more than the few hours. In this case consideration should be given to how best to protect and reassure the public. More extensive protective actions, such as sheltering for longer or further out from the incident site, a wider distribution of stable iodine and evacuation of the inner downwind areas, could all be considered. All of these raise communication and organisation issues and will lead to more people seeking support from the local authorities and health authorities.

The 2011 HIREs for the AGR fleet (see Ref. 31 for an example) reported that "systematic analysis has identified that various internal and external events could result in a radiation accident leading to releases of radioactivity and potential radiation emergencies. These initiating events include failed fuel, failure in containment, fires and fuel movement operations". Faults could run for a few seconds or for several days depending on the circumstances. The big releases are likely to take several hours. They went on to say that the lower ERLs could be exceeded out to < 800m (shelter, whole body), < 1000 m (shelter, organ), < 200 m evacuation (whole body and organ) and < 1000 m stable iodine. At that stage EDF would have recommended a 1 km DEPZ but some sites, particularly those next to a Magnox site have larger DEPZs.

In 2017 the AGR HIREs were revised giving more details of the types of faults considered and slightly changing the dose model used. However, the reports concluded that the previous reference accident source term remained fit for purpose.

They concluded that the 5 mSv dose contour was at 870 m, the sheltering ERL was potentially exceeded out to 680 m under best estimate dispersion and 1,600 m under pessimistic dispersion estimates; the lower ERL for evacuation could be exceeded to 180 m. Food controls (based on 500 Bq/kg Council Food Intervention Level in milk) might extend 43 km downwind in average conditions.

The 2020 consequence report for Hunterston B (Ref. 32) (for example) states that "The assessments required under REPPIR indicate detailed planning is justified for the urgent protective actions of administration of stable iodine and implementation of sheltering within a distance of ~ 2 km from the site for protection of the public. The protective actions should be capable of being enacted as soon as is practical after the declaration of a Radiation Emergency has occurred or before a release starts to maximise the averting of dose. Stable iodine can be administered up to 5-8 hours following exposure as averting iodine inhalation dose of ~ 50% is still possible". The increase in distance compared to the previous assessments seems to be due to a switch in the weather assumption from average weather to pessimistic weather (in the sense that it is a weather pattern that does not encourage the plume to spread upwards and outwards as it travels downwind).

The iodine prophylaxis distance is estimated at about 2,000 m; the shelter distance as about 950 m and evacuation at 300 m. EDF then recommend shelter and stable iodine to 2,000 m and that this be the minimum radius of the DEPZ.

2.2.5. Commercial PWRs in the UK (Sizewell B)

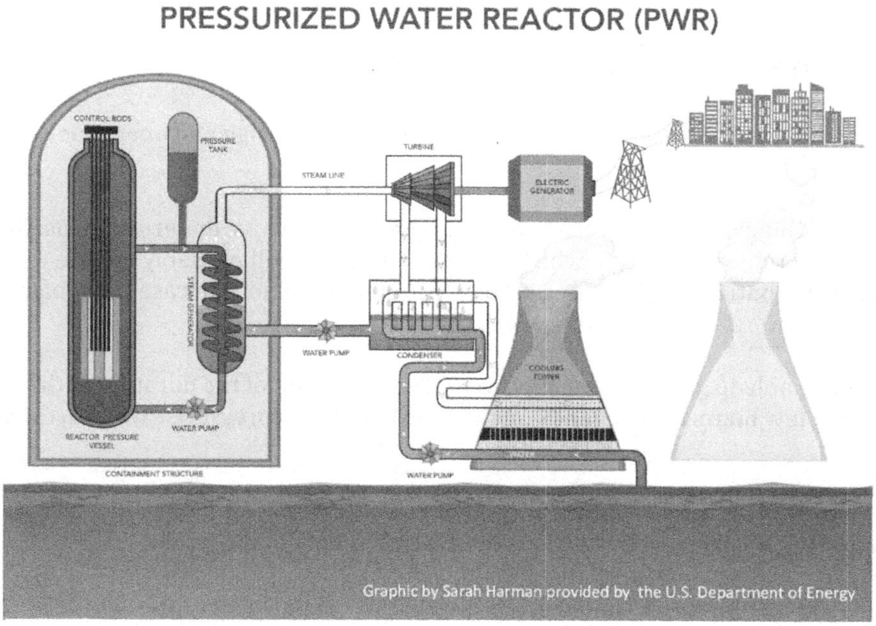

Figure 8 Schematic of a PWR

The pressurised water reactor is a much smaller reactor than the AGR with a higher power density. Fuel is enriched ^{235}U as uranium dioxide pellets and clad in stainless steel. It is cooled by water under high enough pressure to ensure that it remains as a liquid within the primary circuit and to ensure that steam films do not build up on the fuel surfaces reducing heat transfer. Water is pumped through the reactor and into one of a number of steam generators in which the heat is transferred to a secondary system water feed producing steam to drive turbines.

Sizewell B is Britain's only currently operating commercial pressurised water reactor. It is situated on the east coast, near Sizewell in Suffolk, around 35 km north-east of Ipswich. It is a four-loop reactor rated at 1188 MWe (meaning that at full power it can generate 1188 MW of electricity). Hot, pressurised water from the reactor is pumped via four pipe loops to four steam generators (only one loop is shown in the diagram). The steam generators transfer the heat in the water they receive to a separate water/ steam circuit (the secondary circuit). This steam is then used to drive the turbogenerators which produce electricity.

Sizewell B's pressure vessel has an inside diameter of approximately 4.4m, a thickness of 0.21m and an overall height of 13.6m. The fuel is slightly enriched uranium dioxide in a zirconium clad. The coolant is water, which is also the moderator, it is kept at high pressure to prevent it turning to steam.

The pressure is controlled by the pressurizer ("pressure tank" in diagram), which is a large vessel with water at the bottom and steam at the top. The water pressure in the primary circuit is controlled by either heating the pressurizer with electric heaters or cooling it with a water spray to control the size of the "bubble". More heat results in a bigger bubble, compressing the water and producing higher pressure. Spray cooling condenses some of the steam, reducing the size of the bubble and reducing the water pressure. There is one pressurizer. Pressuriser water level is an important parameter in many accident sequences as it is the best indicator of the amount of water in the system, important when dealing with a leak. It also provides a buffer for pressure transients in the primary circuit as the steam bubble is more compressible than liquid water.

Four loops are used to pump water over the hot fuel and into steam generators where the heat is passed to the secondary circuit where it turns feedwater into steam to drive turbines and generate electricity.

There are diverse methods of removing heat from the reactor under a range of circumstances including the emergency core cooling system which is specifically designed to mitigate the consequences of loss of coolant accidents. The principal protection required following a loss of coolant accident is to ensure adequate shutdown margin (a measure of the amount of reactivity below critical) and that the core is adequately cooled in order to limit fuel damage, minimise the release of fission products from the fuel cladding and prevent a loss of core geometry.

Reactor, primary circuits and steam generators are all held within a robust containment structure of concrete and steel designed to be a final barrier against fission product release.

Criticality is controlled by:

- "burnable" poisons (neutron absorbers that are depleted during operation) within the new fuel assemblies to reduce the reactivity of new fuel, used to balance changes in core reactivity as the fuel is used up and so permit power operation to take place at an appropriate temperature;
- Longer term control of reactivity is maintained by levels of boron-10 (a neutron absorber) in the coolant water;
- Control rods made of a neutron absorber provide reactor control and start-up, load following and shutdown capabilities. At Sizewell B there are 53 control rod clusters divided into "control" and "shutdown" banks. The latter set being held above the reactor while it is critical ready to drop under gravity should the need to shutdown arise;
- An emergency boronation system can introduce higher concentrations of boron into the circuit to shut down and hold down the reactor in the event of a control rod malfunction.

The four main barriers against fission produce release are:

1. The fuel matrix;
2. The fuel clad;
3. The primary circuit (pressure vessel); and
4. Containment

A detailed discussion of potential faults for PWRs can be found in Ref. 33 these include:

- Criticality Control Faults
  - Control rod ejection/withdrawal/malfunction;
  - Incorrect connection of an inactive reactor coolant system loop (by adding cold water you make the reactor more reactive);
  - Boron dilution issues;
  - Inadvertent loading of a fuel assembly into an improper position;
- Coolant flow issues
  - Single or multiple main coolant pump trip;
  - Inadvertent closure of a main isolation valve in a reactor coolant system loop;
  - Seizure of one main circulation pump or shaft break of one main circulation pump;
  - Coolant flow blockage in a fuel assembly;
  - Primary to secondary leakage;
- Heat removal perturbations
  - Steam line breaks;
  - Inadvertent opening of steam relief valves;
  - Secondary pressure control malfunction with an increase of steam flow;
  - Feedwater system malfunction/line break/pump trip;
  - Reduction of steam flow from steam generator for various reasons;
  - Loss of Coolant Accidents (LOCA)

In addition to the reactor the site also has stocks of used fuel. Sizewell B does not allow on-load refuelling. Every 18 months or so it is shutdown, opened up and some of the fuel (typically about 40%) removed and replaced with fresh fuel. The removed fuel is transferred to the Fuel Building which contains the stainless steel lined Fuel Storage Pond within which are racks for holding both new and irradiated fuel. Fuel is always handled under borated water. The water removes the decay heat from irradiated fuel and provides an effective and inexpensive radiation shield. The boric acid in the water ensures the fuel assemblies cannot become critical in any circumstances.

Fuel is transferred to on-site dry storage after a number of years in the Fuel Storage Pond when the decay heat has reduced to a level that can be safely managed in dry store.

The 2017 HIRE (Ref. 34) identified the following classes of design basis faults:

- Loss of Coolant Accidents (LOCA) affecting the primary circuit during power operation and at shutdown and which may result in fuel damage;
- Faults where the primary circuit is initially intact during power operation and at shutdown and which may result in fuel damage;
- Faults in interfacing systems in the Auxiliary Building during power operation and at shutdown;
- Faults during fuel handling operations and irradiated fuel storage;
- Faults affecting the radioactive waste treatment plant.

The HIRE then identified 5 candidate individual or bounding design basis fault sequences of which two had higher off-site doses so were analysed further. These were loss of coolant following shutdown bypassing containment and a set of faults involving intermediate loss of coolant accident with fuel failures and with iodine filters not working. The lower ERL for shelter could be exceeded out to 320 m for one fault and the lower ERL for stable iodine out to 527 m for the other (they release different isotopic mixtures).

The 2020 consequence report for Sizewell B (Ref. 35) states that "*EDF Energy has considered a wide range of accident scenarios in the hazard evaluation process and selected a candidate release as the basis of the consequences assessment. The candidate release assumes the most pessimistic attributes from a number of fault sequences in terms of time to release and quantity of activity released it, therefore, does not correspond to the release from a specific individual fault*". This use of a compound source term loses some of the detail that an emergency planner would like to know. What range of severity of faults are plausible? Is there likely to be a warning time? (EDF hint that there may but suggest assuming that there will not be), how will we know on the day what magnitude of fault we are facing? What might the duration of the fault be? (The compound source term is released over 5 hours after which, they claim, the system will run out of pressure to eject radioactivity or they will have sealed the breach.

Stable iodine is indicated to 1350 m, shelter to 500 m and evacuation to 250 m.

"*Assessments indicate that the radiation concentrations in milk under likely dispersion conditions would exceed the Euratom Maximum Permitted Levels (MPL) to a distance of ~ 32 km and concentrations in unprocessed leafy green vegetables would exceed the MPLs to a distance of ~33 km.*"

Further reading: USNRC Technical Training Center, Reactor Concepts Manual, Pressurized Water Reactor (PWR) Systems (Ref. 36).

2.2.6. Future PWRs

EPR (Hinkley Point C and Sizewell C)

As PWRs have developed there have been two competing themes. The first is to add more systems to provide ever more diverse, redundant, segregated and resilient safety systems and the second is to provide simplified systems that are less complex to build, easier to operate and more inherently safe. The UK-EPR (the Hinkley Point C and Sizewell C design) is an example of a complex reactor system whereas the AP600 (Ref. 37) is a simplified design.

Reactors being built at Hinkley Point C and proposed for Sizewell C are UK-EPRs. In the past EPR has meant both "European Pressurised Reactor" and "Evolutionary Power Reactor", but it is now just known as the EPR. The reactor design is dubbed a "third generation design" and has evolved from the Framatome N4 and Siemens Power Generation Division Konvoi reactors. As such there is a lot of experience behind it. Detailed documentation associated with the Generic Design Assessment can be found at reference 38.

The system is designed with safety in mind. The initial response to most initiating events is automatic and for many transients will not lead to a reactor scram. The system is designed with the intent that an operator action should not be required in the control room for the first 30 minutes, nor elsewhere on plant (called a "local action") for an hour after the first significant information is given to the operator.

The UK EPR design has control rods managed by the Control Rod Drive System (CRDS) which can react quickly enough to quench the chain reaction and keep it sub-critical, preventing reactor damage, until other systems such as boron injection can be initiated. The Emergency Boronation System (EBS), which is independent of the normal operational system, provides long term sub-criticality for a wide range of faults.

The reactor building, the spent fuel building and two of the safeguard buildings are heavily reinforced; designed to survive the impact of an aircraft crash. Two further safeguard buildings are on opposite sides of the site with the intention that at least one would survive any impacts. Each Safeguard building is able to totally fulfil one of the two essential safety functions (stopping the nuclear reaction and cooling the reactor). The key buildings float on a massive reinforced concrete raft providing a high degree of protection from seismic activity. In case of a reactor melt down there is a carefully designed core catcher which will spread the molten reactor to avoid criticality and to cool it down. The double containment building is designed to reduce and delay the release of radioactive material even in the most extreme fault.

The reactor's ability to absorbed severe accident sequences is truly impressive. The radiological implications of design basis accidents are analysed in Reference. 39 in which table 38 shows that the doses resulting from a range of accidents with single initiating events are up to 5.5 mSv over seven days at 500m. Under REPPIR-19 these would not quite justify detailed emergency planning.

HPR 1000 (Bradwell B)

There is an outline plan to build a HRP 1000 at Bradwell. This reactor design is currently in the later stages of the Generic Design Assessment (GDA) due to finish in late 2021 (Ref. 40). The aim of the current process (step 4) is to:

- confirm that the higher-level claims and arguments are properly justified;
- progress the resolution of issues identified during Step 3;
- complete sufficient detailed assessment to allow ONR to decide whether Design Acceptance Confirmation (DAC) or Statement of Design Acceptance (SoDA) can be issued.

The objective for GDA is to provide confidence that the proposed design is capable of being constructed, operated and decommissioned in GB in accordance with the standards of safety, security and environment protection required.

The papers submitted to GDA thus far can be found on their website (Ref. 41). The emergency planning section is (Ref. 42). This makes the claim that "*Emergency arrangements will be in place, prior to commissioning, that will be in accordance with up-to-date standards in the event of a release of radioactive substances*" (Claim 3.2.5).

Under this claim they have proposed a number of "arguments"

- Emergency arrangements adequately consider current safety analyses and accident mitigation studies;
- Emergency arrangements include adequate on-site and off-site equipment and facilities, including the provision of technical and operational support centres and their communication systems;
- Emergency arrangements include effective interaction with relevant off-site organisations;
- Emergency arrangements are tested and reviewed regularly and comprehensively, and lessons learned, are adopted in order to be effective when operations commence and maintained during the life of the plant;
- Emergency arrangements are in accordance with current standards, methods and good practices.

2.2.7. Small Modular Reactors

According to the IAEA (Ref. 43) "Small Modular Reactors (SMR) are newer generation reactors designed to generate electric power typically up to 300 MW, whose components and systems can be shop fabricated and then transported as modules to the sites for installation as demand arises. Most of the SMR designs adopt advanced or even inherent safety features and are deployable either as a single or multi-module plant. SMRs are under development for all principal reactor lines: water cooled reactors, high temperature gas cooled reactors, liquid metal, sodium and gas-cooled reactors with fast neutron spectrum, molten salt reactors, and most recently micro-reactors. The key driving forces of SMR development are fulfilling the need for flexible power generation for a wider range of users and applications, replacing ageing fossil-fired units, enhancing safety performance, and offering better economic affordability".

The same report claims that there are more than seventy SMR designs under development around the world and gives an overview of many of them. This includes the UK SMR being championed by Rolls Royce (Ref. 44) which is at the "conceptual stage". This is a three loop Pressurised Water Reactor (PWR) providing about 400-450 MWe from 1200-1350 MWth using industry standard UO_2 fuel enriched up to 4.95%, clad with a zirconium alloy and arranged in a 17x17 assembly. The core contains 121 fuel assemblies (compared to 193 at Sizewell B) and has an active fuelled length of 2.8 m (compared to 3.9 m for Sizewell B), delivering a thermal power of 1276 MW(t). Each fuel assembly contains 40 poisoned fuel pins, with the remaining 224 fuel pins being unpoisoned. The poison used is distributed Gd2O3 (containing natural gadolinium) at 8 wt%.

Reactivity control provided through movement of control rods and use of the negative moderator temperature coefficient inherent to PWRs. This eliminates the complexities of boron chemical control present in current PWRs. Boron is available to provide a diverse system for emergency shutdown and holddown.

The RPV diameter is constrained to be less than 4.5 m to ensure that the UK road transport height of 4.95m is not exceeded.

Containing less fuel than a full size reactor and with a range of passive and active safety devices, this reactor is designed to exceed current safety design expectations. It will be interesting to see what emergency plans they have if one is ever built. The IAEA report that the development plan has first operation in 2030.

The Canadian, Russian and Chinese programmes to develop Small Modular Reactor are more advanced than the UK programme. Canada and Russia benefit from obvious domestic markets while the Chinese investment in nuclear research and development exceeds that of other countries.

2.2.8. Micro Reactors

There is another distinct class of small reactor system being considered. Typically, of under 30MWe and 100MWt output classed as Micro Nuclear Reactors (MNRs). These are smaller, factory-built systems that could be easily transported by road, sea, air or rail. Several designs have been suggested including High Temperature Reactors, Molten Salt reactors and Liquid Metal Reactors as well as a number of "evolutionary" designs which are shrunken versions of the PWR and BWR families.

These reactors could, in principle, be deployed to remote areas or areas where distribution grids are unreliable or damaged and provide reliable hat and power with little on-site work. The sales pitch suggests that "*these systems could provide up to 10,0000 homes with clean power–24 hours a day, 7 days a week – for 10 years without stopping*" (Ref. 45).

Reference 46 is a report commissioned by the UK government which concludes that "Overall, a potential global accessible market of up to 2850MW (equivalent to 570 units of 5MW each) has been estimated by around 2030-2035. This covers both the UK market and internationally accessible markets (notably remote islands and desalination plant). It should be noted that this capacity is only equivalent to less than two new large reactors, illustrating the relatively small size of the short-term market".

It also makes the statement "Based on core thermal power it is estimated that MNRs could have a detailed emergency planning zone as little as 25m radius, but unlikely to exceed 250m". That would be interesting to see.

2.2.9. Submarine PWRs

The Royal Navy operate a number of nuclear powered submarines and it is possible that UK waters will be visited by nuclear powered vessels from other countries. British submarines use small Pressurised Water Reactors (PWR).

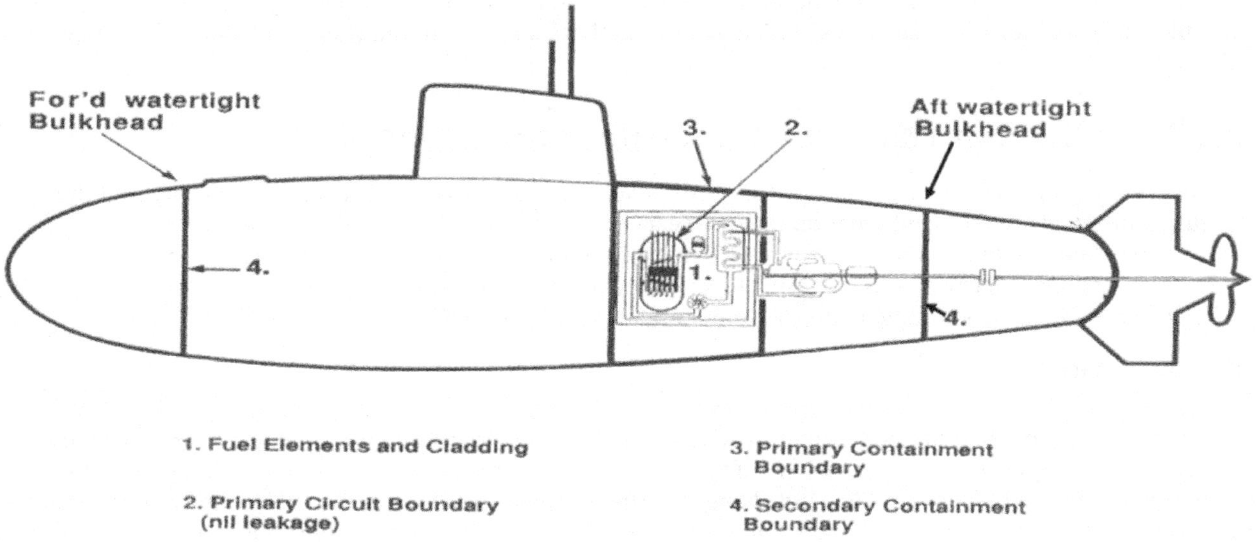

Figure 9 - Simple schematic of a nuclear submarine

The nuclear reactors are mounted in the middle of the vessel. The layers of protection keeping the fission products inside the vessel are shown in Figure 9. They are the fuel elements and cladding, the primary circuit boundary, the primary containment boundary, which is the reinforced room in which the reactor is housed, and the secondary containment boundary, which is the pressure vessel in which the crew live. To escape to atmosphere the fission products have to pass through all of these boundaries.

Detailed safety cases exist for the submarines but, for obvious reasons, these are highly confidential. Local authorities have been given outline descriptions of the faults that are possible and a "representative Accident", which assumes several concurrent failures of the safety barriers which is described thus (Ref. 47):

1. A number of cautious assumptions are made about the radioactive material inventory and other characteristics of the reactor;
2. A leak occurs in the primary cooling circuit of the reactor which cannot be isolated and which is beyond the capacity of the coolant make up systems;
3. A series of extremely unlikely engineering and other failures also occur;
4. The primary coolant leak coupled with the engineering and other failures lead to damage to the fuel within the reactor after >3 hours, resulting in elevated gamma radiation levels around the reactor;
5. The fuel damage in turn releases some radioactive material from the reactor. This is largely contained within the submarine but a small proportion may be released to the environment over the following 1 to 2 days;
6. The radioactive material would be carried downwind and would therefore present a hazard in the downwind sector only. This hazard would arise principally via inhalation initially.

The Representative Accident analysis determines a period of approximately 3.5 hours from the onset of the initial event (a 15mm non-isolable leak) to the onset of radiological hazard. Other, less likely, fault sequences could reduce the warning time dramatically.

The emergency plans for submarine berths are designed to make best use of the expected time between the initiating event and any release of radioactivity by getting people, including the submarine crew, away from the immediate vicinity, sheltering those further out and providing stable iodine tablets (denoted SITS in some navy documents but often as PITs elsewhere) to those who may benefit from them. The submarine crew evacuate after they have done all that is possible to achieve containment to minimise the release of fission products.

The main concerns in a nuclear submarine accident involving core damage are the intense gamma shine around the hull, particularly perpendicular to the vessel (fore and aft the radiation has to pass through large bulkheads which provide significant attenuation) and the potential for a release of fission products. The latter is considered to be less likely as submariners have a lot of experience of establishing containment (Ref. 48).

Another concern with submarines is the potential for a significant release of radioactivity to the water it is sitting in. This would cause a different set of issues, with control of activities on and in the water and along shorelines to be considered.

2.2.10. Nuclear reactor - main safety functions

For events relating to operating nuclear reactors (AGR, PWR and naval PWR) it is useful to bear in mind the three main safety functions and listen for confirmation that the reactor is:

1. safely held down (no likelihood of re-criticality as xenon reduces or temperatures rise);
2. safely cooled (decay heat is not going to lead to further damage); and
3. contained (the radioactive material cannot escape in large quantities).

Criticality control

Criticality control is about stopping the chain reaction and reducing the heat generation in the core. In an accident, core criticality can be complicated as the reactivity of the core changes with changes in temperature, changes in the levels of short lived poisons (such as xenon-135) and changes in the composition of the gas in the cooling circuit. There is a possibility of a situation occurring where a damaged reactor is sub-critical (shut down) but if current trends continue (burn off of Xe-135, rises in temperature etc.) it could restart itself and cause more problems.

Neutron Poison: a substance (other than fissionable material) that has a large capacity for absorbing neutrons in the vicinity of the reactor core. This effect may be undesirable in some reactor applications because it may prevent or disrupt the fission chain reaction, thereby affecting normal operation. However, neutron-absorbing materials (commonly known as "poisons") are intentionally inserted into some types of reactors to decrease the reactivity of their initial fresh fuel load. (Adding poisons, such as control rods or boron, is described as adding "negative reactivity" to the reactor.)

(US NRC)

AGR operators have three diverse ways of absorbing extra neutrons and stopping the chain reaction:

1. the insertion of the control rods;
2. nitrogen injection; and
3. boron bead injection.

In water cooled reactors control rods and boron injection are available to shut down the reactor.

In an emergency listen out for conversations that discuss whether the reactor is safely "held down" as there are some circumstances where the reactor can be sub-critical but changes, such as rising temperature, threaten to restart the chain reaction.

Cooling

Cooling the reactor requires a cooling fluid to move through the reactor taking off the heat and for that fluid to be cooled in turn. There are several ways in which heat can be removed from an AGR reactor:

1. There are 4 main boilers which can remove heat from the primary circuit;
2. There are 4 decay heat boilers (smaller, simpler and able to take poor quality cooling water).

There are multiple ways to provide water to the boilers.

The coolant in the primary circuit is usually CO_2 but it can be air (it is air during maintenance shutdowns).

Commercial PWRs have diverse ways of providing cooling to intact cores, to damaged cores and to containment buildings.

Listen out for discussions about core temperature, particularly core temperature rises. If temperatures are rising ask about the temperatures at which things start to get dangerous, for predictions about how long it will take to get to those temperatures under current conditions and for the operator's plan to provide cooling.

Barriers to release (containment and confinement)

There are three main barriers preventing the fission products from being released:

1. Fuel matrix;
2. Fuel cladding;
3. Primary circuit.

Submarines also have the vessel's hull as another line of defence.

Listen out for news about leaks in the primary circuit and what the company is doing, or planning to do about it. On AGRs they may be "blowing down" the reactor (deliberately depressurising it to allow for a repair) and may be doing this through "blow down beds" or "iodine beds" which are filters designed to remove particulates and iodine from releases. Time to depressurise might be important. How long will it take and what are the consequences for atmospheric release? A blow down might give a spike of higher discharges while it is happening, but give the operators the ability to prevent greater releases later.

Also listen out for news of the fuel integrity. If there are concerns about damaged fuel then the release could get worse.

In a depressurisation the operators may also have concerns about reactor cooling and about the composition of the gas in the primary circuit, but would normally be confident that they have these under control.

For PWRs, including submarines, listen out for conversations about the tightness of containment. If the operators are monitoring the pressure inside containment they will be able to give an estimate about the change of leakage rate as a function of time going forward.

2.3. Nuclear fuel storage and reprocessing (Sellafield)

Sellafield is a large and complex nuclear fuel reprocessing site situated in Cumbria. The site has been in existence since 1947 and is where the spent nuclear fuel from the UK nuclear power plants is reprocessed and stored. Legacy buildings hold large inventories of nuclear wastes in facilities that fall short of modern standards. Major projects are underway to recover this waste, process it and store it in safer forms and in modern facilities.

The work at Sellafield can be considered under a number of headings:

- Retrievals of nuclear waste, fuel and sludge that are stored inside legacy ponds and silos, the highest risks and hazards at Sellafield;
- Remediation. As well as the legacy ponds and silos, the site has hundreds of nuclear and non-nuclear facilities that need to be cleaned-up;
- Spent nuclear fuel management. The site used to reprocess spent nuclear fuel. That means taking the fuel that has been used inside a nuclear power station and extracting the individual component parts of plutonium (1%), uranium (96%) and waste (3%). (From Ref. 49). The Magnox Reprocessing Plant is due to close in 2021. Thorp (the Thermal Oxide Reprocessing Plant), which reprocesses AGR and other fuels, ceased operation in 2018.

See reference 50 for an informative presentation about Magnox reprocessing outlining the process and the improvements made over the years. See reference 51 for background and detailed discussion of the chemistry involved in fuel reprocessing.

The Highly Active Liquor Evaporation and Storage (HALES) facility supported both Thermal Oxide Reprocessing Plant (THORP) operations and Magnox reprocessing operations at Sellafield by receiving Highly Active Raffinate[5] (HAR) and concentrating it through evaporation. The result is concentrated liquor known as Highly Active Liquor (HAL). The HAL

[5] Raffinate definition: the liquid left after a solute has been extracted by solvent extraction

is held in interim storage within HALES in Highly Active Storage Tanks (HASTs) prior to it being transferred to another facility, the Waste Vitrification Plant (WVP), where it is processed into glass blocks suitable for long term storage.

Sellafield is also the site of the Sellafield MOX Plant (SMP) which made mixed oxide fuel by blending plutonium from reprocessing with uranium. This plant, which had a troubled history of technical difficulties and poor production levels, was closed abruptly following the collapse of its market in the wake of the Fukushima event.

According to the last REPPIR-01 HIRE, (Ref. 52) the main parts of the sites radioactive material inventory (as of 2010) were as follows:

- High Level Waste (HLW) - 1,620m3
- Intermediate Level Waste (ILW) - 67,170m3 (inc Windscale)
- Separated plutonium - 113 Te

The important conclusions from the HIRE were that:

- A Radiation Emergency is reasonably foreseeable at the Sellafield site;
- The maximum radiation dose to a member of the public from a reasonably foreseeable radiation emergency at Sellafield has been calculated to be less than 320 milli-sieverts (mSv);
- The bounding 5 mSv dose contour from a reasonably foreseeable radiation emergency at Sellafield extends to a distance of 6 km from the centre of the current Detailed Emergency Planning Zone around the site.

The reference accident is based upon seismic damage to a number of facilities leading to the release of radioactive material but detail is lacking due to security measures.

The 2019 REPPIR-19 Consequence Report (Ref. 53) reports the analysis of 10 different atmospheric release scenarios. It concludes that shelter would be appropriate out to 2.7 km in average weather conditions and 5.3 km in pessimistic weather conditions based on damage that might be sustained in a 1 in 1000 year earthquake. For evacuation the distances are 0.8 km and 1.8 km, with the recommendation that plans are capable of completing evacuation within 2 hours of the alert.

Food controls may be needed out to 72.5 km for milk and 137.5 km for green vegetables.

2.4. Recycling and waste services

A lot of the material being taken out of decommissioning nuclear sites is mildly radioactive, not necessarily enough to hurt someone but enough to cause anxiety. Sites such as Lillyhall in Cumbria are specialists in handling contaminated metals. It is a small low hazard facility that undertakes sorting and segregation, size and volume reduction and shot-blasting allowing around 95% of the wastes to be recycled (Ref. 54). There is also a landfill site there, currently limited to a maximum average activity of 4 Bq/g, or 40 Bq/g for tritium but which has applied to increase these permitted levels (Ref. 55).

Inutec Limited, Winfrith have recently taken a nuclear site licence for part of the Winfrith site offering waste characterisation and disposal services (Ref. 56).

2.5. Dounreay

Dounreay was the UK's centre of fast reactor research and development from 1955 until 1994. Its history includes three nuclear reactors, two fuel reprocessing facilities and the production of enriched uranium fuel elements. Dounreay is now a complex nuclear decommissioning site. There are also a variety of radioactive waste storage and disposal facilities.

The principal operations remaining on site include:

- storage of irradiated fuel;
- storage of unirradiated fuel;
- storage of intermediate level liquid waste;
- treatment and storage of liquid and solid intermediate level waste;

- treatment and storage of solid low level waste;
- conversion of enriched uranium stocks to a form suitable for long-term storage.

According to their August 2018 emergency plan (Ref. 57), "the main across-site or off-site hazards to on-site workers and the public identified from the referenced safety cases are:

- aircraft crash: Inhalation of contamination arising from an aircraft impact resulting in a large fire and loss of containment;
- security Related Events: Inhalation of contamination arising from a malicious aircraft impact resulting in a large fire and loss of containment (bounding case for security related events on site);
- external Events: Inhalation of contamination arising from a seismic event;
- fissile Material: Radiation and airborne release as a result of a criticality;
- liquid Alkali metal: Inhalation of contamination arising from an Alkali metal fire;
- RAM Transport Accidents: On site vehicle collision resulting in loss of radiological containment and coincident fuel fire".

Under REPPIR-01 Dounreay had a DEPZ of 1.15 km.

2.6. The nuclear deterrent

The UK's continuous at sea deterrent (CASD) is a policy of having one nuclear armed, nuclear powered submarine at sea at all times. It is currently based on the Vanguard Class Submarines each carrying about 40 warheads. The new Dreadnought class is being developed to replace the Vanguards (Ref. 58).

Key facilities are the AWE sites near Reading which design, build and maintain the warheads, the Rolls Royce site in Derby where reactor cores are designed and built, Barrow, where the submarines are built and where the reactors are first tested, the dockyards at Devonport and Rosyth where the submarines and based and maintained, submarine berths dotted around the UK and Vulcan, a research and development site near Dounreay. There are also a number of legacy facilities dotted around the UK left over from the rapid development of the UK deterrent.

The Ministry of Defence (MOD) have summarised information for the emergency services (Police, Fire and Rescue and Ambulance) and Local and Health Authorities on contingency arrangements to be implemented in the unlikely event of an emergency, including those arising out of terrorist acts, during the transport of nuclear weapons and Special Nuclear Material (SNM) in Ref. 59.

2.6.1. Nuclear warheads

The atomic weapons establishment (AWE) operates two sites near Reading; Aldermaston and Burghfield. Their role is to "Deliver and maintain a safe and secure Nuclear Warhead Stockpile" which includes: initial research and design through to development, manufacture, maintenance and finally decommissioning and recycling or disposal, providing continual assurance that the warheads are safe and fit for purpose.

Parts of this process bring high explosives close to plutonium, uranium and tritium assemblies.

The 2019 off site plan for AWE (Ref. 60) states that the reference accident for Aldermaston was '*a seismic event leading to consequential fires causing simultaneous loss of containment in multiple facilities with an unfavourable wind direction that would cause cumulative doses from two separate facilities*'. While for Burghfield it has '*a seismic event leading to catastrophic building collapse and consequential explosions causing simultaneous loss of containment*'.

AWE have recently produced Consequence Reports for Burghfield (Ref. 61) and Aldermaston (Ref. 62). Both Consequence Reports recommend shelter as the prompt protective action of choice but the Aldermaston report admits that there will not be enough time to implement this to avert any radiation dose and the Burghfield report comes very close to reaching the same conclusion.

The consequence report for Aldermaston is a bit coy about the types of accident considered although it does admit to criticality events with no off-site release and to fires leading to a release of tritium or plutonium. It concludes that the

7.5 mSv dose contour (where shelter ceases to be indicated) is at 1540 m around the site centre location when using an unlikely set of weather conditions to minimise the sideways spreading of the plume.

For Burghfield the "most likely predicted accidents would spread material by explosive distribution, these are non-fission incidents, where the material that would dominate in this type of release will be plutonium (which is an alpha emitting actinide) in an inhalable particulate form". This fault gives no warning and has a very short duration. First pass inhalation dose dominates. This combination of circumstances makes the averting of any dose at all extremely difficult. None the less AWE recommend a proposed minimum geographical extent to be covered by the Local Authorities Off-Site Emergency plan extending to a radial distance of 3,160 m from the Burghfield Site centre location.

We are told that it is inconceivable that an accidental nuclear explosion can occur.

2.6.2. The nuclear submarine programme

Nuclear core production

The Neptune Reactor and Nuclear Fuel Production Plant (NFPP) licensed sites are both situated within the boundary of the Rolls Royce Marine Power Operations Limited (RRMPOL) Raynesway site which is approximately 3 kilometres south-east of Derby city centre.

The principal nuclear facility is the Neptune test reactor which is a low power reactor having a designed thermal power output not exceeding 300 watts. This is used to conduct experiments on the reactor cores manufactured at the nearby Manufacturing Site in order to confirm the design data.

At the Manufacturing Site, uranium fuel is processed and nuclear reactor cores are fabricated and assembled for the Royal Navy's submarines. The uranium is received from the Ministry of Defence and its form and shape are then changed by the manufacturing process. Much of the work involves conventional manufacturing processes common to any engineering industry. At the end of this process, all the components are brought together and the finished reactor core is trial assembled.

Under REPPIR-01 RRMPOL's assessment for the Neptune Reactor site showed that it is not reasonably foreseeable that a radiation emergency could arise at the facility. The NFPP HIRE identified a number of accident scenarios in which an unintended criticality (a self-sustaining nuclear chain reaction) event could theoretically occur and result in a reasonably foreseeable radiation emergency.

A cyclic criticality event was chosen as the reference accident. A 310 m radius was estimated for the 5 mSv contour.

The 2020 consequence report estimated a distance of 360 m to "protect against the direct neutron and gamma radiation hazards from the [criticality] incident, in accordance with the lower ERL for shelter of 3 mSv" with an OPZ of 1 km being set by the Secretary of State for Defence.

Nuclear submarine construction and initial testing (Barrow)

Barrow BAE Systems Marine Ltd (BAESM) constructs, tests and commissions nuclear powered submarines under contract from the Ministry of Defence (MOD) at the Barrow GB nuclear site. The actual operation of the nuclear reactor, at this location, takes place only for a short period of time at the end of each submarine build (i.e. it is an occasional operation only). The submarines are subsequently prepared for sea at the Ramsden Dock Basin (RDB), which is a nuclear warship site. The operator of this latter site is the MOD. The RDB is a holding berth, and submarines are intended to spend a limited number of days (or perhaps hours given favourable weather and tidal conditions) at this location prior to their transit to sea.

Up until 2014 the Barrow DEPZ was based on 2 km circles centred on the two sites. The outcome of the risk assessment process determined 1. 5 km in 2008 based on the change in class of submarine being built and tested but the DEPZ was not changed on the ground. The ONR agreed in 2014 with an assessment of 0.690 km but, to cope with the geography of the Walney Island the DEPZ extends for up to 5 km.

Nuclear Dockyard (Plymouth & Rosyth)

Within the dockyard Port of Plymouth the MOD controls activities at the HMNB Devonport Berths and anchorages (an authorised site[6]) and Babcock operate an adjoined licensed site for the refitting and refuel/defuel of nuclear-powered submarines, where submarines undergo major refits and upgrades.

According to the 2017 HIRE for the site (Ref. 63) the activities that take place in Devonport are:

- Operation of Nuclear Steam Raising Plant (NSRP) (i.e. nuclear reactors);
- Movement and berthing of vessels;
- Testing, maintenance and repair of the NSRP and associated systems;
- Lay up of decommissioned submarines.

The site has a number of X berths, where nuclear maintenance, repair and ammunitioning is permitted and Z berths for which prior justification and approval are required for these operations.

The site also hosts a number of nuclear submarines that have reached the end of their service life and are awaiting decommissioning (long term berthing). Some of the submarines that are laid up in the docks have had all of the nuclear fuel removed while others have not. Risk assessments have shown that these vessels do have reasonably foreseeable accidents above the REPPIR-01 threshold.

The HIRE identified the reference accident, for an operating reactor, as a Loss of Coolant Accident (LOCA) where core damage occurs but the containment remains intact therefore there is only a very small release of radioactive material into the environment. For this accident sequence, the consequence is dominated by gamma shine from the submarine hull, with a 5mSv dose contour (unmitigated) extending approximately 870m.

Lower probability faults have also been identified. The bounding case sensitivity study has a slightly greater atmospheric release of radioactive material with a 5mSv dose contour (unmitigated) extending to 1170m. In this accident sequence, the radiation dose beyond 500m is dominated by the release of radioactive fission products. In the past ONR have used this accident sequence to determine the Detailed Emergency Planning Zone (DEPZ) for HMNB Devonport berths.

Note: Rosyth Dockyard (in Fife) was used for the refit, repair and maintenance of operational nuclear submarines from the 1960s until 2003. As each submarine reached the end of its operational life, it was defueled and decommissioned onsite. A MOD project for the final dismantling of the laid-up submarines is currently at the planning stage. ONR have agreed with the site operator's assessment that there are no credible faults that would lead to a nuclear emergency response organisation being required (Ref. 64).

Nuclear submarine berths

Several local authorities play host to nuclear submarines. These range from facilities where submarines are based, maintained and/or tested, berths that are visited relatively often and berths that are occasionally visited by nuclear submarines. There are a hierarchy of rules restricting, but not banning, the operation of nuclear reactors in these facilities.

A Z-berth is a berth designated for occasional operational or recreational visits by nuclear-powered warships. Its safety arrangements will include a public safety scheme prepared after discussion with the local authorities and civil emergency services, which will provide for a Royal Naval incident commander and a naval emergency monitoring team in the area to stand by throughout any such visit to respond to an emergency. The monitoring team has access to a supply of potassium iodate tablets, and carries out monitoring during the visit.

---

[6] HMNB Devonport, as a MOD establishment, is not subject to licensing under the Nuclear Installations Act 1965. The MOD operates an internal nuclear regulatory regime that mirrors the Office for Nuclear Regulation (ONR) licensing approach.

An X-berth is a berth designated for frequent and regular use by nuclear-powered warships, or a berth in a dockyard or naval base which is suitable for the building, refit, repair or maintenance of nuclear-powered warships. All X-berths have as an integral part of their safety arrangements a permanent health physics department, a local emergency monitoring organisation and a local safety plan prepared under the auspices of a local liaison committee. The United Kingdom X-berths are at Barrow-in-Furness, Devonport, Faslane and Rosyth. (Ref. 65).

The hazards and emergency planning needs of nuclear submarines are covered in Section 2.2.9.

The consequence report for Portland (Ref. 66), which is near identical to all the other berth consequence reports contains the following advice and rational with regard to prompt protective actions:

1 Controlled evacuation of the immediate area around the submarine to 200 m in all directions is to protect against the direct gamma radiation hazards from the submarine, in accordance with the lower emergency reference level (ERL) for evacuation of 30 mSv;
2 Sheltering indoors up to 400 m in all directions from the submarine is to protect against the direct gamma radiation hazards from the submarine, in accordance with the lower ERL for shelter of 3 mSv;
3 Consumption of SITs in the 1.5 km downwind zone is to protect against an uptake of radioactive iodine to the thyroid, in accordance with the lower ERL for stable iodine of 30 mSv;
4 Sheltering within the 1.5 km downwind zone is to protect against contamination following a release of radioactive material, in accordance with the lower ERL for shelter of 3 mSv.

The recommendation to shelter and consume SITs out to 1.5 km in the downwind sector differs from the previous recommendation of 1.2 km due to the consideration of a wider range of weather conditions in accordance with Schedule 3. This is in line with the current emergency arrangements, and ensures the public are protected from a full range of possible consequences identified.

The consequence reports for all operation berths are near identical, differing only in the identification of the site. All operational berths have a 1.5 km radius DEPZ, although some have "lumps and bumps" projecting beyond this distance to encompass groups of people just over the boundary.

Naval Reactor Test Establishment (Vulcan)

The Vulcan Naval Reactor Test Establishment (NRTE) is a Ministry of Defence (MoD) establishment housing the prototype nuclear propulsion plants of the type operated by the Royal Navy in its nuclear submarine fleet. Construction began on site in 1957 with the first reactor operational in 1965. Vulcan is a defence authorised site.

It now contains the Shore Test Facility (which is a PWR (shutdown)), some used fuel storage ponds and Decontamination Waste Treatment Facility (DWTF), which is in care and maintenance.

An analysis of the safety case for the Shore Test Facility concludes "In the improbable event of a radiation emergency, the likely exposures to those members of the public within the zone extending 2 km from the location of the plant could exceed 5 mSv. It is very unlikely that exposures in excess of 5 mSv could be received beyond this zone". (Quoted in the Dounreay and Vulcan NRTE Emergency Plan, Version 7.0 - Public Version (Ref. 67).

Within the 550 m automatic protective action zone personnel go to shelter stations prior to evacuation from the site. There is a 2 km DEPZ. In the event of an off-site nuclear emergency, members of the public within the DEPZ but outside the automatic protective action zone will be recommended to shelter and consideration will be given to using any warning time to evacuate some or all of them from the area before a release begins.

Stable iodine was removed as a protective action in March 2018 following changes in the operations on-site (i.e. a rather slow response to the final shutdown of the nuclear reactor in 2015).

2.7. Non-reactor site emergencies

This section takes a brief look at a few other scenarios that could present the local authority with a radiological emergency or clean-up operation to manage.

Nuclear accidents are not the only situations in which the public can be exposed to additional doses of ionising radiation. This section briefly looks at the consequences of:

1. Transport accident;
2. Dirty Bomb - a device left by terrorists which use convention explosives to spread a radioactive material;
3. Orphan source - where a radioactive source escapes control and exposes people as they go about their daily life;
4. Industrial accident;
5. Nuclear satellite re-entry;
6. Nuclear bomb; and
7. Criticality accident.

In all these cases protection of the public relies on the controls of time, distance and shielding in the immediate event and decontamination as the event progresses. A tabulation of recorded accidents is given in appendix A.

2.7.1. Transport accident

It can be seen from Sections 2.1 to 2.6 that there are a number of sites across the UK associated with the nuclear industry or nuclear defence. Material is routinely transported from site to site. This material includes new nuclear fuel, used nuclear fuel, nuclear wastes and nuclear weapons (in components or complete). In addition radioactive materials are transported for the health sector and for industrial purposes such as non-destructive testing. Transport can be by road, rail, sea or air. Up to half a million packages containing radioactive materials are transported to, from and within the UK every year and accidents are rare but do happen. In the UK details of accidents and incidents involving the transport of radioactive materials are entered into the Radioactive Materials Transport Event Database (RAMTED), which contained information on 1034 events that were known to have occurred between 1958 and 2014 (Ref. 68).

Under IRR-17 these activities must be risk assessed, all reasonably practical steps taken to reduce the risk and contingency plans be put in place if the residual risk merits them.

The main source advice for transport emergency planning is an ONR publication (Ref. 69). The relevant regulations are the Carriage of Dangerous Goods Regulations and Use of Transportable Pressure Equipment Regulations 2009 (as amended on 21st April 2019 by the Carriage of Dangerous Goods Amendment Regulations 2019) CDG09(19) (Ref. 70) which implements parts of the EU Basic Safety Standards Directive.

CDG09(19) require duty holders (both the consignor and the carrier) to have a plan where they have reached the conclusion that a radiation emergency might occur. The emergency plan must detail the arrangements to restrict, so far as is reasonably practicable, the radiation exposure of any person that may be affected by a radiation emergency, before the carriage of radioactive material takes place. This includes the vehicle crew, the public, attending emergency services and any persons exposed to ionising radiation as a result of a loss of radiation shielding, release of all or part of the contents of a package or an uncontrolled criticality when transporting radioactive material. ONR expectations are clearly and succinctly outlined in Ref. 71.

In a similar fashion as REPPIR-19, to be considered adequate, an emergency plan must contain the information in CDG09(19) Schedule 2, Part 1, paragraph 3 and be effective in meeting the principles and purposes in CDG09(19) Schedule 2, Part 1, paragraph 4.

A useful US training video for emergency responders called to a transport accident involving radiation can be found at reference 72 with notes provided as reference 73.

2.7.2. Dirty bomb (RDD)

A Dirty Bomb or radiological dispersal device (RDD) is a device that combines conventional explosives with radioactive material. It is not a nuclear explosion. The intention might be to spread alarm, based on the well-known fear of radiation, and to make recovery from the conventional bomb more difficult i.e. to deny access to an area for a prolonged period causing economic and social disruption. It is important to understand that the immediate health effects from exposure to the low radiation levels expected from an RDD would likely be minimal.

A dirty bomb is not the only conceivable form of a radiological dispersal device. Sprays of radioactive materials from vehicles or aircraft, for example, is also possible.

To detonate a dirty bomb in a public space terrorists would need the technology to deliver and initiate a conventional bomb and they would also need to obtain a radioactive source without coming to the attention of the authorities. There are many radioactive sources in use in hospitals and in industry but they are subject to tight control. They also tend to be in non-dispersible form, such as metals or ceramics, or tightly encapsulated. The stronger sources would be powerful enough to make anyone tampering with them or carrying them outside their heavy shielding ill quite quickly. The terrorist group would have to find a way to handle the material and to make it dispersible in order to make a successful dispersal device as opposed to just a high profile news story.

The immediate danger would be the bomb blast and priority should be given to treating those with serious injuries and providing support to the other injured people and worried bystanders. At this stage no one would know the bomb contained radioactive material unless tipped off by the terrorists.

The first responders' radiation detectors would pick up any significant radiological hazards provided they were being used correctly. You could consider that radiation had been confirmed if two separate detectors showed its presence in different places. (This avoids false alarms caused by faulty detectors or legitimate radiation sources[7]). Thereafter the dust produced by the explosion should be considered to be hazardous and there will be the potential for "hot particles" which are small pieces of highly radioactive material which are hazardous to approach and very dangerous to touch. Almost all of these hot particles will be within the ballistic range of the explosion although some may have been transported further on people or vehicles leaving the area.

On alert of an explosion the local police control room will facilitate the deployment of first responders, gather and share information, promulgate the alarm, ensure coordination with the other emergency services at control room level and at the scene.

The first responders' priority on arriving at the scene of a bomb blast would be to consider their own safety, to assess the scene (The Emergency Services in the UK use the METHANE prompt - Figure 10), to set up rendezvous points for incoming responders and a forward command post some distance upwind of the event. Working together the police will seek to regain control of the area, setting up cordons as appropriate; the fire and rescue service will concentrate on aiding victims and the ambulance service will provide first aid skills and transport to hospitals.

In the UK the Ambulance Service have a strategic obligation to protect the health, safety and welfare of all health service personnel on scene from the hazards posed although that in no way reduces the obligations of the individual services or individuals to take care for themselves and others.

The early intention would be to move people away from the scene, giving priority to moving people from the downwind sector. People in undamaged buildings near the scene of a blast may be best advised to stay there with the doors and windows shut, air conditioning off (i.e. shelter) until more consideration can be given to moving them safely. Care should be taken as experiments performed on mock dirty bombs (Ref. 74) show that, under some circumstances the airborne levels can be significantly elevated for tens of minutes. Respiratory protection, as well as limited stay-times, should be considered.

In the longer term the local authority is likely to have to manage a cordoned off area for an extended period of time with people displaced from their homes and businesses. The response may be similar to the Salisbury nerve agent incident (2018). See the government summary of first two months of this event in Ref. 75 and notes on the decontamination in Ref. 76.

---

[7] The US advice (Ref. 66) suggests readings 50m apart and at least 3 to 5 times the natural background levels.

M	Major Incident Declared?	Has a major incident or standby been declared? (Yes / No - if no, then complete ETHANE message)
E	Exact location	What is the exact location or geographical area of the incident?
T	Type of Incident	What kind of incident is it?
H	Hazards Present or suspected	What hazards or potential hazards can be identified?
A	Access - routes that are safe to use	What are the best routes for access and egress?
N	Number, type, severity of casualties	How many casualties are there, and what condition are they in?
E	Emergency services present and those required	Which and how many, emergency responder assets/personnel are required or are already on-scene?

Figure 10 The METHANE Situation report prompt

The US NCR has the following advice for people caught up in a dirty bomb (Ref. 77): "*Move away from the immediate area - at least several blocks from the explosion - and go inside. This will reduce exposure to any radioactive airborne dust.*

- *Turn on local radio or TV channels for advisories from emergency response and health authorities.*
- *If facilities are available, remove clothes and place them in a sealed plastic bag. Saving contaminated clothing will allow testing for radiation exposure.*
- *Take a shower to wash off dust and dirt. This will reduce total radiation exposure, if the explosive device contained radioactive material.*
- *If radioactive material was released, local news broadcasts will advise people where to report for radiation monitoring and blood and other tests to determine whether they were exposed and what steps to take to protect their health."*

The US Department of Homeland Security has published a planning document showing what an emergency plan and response for this type of event could look like (Ref. 78). It describes a response over the first 100 minutes (Table 1) that includes recognition (by first responders equipped with radiation detectors), promulgation of alert, initiation of a multiagency response including evacuation of those nearby and the advice to shelter in place for those further out, the measuring and mapping of radiation levels around the scene and downwind, completion of evacuation and finally clean-up via scene of crime investigations. This would be a useful starting point for any planning team looking at dirty bomb preparedness.

The report warns that "Many of the public safety recommendations presented in this guidance are time sensitive and it is important for local agencies to plan for both how the operation is achieved and when it must be completed to have the desired impact. A complete local RDD response protocol will include an achievable, detailed timeline of how Tactics and co-ordination are executed."

The YouTube video (Ref. 79) provided with this document is well worth 20 minutes to watch.

Advice to individuals affected by a dirty bomb can be found in a US Federal Emergency Management Agency (FEMA) factsheet (Ref. 80) and in a recent IAEA publication aimed at medical physicist responders (Ref. 81).

2.7.1. Orphan source and radiation exposure device

An orphan source is an amount of radioactive material that has escaped control. These can be very powerful sources, although the term does also apply to weak sources. There have been a number of cases where sources have been lost and gone on to cause significant harm. The IAEA have recorded 68 such incidents in total (Ref. 82).

A radiation exposure device is consists of a large quantity of radioactive material clandestinely placed to expose people to ionizing radiation.

Mission	Tactic	Activities
Recognise that radiation is present at the scene of an explosion (0 -5 Minutes)	Initial response & on-scene recognition	
	Confirm the presence of radiation	At least two readings sufficiently above background from different locations.
Inform (5 - 10 minutes)	On scene commander or designate notifies control.	
	Issue protective action advice to the public	Move away from scene/ shelter in place. The initial message could be a prepared generic statement. Plan should outline how this would be promulgated.
	Promulgate alert to other emergency, health and specialist services.	
Initiate a multiagency response (5 - 40 minutes)	Initiate lifesaving operations	Search and rescue, fire suppression, medical triage and treatment.
	Secure and manage the scene	Determine and manage hazards. Establish Command and Control foci and joint working. Establish initial public safety boundaries and crime scene area
Measure and map (15 - 90 minutes)	Conduct an initial characterisation of radiological contamination, locate non-uniform high radiation areas (hot spots). Provide data for mapping.	First responders should determine the direction of travel of the initial smoke from the event to provide priority direction for survey.
Evacuate and monitor (> 70 minutes)	Commence phased evacuations	Establish evacuation routes based on situation (radiological measurements and wind direction).
	Monitor and decontaminate	Undertake quick screening of individuals at exits from the hot zone to the extent practical without slowing the evacuation.

Table 1 Dirty Bomb: Outline response plan

The worst orphan source incident to date (written in September 2020) was the accident in Goiania (capital and largest city of the Brazilian state of Goiás) in 1985 which resulted from the theft of a 50.9 TBq caesium-137 source from a partly demolished private medical facility (Ref. 83). This was dismantled by people unaware of the risks. The glittering powder attracted interest and several people used it on their skin as decoration. A few weeks later, after a doctor recognised the effects of radiation in patients presenting with burns and sickness, the area affected was monitored and evacuated. People sheltered at a sports facility for several days awaiting monitoring. About 110,000 persons asked to be monitored (about 10% of the local population). Of these, 249 were shown to be contaminated, 150 of which required further medical care. 14 required intensive care, 4 died. The decontamination of the area took six months, some flats were evacuated pending decontamination and some demolished. Forty five public places (including streets, squares and shops) were decontaminated.

Another example of an orphan source emergency, this time a by-product of an assassination, was the Polonium-210 incident in London, 2006. Alexander Litvinenko was admitted to hospital on 3rd November with vomiting, diarrhoea and abdominal pain. After some delay it was determined that he had been poisoned with about 1 GBq of Po-210, an isotope that emits only alpha particles and thus is not detected by many types of radiation detector in common use. The identification of exposed individuals and identification and decontamination of contaminated places was a major undertaking. 3837 people reported concerns following a media campaign. After a questionnaire 1844 were followed up and 753 people provided urine samples. 86 people were assessed to have had a dose less than 1 mSv, 36 between 1 and 6 mSv and 17 over 6 mSv (Ref. 84).

A third example (although it might be considered to be an industrial accident or an accidental radiation exposure device) is the Cochabamba bus incident of 2002 (Ref. 85) in which an industrial radiography source (0.67 TBq) was transported in the luggage hold of a civilian bus for a six hour journey in which time the bus had a changing number of passengers. Unfortunately the source had not been correctly wound back into the shielding. Doses to three workers were estimated by dose reconstruction and by taking blood samples. Doses to passengers were estimated from their position in the bus, by interviews and by medical examination. In this case none of the persons examined had had early symptoms that could be associated with radiation. This incident was used as the scenario for a European dose assessment inter-comparison exercise (Ref. 86).

Radiation sources are used for a range of purposes, a number of different isotopes are used and they come in a wide range of source strengths. Responders need to be aware of the nature of the source they are approaching. Table 2 shows the activity and dose rates associated with typical sources used in industry and medicine. Some of them are very powerful and used incorrectly or maliciously could cause a lot of harm (Ref. 87).

In the UK the Ionising Radiation (Basic Safety Standards) (Miscellaneous Provisions) Regulations 2018 (Ref. 88) require "*The Scottish Ministers, the Welsh Ministers, a Northern Ireland department and the Secretary of State must each exercise their functions to ensure that arrangements are made for—*
(a) raising general awareness of the possible occurrence of orphan sources and potential hazards; and
(b) issuing guidance for persons who suspect, or have knowledge, of the presence of an orphan source on informing the relevant minister and on the actions to be taken".

They also require the same roles to "*ensure that an appropriate course of action is organised to recover orphan sources left behind from past practices*".

Figure 11 radiation warning signs

The best defences from orphan sources are time, distance and shielding. Recognise the radiation warning signs (Figure 11) and if in doubt stay away.

If an intact source were to be left in a public area then people near that source would be exposed. The effects of that exposure would depend on the strength of the source, how near people got to it and how long they stayed near it. This is illustrated by Figure 12 which shows the doses received at different times and distances from an unshielded 50 TBq Cs-137 source (Gamma ray dose constant of 1.02E-04 mSv/hr^{-1} per MBq at 1m) (Ref. 89). The exposure time on the vertical axis ranges from 5 minutes to 2 hours and the distance on the horizontal axis from 1 to 10 metres. Doses in the top left hand corner (long exposure and close to the source) are over 5 Sv and could well be fatal. Doses in the bottom right corner are below 100 mSv and people is this range would probably not show any symptoms. The two middle blocks correspond to 1 - 4 Sv (likely to be ill, could be fatal at high end) and 100 mSv to 1 Sv (could show some symptoms but likely to recover with medical care).

This sort of table can be used by the emergency services and other responders to estimate how long they can stay near the source for a particular dose limit although it is better to use measurements of the dose rates rather than estimates. Their stay-time could be increased if they could put some shielding between themselves and the source. The first signs of a powerful source being left in a public area might be people presenting at hospitals, doctors surgeries or calling for an ambulance with symptoms of acute radiation sickness or unexplained burns. Importantly radiation sources for commercial use tend to be made with very robust encapsulation to prevent the spread of contamination. If such a source was undamaged then no one would be contaminated and the casualties would pose no hazard to those helping them or to any medical facility they were taken to.

Radiation sources are controlled by Council Directive 2003/122/EURATOM of 22 December 2003 (Ref. 90) (the HASS directive) which states that "*the purpose of this Directive is to prevent exposure of workers and the public to ionising radiation arising from inadequate control of high-activity sealed radioactive sources and orphan sources and to harmonise controls in place in the Member States by defining specific requirements ensuring that each such source is kept under control*".

Practice or application	Radionuclide	Typical activity	Dose rate at 1 m (mSv/hr) (See note)	Time at 1m to exceed 1mSv
Sealed sources used in medicine				
Bone densitometry	Am-241	1 - 10 GBq	3E-02	30 h
	Gd-153	1 - 40 GBq	4.E-01	2 h
	I-125	1 - 10 GBq	6.E-02	20 h
Manual brachytherapy	Cs-137	50 - 500 MBq	3.E-02	30 h
	Ra-226	30 - 300 MBq	2.E-04	200 d
	Co-60	50 - 500 MBq	1.E-01	8 h
	I-125	50 - 1500 MBq	9.E-03	5 d
	Ir-192	200 - 1500 MBq	1.E-01	8 h
	Cf-252	50 - 1500 kBq	3.E-07	400 yr
Remote loading brachytherapy	Co-60	About 10 GBq	3.E+00	20 min
	Cs-137	0.03 - 10 MBq	6.E-04	70 d
	Ir-192	About 400 GBq	3.E+01	2 min
Teletherapy	Co-60	50 - 1000 TBq	3.E+05	< 1 s
	Cs-137	500 TBq	3.E+04	< 1 s
Blood irradiation	Cs-137	2 - 100 TBq	6.E+03	1 s
Sealed sources used in industry				
Industrial Radiography	Ir-192	0.1 - 5 TBq	4.E+02	9 s
	Co-60	0.1 - 5 TBq	1.E+03	3 s
Well Logging	Cs-137	1 - 100 GBq	6.E+00	10 s
	Am-241/Be	1 - 800 GBq	2.E+00	20 s
Smoke Detectors	Am-241	0.02 - 3 MBq	9.E-06	10 yr
Lightning rods	Am-241	50 - 500 MBq	2.E-3	30 d
	Co-60	4 - 8 GBq	2.E+00	30 min
	Eu-152/154	7 - 40 GBq	5.E+00	10 min
Moisture/density detector	Am-241/Be	0.1 - 2 GBq	6.E-03	7 d
	Cs-137	400 MBq	2.E-02	2 d
	Cf-252	3 GBq	6.E-04	70 d
Static electricity eliminator	Am-241	1 - 4 GBq	1.E-02	3 d
	Po-210	1 - 4 GBq	4.E-06	30 yr
X-ray fluorescence analyser	Fe-55	0.1 - 5 GBq	2.E-03	30 d
	Cd-109	1 - 8 GBq	1.E+00	50 min
Sterilization and food preservation	Co-60	0.1 - 400 PBq	1.E+08	< 1 s
	Cs-137	0.1 - 400 PBq	2.E+07	< 1 s
Calibration facilities	Co-60	1 - 100 TBq	3.E+04	< 1 s
	Cs-137	1 - 100 TBq	6.E+03	< 1 s
Level gauge	Cs-137	0.1 - 20 GBq	1.E+00	50 min
	Co-60	0.1 - 10 GBq	3.E+00	20 min
	Am-242	4 GBq	1.E-02	3 d
Thickness gauge	Cs-137	1 TBq	6.E+01	1 min
	Kr-85	0.1 - 50 GBq	1.E-2	4 d
	Ti-204	40 GBq	4.E-3	10d
Density gauge	Cs-137	1 - 20 GBq	1.E+00	50 min
	Am-241	1 - 10 GBq	3.E-02	1 d
Sealed sources used in research				
Irradiators	Co-60	1 - 1000 TBq	3.E+05	< 1 s
	Cs-137			< 1 s
Calibration facilities	Cs-137	< 100 TBq	6.E+03	< 1 s
	Co-60	< 100 TBq	3.E+04	< 1s
	Cf-252	< 10 GBq	2.E-03	20 d

Table 2 Some typical source strengths and dose rates

Note: The notation 1.E-01 means 1×10^{-1}

Dose in Sv staying at distance on x-axis for time on y-axis

Staytime in minutes	1	2	3	4	5	6	7	8	9	10
120	10.170	2.543	1.130	.636	.407	.283	.208	.159	.126	.102
115	9.746	2.437	1.083	.609	.390	.271	.199	.152	.120	.097
110	9.323	2.331	1.036	.583	.373	.259	.190	.146	.115	.093
105	8.899	2.225	.989	.556	.356	.247	.182	.139	.110	.089
100	8.475	2.119	.942	.530	.339	.235	.173	.132	.105	.085
95	8.051	2.013	.895	.503	.322	.224	.164	.126	.099	.081
90	7.628	1.907	.848	.477	.305	.212	.156	.119	.094	.076
85	7.204	1.801	.800	.450	.288	.200	.147	.113	.089	.072
80	6.780	1.695	.753	.424	.271	.188	.138	.106	.084	.068
75	6.356	1.589	.706	.397	.254	.177	.130	.099	.078	.064
70	5.933	1.483	.659	.371	.237	.165	.121	.093	.073	.059
65	5.509	1.377	.612	.344	.220	.153	.112	.086	.068	.055
60	5.085	1.271	.565	.318	.203	.141	.104	.079	.063	.051
55	4.661	1.165	.518	.291	.186	.129	.095	.073	.058	.047
50	4.238	1.059	.471	.265	.170	.118	.086	.066	.052	.042
45	3.814	.953	.424	.238	.153	.106	.078	.060	.047	.038
40	3.390	.848	.377	.212	.136	.094	.069	.053	.042	.034
35	2.966	.742	.330	.185	.119	.082	.061	.046	.037	.030
30	2.543	.636	.283	.159	.102	.071	.052	.040	.031	.025
25	2.119	.530	.235	.132	.085	.059	.043	.033	.026	.021
20	1.695	.424	.188	.106	.068	.047	.035	.026	.021	.017
15	1.271	.318	.141	.079	.051	.035	.026	.020	.016	.013
10	.848	.212	.094	.053	.034	.024	.017	.013	.010	.008
5	.424	.106	.047	.026	.017	.012	.009	.007	.005	.004

Distance from source in metres

Figure 12 example total dose for distance and time table for 50 TBq Cs-137

2.7.2. Industrial accident

X-ray generators and sealed radioactive sources are used in non-destructive testing in factories, industrial premises and in building sites across the UK. These use the penetrating power of radiation to look for changes in thickness of metals or quality of welds in the same manner that medical x-rays look for broken bones.

Used correctly there is no danger to the user or people in the vicinity but accidents can and do happen (see Section A.9). If the equipment is used incorrectly then people, usually the operator, can get large radiation exposures in a short period of time and experience acute radiation sickness and long term injury.

If such an event happened in the UK there would be very little for the responders to achieve other than making the scene safe which involves switching off X-ray machines or recovering radioactive sources. Sources tend to be well sealed so there should be no contamination issues.

There may be a role for the police and local authority in reassuring members of the public that there is no remaining radiological hazard. Media and public interest may be prolonged as the resulting radiation illness can take many months to run its course.

2.7.3. Nuclear satellite re-entry

Between 1967 and 1988 the Soviet Union launched over 30 fission reactors into space while the USA launched one (Ref. 91). Many of these systems are still in orbit. There are supposedly "parked" in high orbits and unlikely to crash to earth for several hundred years so a radioactive satellite re-entry is inevitable at some time in the future but apparently unlikely in the next few decades at least.

Reassuringly IAEA Safety Series 119 (Ref. 92) states that launched systems are generally designed to fragment and burn-up on re-entry and that "*If the total inventory of a 100 kW(th) reactor* [this is a typical power rating for a reactor in orbit] *is deposited homogeneously over an area of 100 000 km², the net average external dose rate increase in the central part of the footprint would be about 0.05 μSv/h, significantly less than the natural background.*" (Note "If"!).

The experience of the Cosmos 954 satellite re-entry in January 1978 over the Northwest Territories of Canada gives an idea of the issues that might arise. It spread debris over 124,000 square kilometres. Some of this debris was sufficiently radioactive to be hazardous to health. The Canadian Nuclear Safety Commission reports (Ref. 93) that "by October 1978, over 4,000 flakes of radioactive debris were recovered from the region, including core fragments. It took nine months, 4,500 hours of flying time and $13,970,143 (Canadian[8]) to recover from the crash of the COSMOS 954".

We can conclude that a nuclear satellite crashing on the UK would be a wide area, long term emergency, would need some clear communication of the risks to the public and a major radionuclide survey of the affected area, and probably beyond.

In addition to reactors there have been a number of radioisotope thermoelectric generators (RTGs) launched. These use the heat generated by the decay of radionuclides (generally Plutonium-238 (Pu-238) or Americium-241 (Am-241)) to provide power. NASA is also using a new generation of Radioisotope Thermoelectric Generator (RTG) with the Multi-Mission Radioisotope Thermoelectric Generator (MMRTG) designed to generate 110W at launch which powers the Curiosity Mars rover and is planned for future Mars missions. NASA explain that radionuclides are safely contained in "the ceramic form of the plutonium dioxide heat source material, iridium metal cladding, graphite sleeves that protect the fuel clads, and the rugged carbon-fibre material that forms the aeroshell". They also report a predecessor being aboard a failed launch in 1968 and the nuclear material being recovered from the ocean floor and reused. (Ref. 94). The US safety targets for the use of nuclear system in space are given in (Ref. 95).

2.7.4. Nuclear bomb

> *"The energy liberated in the explosion of such a super-bomb is about the same as that produced by the explosion of 1,000 tons of dynamite. This energy is liberated in a small volume, in which it will, for an instant, produce a temperature comparable to that in the interior of the sun. The blast from such an explosion would destroy life in a wide area. The size of this area is difficult to estimate, but it will probably cover the center of a big city.*
>
> *In addition, some part of the energy set free by the bomb goes to produce radioactive substances, and these will emit very powerful and dangerous radiations. The effects of these radiations is greatest immediately after the explosion, but it decays only gradually and even for days after the explosion any person entering the affected area will be killed.*
>
> *Some of this radioactivity will be carried along with the wind and will spread the contamination; several miles downwind this may kill people".* Frisch-Peierls Memorandum, March 1940 (Ref. 96)

[8] On 27/1/21 the Canadian Dollar was worth £ 0.57. Inflation since the reference was written has not been taken into account.

A nuclear bomb exploding anywhere within the UK would be an event which would require national, if not, international support and co-ordination. This section looks at the range of damage that is possible with known bomb designs.

A nuclear bomb detonation in the UK is considered to be a very low probability. The 2020 National Risk Register (Ref. 98) mentions the use of an improvised nuclear device by terrorists (this is a nuclear explosion as opposed to a dirty bomb dispersing a radioactive source) "*resulting in much greater numbers of casualties and widespread, long-term impacts of a magnitude above all other terrorist attacks*". The Chemical, biological, radiological and nuclear (CBRN) response discussed on the Centre for Protection of National Infrastructure website is basic (Ref. 99). The UK emergency services response to a CBRN event is outlined in Reference 100.

The consequences of such a detonation would depend on the nature of the attack. A single improvised bomb is likely to have a relatively low explosive yield, a nuclear device stolen from one of the nuclear powers would pose a far greater threat and an all-out attack from a superpower would be another matter entirely.

The explosive power of a nuclear bomb is measured in terms of the amount of TNT that would be required to give a similar size explosion. (A ton of TNT releases about 4.184×10^9 joules) There are a wide range of explosive yields available from nuclear devices (see Table 3 adapted from Wikipedia (Ref. 101)).

Name	Est. yield (kT)	Details
David Crocket	0.01	US tactical nuclear weapon
Hiroshima (Little boy)	13 - 18	1st atomic bomb attack - 6/8/1945
Nagasaki (Fat man)	18 - 23	2nd atomic bomb attack - 9/8/1945
Hurricane	25	1st British test - 3/10/1952
Castle Bravo	15,000	Largest test by USA 1/3/1954
Tsar Bomba	50,000	Largest ever nuclear explosion (USSR) 31/10/61 You can see a Russian documentary, with subtitles, about this (Ref. 102)
Test No. 6	3,300	China 17/6/67
Canopus	2,600	France 24/8/1968
Pokhran-II	45 - 50	India 11/5/1998
Chagai-I	40	Pakistan 28/5/1998
North Korea	200 - 300	Test 3/9/2017

Table 3 Notable nuclear explosions

A nuclear explosion has a number of identifiable steps:

- Fireball - caused by the release of energy;
- A wave of thermal radiation - very fast moving and capable of blinding, burning or killing outright. First degree burns (similar to bad sunburn) may be expected out to a distance of 3.2 km from a 10 kT bomb; second degree burns (blistering, possible shock and death if extensive and untreated) out to 2.4 km and third degree burns (destruction of tissue, shock and death without specialist treatment) out to 1 km (FEMA Ref. 103);
- A shock wave that can cause considerable physical damage to anything in its way. At 0.4 km from a 10 kT explosion the wind could be 160 miles an hour, sufficient to suck people out of buildings;
- An electromagnetic pulse - this can damage electricity supply and telephone systems as well as computers and other electrical systems. Wireless communications may be disrupted for a while even if the hardware is unaffected. A blast high in the atmosphere could damage any satellites within range. This is considered to be a tactical use of nuclear weapons, and one to be concerned about as it greatly damages the ability of a modern society to function without causing so much longer term fallout;
- A mushroom cloud as everything within the fireball is vaporised and carried upward. This can include great quantities of dust if the blast is low enough to hit the ground and produce a crater;
- Fallout - caused by the dust and vapour produced cooling, condensing and returning to ground level. This fallout can be intensely radioactive although the levels of radioactivity will drop rapidly, at least at first;

To survive a nuclear bomb it is necessary to survive a number of deadly effects:

- The air blast (shock wave) caused by the energy released locally heating up the air and causing it to expand rapidly;
- The thermal radiation produced in the explosion and the interactions with the atmosphere;

- The neutron radiation produced in the explosion;
- The gamma radiation produced in the explosion;
- Any fire storm resulting from the blast;
- The beta/gamma radiation from decay of fission products and neutron-induced radioactivity within the fallout.

The initial pulse of radiation may be sufficient to kill, or at least blind, people beyond the zone that is devastated by the blast. Survival is dependent on being far enough away from the bomb or well shielded. Those looking towards the bomb are obviously more likely to have their eyesight damaged, although this may be temporary for those "lucky" enough to be further away.

The effects of the blast travel more slowly than the flash. They can destroy buildings and kill people over an extensive range. Again survival is about distance and sheltering. Shelters would have to be robust to survive the blast. Underground would be better.

The blast can cause changes in air pressure, both sudden increases and decreases, which can be sufficient to kill. The blast and resulting fire can deplete the oxygen supply leading to death.

For the unprepared and unwarned survival of the blast is a matter of luck - not being too close to the bomb when it goes off or fortuitously being shielded. Thereafter survival of the next few minutes could be enhanced by quickly (very quickly!) finding better shelter from the thermal and radiation blasts and their effects. The advice given is "*if you can't find shelter within a few seconds then lie on the ground and cover your head if possible. Do not look at the flash or fireball if you can help it*".

The next issue is fallout - the radioactive dust and gases produced by the explosion. Survival is about putting as much distance or shielding between individuals and the fallout as possible. Evacuations should be away from the fallout. Any survivors in shelter will need advice, reassurance, and probably, medical attention. A good source of such advice is the USA Ready Government web-site (Ref. 104).

For the few days or weeks after the explosion avoidance of the fallout is necessary to prevent the exposure to a large radiation dose. Evacuation away from the dust falling from the sky followed by decontamination is the quick route out. Staying in a fallout shelter until the radiation levels have dropped low enough not to be a great hazard is the slow way out. Those exposed to high doses of radiation or injured in the blast may be unlikely to survive without prompt medical care.

FEMA report a 7-10 rule (Ref. 105): The 7:10 Rule of Thumb states that for every 7-fold increase in time after detonation, there is a 10-fold decrease in the exposure rate. In other words, when the amount of time is multiplied by 7, the exposure rate is divided by 10. For example, let's say that 2 hours after detonation the exposure rate is 4 Sv/hr. After 14 hours, the exposure rate will be 1/10 as much, or 0.4 Sv/hr.

Recently published advice to potential responders to a nuclear detonation in the USA can be found on the internet (Ref. 106), a fuller health and safety guide is also available (Ref. 107). Table 4 shows the range out to which significant effects might be seen. In this table significant effects are 50% mortality from shockwave and heat and a radiation dose of 4 Sv.

Yield (kT)	Shockwave (km)	Heat (km)	Initial Radiation (km)	Fallout Radiation (downwind km)
1	0.3	0.6	0.8	Up to 6
10	0.6	1.8	1.3	Up to 10

Table 4 Range of significant effects from atomic bomb (km) from Ref. 108.

Figure 13 shows predicted bomb damage ranges for a 10 kT, ground-burst nuclear explosion in an urban environment. The Moderate Damage zone from about 0.5 - 1.0 miles is described as "*Significant building damage and rubble, downed utility poles, overturned vehicles, fires, many serious injuries, greatest lifesaving opportunities*" (Ref. 106).

The film "War Game" (1960) is available on the internet (Ref. 109). This shows a fictional docudrama depicting a hypothetical nuclear attack on Britain and shows what the local authorities would be up against. Dated, informative, occasionally accidently amusing, and terrifying, it is well worth watching.

A quick read of the "*planning guidance for response to a nuclear detonation*" prepared by the US Homeland Security Council (Ref. 110) should be enough to convince us that the situation facing responders would be terrible. This publication and the quick guide that accompanies it (Ref. 106) would be useful additions to any responders' library.

Some brief information about the UK nuclear bomb tests can be found at reference 111.

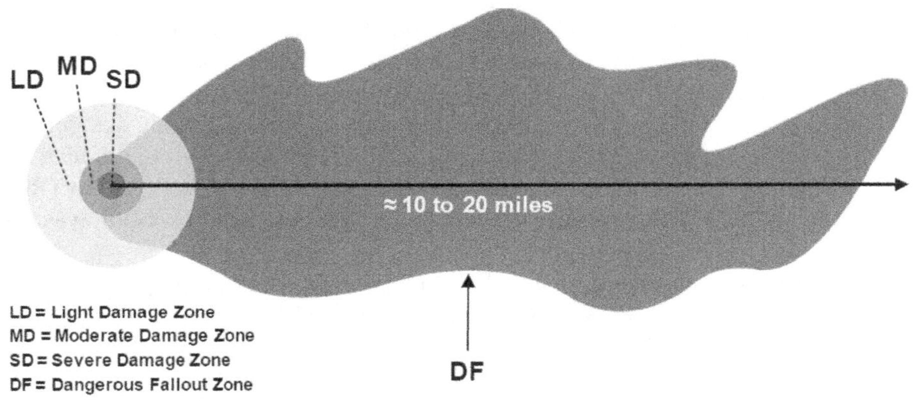

https://www.dhs.gov/sites/default/files/publications/IND%20Health%20Safety%20Planners%20Guide%20Final.pdf

Figure 13 Atomic bomb effects ranges

2.7.5. Criticality accidents

A critical accident or criticality excursion occurs when mishandled fissile materials are put into a configuration that results in a supercritical configuration leading to an uncontrolled divergent (i.e. increasing in rate) chain reaction. They have been known to occur in process plants where enriched fissile material is being processed to make fuel and in research situations where experiments have been mishandled. See reference 112 for a detailed review of historic criticality accidents.

Criticality excursions tend to be very quick events with the energy generated destroying the assembly instantly although a few have continued in a steady state or cyclical pattern for several hours (the record seems to be 37.5 hours for a plutonium processing accident at Hanford Works, USA, 7 April 1962). These accidents release a large amount of radiation in a short period of time and have often resulted in fatal doses to those nearby. They do not tend to release sufficient radioactive material to the atmosphere to warrant extensive protective actions off site.

For the affected local authority, a criticality accident response is likely to be one of information gathering and reassurance. For the local hospital there may be the need to manage patients with high radiation doses who may, or may not, have severe conventional injuries as well. The treatment of acute radiation syndrome is complex and not often required so little practiced (See Section 4.8.5).

In the meantime, the operator will be mustering personnel, aiding any casualties and identifying those who may have had a high exposure (from dosimetry, by their position at the time of the spike(s) and by the shortness of the delay of onset of any symptoms of ARS). They will be trying to ensure that the assembly is safely sub-critical to avoid further criticality spikes and will be trying to prevent any spread of contamination and toxic chemicals.

3. A concise history of nuclear emergency planning

The extent of emergency preparedness around nuclear sites is as much a political decision as it is a technical one, partly informed by past accidents and the response to them by the operator, the governments and the general public. In this section the history of nuclear emergency planning is discussed to show some of the debates and some of the developments that have lead us to the current situation.

In the USA the Atomic Energy Act of 1946 was concerned with gaining government control of the nuclear industry, imposing secrecy and supporting the weapons programme. It also established the five-member U.S. Atomic Energy Commission (AEC) to manage the programmes. The Act was replaced in 1954 with one aimed at encouraging civil uses of nuclear power whilst protecting public health and safety. It introduced site licensing to the US.

In the introduction a rule of thumb, used in the USA to estimate the exclusion zone required around a power reactor was given. As reactors got more powerful this became too limiting and exceptions were made starting in 1952 and 1953 for reactors that had containment buildings around them to significantly reduce any accidental releases of radionuclides during accidents. This move from isolation to containment continued with the development of the philosophy of accident prevention based on quality assurance and conservatively designed components and then the redundancy, diversity and segregation of safety systems leading to the concept of defence in depth; all designed to progressively reduce both the probability and consequences of accidents. A USA review (Ref. 5) suggests that it was the submarine programme that lead the way with accident reduction given that isolation was not an option.

Redundancy: In this context redundancy means there must be more than one system capable of doing the same job so that if one fails or is under maintenance there is another system available.
Diversity : In this context diversity requires that the different systems are designed in different ways or work on different principles so that if there is a design or build fault on one system it does not affect the effectiveness of the others.
Segregation: Systems must be segregated by either distance or engineered structure in such a way as to greatly reduce the likelihood that they could all be damaged by the same event.

The UK development of reactor safety and siting rules have been reviewed in the literature a number of times. See references 113 - 115. A paper produced in 2006 (Ref. 113) for the Department of Trade and Industry (a UK government department dissolved in 2007) suggests that Harwell was chosen for the UK's first nuclear site on the wonderfully non-scientific criteria that it "*should not be too far from London; there should be easy access to a University; there should be some degree of isolation and lastly the countryside should be pleasant to live in*". Sellafield was then chosen as the site of the first military piles based on American criteria of minimum distances to towns of given sizes. The "Farmer curves", the UK equivalent of this system which define remote, semi-urban and city sites is explained in reference 114. The current siting strategy seems to be the 2011 *National Policy Statement for Nuclear Power Generation (EN-6)* (Ref. 116) which recognises emergency planning as a "flag" for local consideration.

The first nuclear reactors in the UK were built at what was then the Atomic Energy Research Establishment (AERE) at Harwell in Oxfordshire. GLEEP – the 3 kW Graphite Low Energy Experimental Pile–was commissioned in 1946.

On December 12 1952 at the NRX reactor core at Chalk River in Ontario, a series of operator errors and communication failures lead to control rods being removed and a power excursion followed by a hydrogen-oxygen explosion. Part of the core was melted, fission products released to the atmosphere and contaminated water was dumped. This is possibly the earliest recorded reactor accident.

In 1954 the influential US congressional oversight Joint Committee on atomic energy was told that "*no matter how careful anyone in the atomic energy business may try to be, it is possible that accidents may occur*". In 1956 they were told that because of technical uncertainties and limited operating experience, "*the determination that the hazard is acceptably low is a matter of competent judgment if safeguards are included in facility design against all possible accidents having unacceptable consequences, then it could be argued that any site would be acceptable assuming that safeguards would not fail and that some dangerous accidents had not been overlooked.*" An excellent summary.

The initial rules of thumb were based on "maximum credible accidents" or "worst conceivable accidents" but it was obvious that you could not base a viable nuclear industry on them. In 1957 report WASH-740 report entitled "*Theoretical Possibilities and Consequences of Major Accidents in Large Nuclear Power Plants,*" postulated a severe accident leading to 3,400 early fatalities, 43,000 acute injuries and costing 7 billion dollars at 1957 values to clean up. (This was a very pessimistic assessment both in terms of the fraction of the core released and in the level of dose that would prove fatal). This paper lead to state insurance of the nuclear industry now achieved in Europe by the Paris and

Brussels conventions.

The 1957 Kyshtym disaster (29/9/1957), the explosion of an inadequately cooled tank containing liquid radioactive waste, released an estimated 800 PBq (20 MCi) of radioactivity making it the third worst nuclear accident to date behind Chernobyl and Fukushima. This produced a plume which deposited radionuclides, notably Cs-137 and Sr-90, over a wide area up to 350 km downwind, an area now called the East-Ural Radioactive Trace (EURT). There were mass evacuations and crude clean-up attempts following the accident but any potential learning for the west was delayed by decades of cover-up from the soviet authorities including the designation of the area as a wildlife reserve with very limited access.

When the Windscale accident (8/10/1957) (see section A.2) happened in 1957 (only a week after the Kyshtym disaster), there was little in place that we would understand as emergency preparedness but the site personnel improvised a response, put the fire out and put in place activities to reduce radiation doses to the public; notably the destruction of milk produced in the area contaminated by the fall out. It was a very uncomfortable experience. Tom Tuohy, the deputy general manager at the site, is quoted as saying "*Mankind had never faced a situation like this; there's no-one to give you any advice*".

The Windscale fire lead to the Nuclear Installations Act (Ref. 117), first placed in 1959 but updated since, notably in 1965 and by the Energy Act in 2014. The Nuclear Installations Inspectorate (NII), forerunners of the ONR, and nuclear site licensing were consequences of this Act. Site license condition 11 states that "*the licensee shall make and implement adequate arrangements for dealing with any accident or emergency arising on the site and their effects*".

A "deliberately cautious" siting policy was used for the Magnox reactors then in 1968 the Nuclear Safety Advisory Committee agreed that AGRs should be built closer to the population centres. This was supported by a Royal Commission on Environmental Pollution (RCEP) in 1976 which concluded that "*safety of the public was considered to derive more from high standards in the design, construction and operation of nuclear power stations than from remote siting*".

The 1979 Three Mile Island accident led to improved local and national arrangements with the establishment of Local Emergency Centres (LECs) for each licensed site and the central Nuclear Emergency Briefing Room.

The nuclear safety regulator, then HSE NII now ONR, published its first set of Safety Assessment Principles for Nuclear Power Reactors in 1979, these are updated periodically, the latest version being 2014 (Ref. 118). Also relevant to the history of UK nuclear safety assessment is *The Tolerability of Risk from Nuclear Power Stations* (1988, 1992) (Ref. 1) and *Reducing Risks, Protecting People* (2001) (known as R2P2) (Ref. 119).

The Chernobyl accident in 1986 (see Section A.5) was a reminder that severe accidents can occur at nuclear reactors although the west did try to hold the line that it was inferior Soviet design and operation that lead to an accident that "could not happen here" (see box). Post Chernobyl the UK reviewed its plans for managing major nuclear accidents abroad and, as part of that process, installed the RIMNET array of radiation detectors around the UK (see Section 11.12.12).

> "*Vastly larger numbers of people have died from accidents due to fire in the last few years than from accidents due to nuclear power generation, including Chernobyl, which would not and could not have happened here*"
> (Lord James Douglas-Hamilton)
>
> Hansard 1987 (Ref. 120)

A recommendation from the US NRC "Three Mile Island Report to the Commissioners and to the public" (Ref. 121)(1980) states that "*The NRC must adopt a policy that requires reasonable offsite emergency planning, and such planning must consider emergency response to low probability accidents having offsite consequences greater than those analysed as credible in the design review. This policy is an important part of the NRC's defence-in-depth concept, which deserves strong reemphasis*".

In 1981 the IAEA published a major report on emergency preparedness (Ref. 122). In the foreword it states that "*Nuclear facilities are designed, constructed and operated with exceptional consideration for safety. Their sophisticated safety systems are designed to protect not only the nuclear facility and its personnel, but also the public and the environment. In the event of a breakdown or failure within a facility, one or more of these systems will be activated, each of which is designed to help ensure the ultimate integrity of the facility.*

Nonetheless, a potential for a serious accident does arise, even though its probability is minimal, and an effective emergency planning and preparedness programme is a necessary part of the support for any nuclear facility. It is desirable to plan in advance for emergency action to protect both the facility staff and the public".

The reader is warned that "*the publication is neither a collection of rules nor a list of approved steps and actions. Rather it is a discussion of the various philosophies, problems and recommendations involved in the off-site response to an accident in a nuclear facility*". However, the picture painted of emergency arrangements and protective actions is recognisable to the current practitioners except it discusses improvised respiratory protection using such things as handkerchiefs, towels and shirts.

Sir Frank Layfield in his report on the Inquiry for the Sizewell B PWR (between January 1983 and March 1985) is quoted as writing "*I endorse the NII's preference for a detailed plan covering a limited area but capable of extension if necessary*" and added that "*a balance should be struck between ensuring that the plans are sufficiently extensive to cope with serious accidents, and avoiding the waste of resources which could occur if there was over-planning for the most improbable accidents*" (taken from "*Outline Emergency Planning for Licensed Nuclear Power Stations*" (Ref. 123)). The balance point referred to in this statement is ill-defined but is probably driven by society's risk appetite and concern about nuclear accidents. Focus rightly turns on to severe accident planning every time there is a major nuclear accident or scare across the world.

The worst conceivable accident criteria was replaced with the concept of a "reference accident". These were described in the UK in 1994 by HSE (Ref. 124) thus; "*Modern safety analysis techniques allow the identification of a spectrum of accidents which can reasonably be foreseen. These are termed 'design-basis accidents' and the one which gives rise to the most significant off-site consequences is taken as the reference accident.*" The concept being that if you had a set of emergency plans (on-site and off-site plans) that could put up a credible response to the reference accident and had the flexibility to respond to the actual events seen on the day of an accident then the preparedness level was appropriate. The two issues that frequently arose when trying to identify the reference accident was agreeing the probability that represented the threshold of "reasonably foreseeable" and the nagging question about whether or not all potential faults had been identified and analysed correctly. History keeps teaching us that accidents do not follow the script and occasionally there is an accident which, due to a combination of events that would be not considered likely, are more severe than the reference accident.

For early Magnox reactors the event upon which emergency plans were scoped was a no-notice catastrophic failure of a main cooling duct leading the reactor circuit to depressurise in a short period of time. It was assumed that this event would be violent enough to damage some of the cooling channels in which the fuel sits occasionally resulting in one of the channels catching fire. This was a challenging fault for emergency planners, requiring prompt actions on a badly damaged site to reduce the consequences and urgent activity off-site to inform and protect the public. This sequence was considered to be pretty unlikely (prompt depressurisation once every 10,000 years of operation and channel fire once every 100 prompt depressurisations therefore 10^{-6} per annum) but, for the sake of argument, was taken as being "reasonably foreseeable". This resulted in the development and maintenance of a competent set of emergency plans capable of rapidly deploying damage assessment and damage repair teams into a very hostile environment to work on the stricken reactor and initiating an off-site plan to implement automatic protective actions within a few kilometres of the site and to consider wider protective actions before turning attention to recovery.

Doomsday variants of this accident included the suggestion that the initial physical shock could warp the reactor core and so prevent the control rods falling in and shutting down the chain reaction. The response to this included the development of articulated control rods that could worm their way into the warped reactor and boron balls or dust with the same intention.

The emergency plan was considered to be "extendible" in that the many organisations involved in the planning and exercising had a number of people trained and ready to go in each role and had the ability to scale up their response by calling in more of their people and then calling on mutual aid arrangements with other industry players or neighbouring authorities. It was considered that this would allow the response to manage a longer term event or one that was more severe. What this "extendibility" did not necessarily deliver was a credible detailed plan to distribute stable iodine to areas beyond the DEPZ nor to evacuate areas near the site should it prove necessary.

Importantly Magnox sites had, and existing UK sites have, a comprehensive set of plans that could, in principle, be extended to manage more severe outcomes - within limits.

For AGRs, which are an evolution from the Magnox reactors, the maximum credible break in the cooling circuit is significantly smaller, reducing the speed and violence of the worst credible depressurisation. The fuel is also considerably more robust. As a result of these developments AGRs are significantly safer to operate than the Magnox fleet was. This resulted in the detailed emergency planning zones being reduced from several kilometres for Magnox reactors to one kilometre for most AGRs, but with similar protective action strategies. The on-site plans, with the capability of coping with the danger to life and damaging effects of a serious high-pressure leakage of hot (both in terms of temperature and radioactivity) carbon-dioxide, were comparable.

By the 1980s/1990s nuclear emergency arrangements in the UK were fairly mature. They were described in the 1994 HSE publication *Arrangements for responding to nuclear emergencies* (Ref. 124). This describes the need for a response to:

- Control the accident at the site;
- Assess the actual and potential accident consequences, and alert the relevant authorities and the public;
- Introduce protective actions to mitigate the consequences as regards
    - Individuals who could be exposed in the short term; and
    - Longer-term effects such as contamination of food supplies, land and adjoining waters;
- Enable the return to normal conditions.

The document describes a detailed emergency planning zone (DEPZ), defined in relation to the maximum size of any accident which can be reasonably foreseen and in which arrangements to protect the public are defined in detail. It also states the expectation that a plan could respond to larger accidents making use of local and national plans prepared to deal with a wide range of emergencies. The document goes on to describe multi-agency co-ordination headed by the police force and outlined the roles of government departments and national agencies.

In 1996 an EU Council Directive (Ref. 125) laid down basic safety standards for the protection of the health of workers and the general public against the dangers from ionising radiations. The emergency planning provisions of this were embedded in the Radiation (Emergency Preparedness and Public Information) Regulation 2001 (Ref. 126) which, together with its guidance, defined the requirements for nuclear emergency planning in the UK. REPPIR-01 placed duties on the operator to have an on-site plan and to cooperate with the off-site bodies, on the local authority to have an off-site plan and public information processes and on the regulator to set the DEPZ and ensure that plans were adequate and were tested.

The Civil Contingencies Secretariat (CCS) was established in 2001, updating the Civil Contingencies Unit, with the stronger remit to make the United Kingdom more effective in planning for, dealing with, and learning lessons from emergencies and disasters (Ref. 127) following issues with the government approach to a number of issues including fuel protests, foot and mouth and flooding. The Civil Contingencies Act (CAA) followed in 2004 defining an emergency as:

a) an event or situation which threatens serious damage to human welfare in a place in the United Kingdom,
b) an event or situation which threatens serious damage to the environment of a place in the United Kingdom, or
c) war, or terrorism, which threatens serious damage to the security of the United Kingdom.

The Cabinet Office "*Dealing with Disaster*" provided the Government's core guidance to emergency planners and local responders - the revised third edition (2003) of this can be found as Ref. 128. This advice was updated in the non-statutory guidance associated with the CCA in 2013 (Ref. 129).

The Fukushima event in 2011 (see Section A.7) lead to a review of the safety of existing reactors in the light of severe external events and focussed emergency planning on severe accidents leading to a systematic review of outline planning at all licensed sites.

REPPIR-19 transferred the setting of the DEPZ from the regulator to the local authority, placing a duty on the Operator to provide the local authority with a "consequence report" recommending a minimum range for prompt protective actions while being rather vague about how it should determine the range. It also introduced the outline planning zone, and defined its radius depending on the type of site.

When the design for the current EPR was developed the aspirations included improvements in equipment and systems used in normal operation to reduce the frequencies of transients and incidents hence limiting the situations that could evolve into an accident; a significant reduction in the probability of core melt; a significant reduction in the potential radioactive releases due to all conceivable accidents including core melt; for accidents without core melt there should be no justification for evacuation or sheltering of the public and core melts with large early releases should be "practically eliminated" (Ref. 130).

The estimation of the risks of the EPR is summarised in *PCSR – Sub-Chapter 14.6 – Radiological consequences of design basis accidents* (2012) (Ref. 131) (This report will have been updated in the meantime but this is the most recent version I can find on the internet). This report indicates the lengths the reactor designers have gone to in order to reduce the probability and consequence of any event.

What does proportionate planning mean for a site that has invested a great deal of effort into reducing the chances of an off-site release of any significance to near zero but cannot totally rule it out and where the personnel responding to events up to an including core melt can expect to walk along well lit, undamaged corridors to the valves they need to open or close in the next half hour or so? This is where politics come into play. Under the written technical rules given in REPPIR-19 a strong argument could probably be made that an off-site plan is not required.

There is some truth in the saying that we are always preparing for our last emergency. After the Windscale accident the NII was established raising the expectations that reactors would be better built, better understood and better prepared for an off-site release (the forerunners of Emergency Reference Levels were created and off-site plans required). After Chernobyl the national plan concentrated on the risk of foreign radioactivity being blown over the UK with the creation of RIMNET and increased focus on measuring off-site releases with the advent of perimeter monitoring systems. Three Mile Island gave us off-site centres and better media handling. After Buncefield lip-service was paid to extendibility planning and Piper Alpha left us with better crisis management processes. Fukushima finally brought home the potential for a severe accident in a modern and developed country. Each major accident was also followed by a rapid reorganisation of the government committee structure that co-ordinates nuclear emergency planning, giving the process a different brief. Post Fukushima, following criticism of the Japanese central response, there was a focus on the government role in a serious accident. What would be more useful than planning for the last emergency would be to plan to respond to the next emergency but we have never proved very good at predicting them.

4. Radiation, contamination and their effects on people

The purpose of this chapter is to equip the reader with sufficient knowledge of radiation and its possible effects on living tissue and human health in order to allow them to make knowledgeable decisions about activities that involve themselves or others being exposed to potential radiation doses.

This is complicated. There are several flavours of radiation that are very different to each other. They interact with matter, including human flesh, in a number of different ways depositing energy in different patterns. This initiates local chemical changes which in turn initiates biological change. This can be repaired by the body, cause local damage, cause interuption in the body's maintenance processes or cause cancer. The harm, and the time to expression, then depends on what part of the body is exposed and how interupting its normal operation affects the health of the individual.

4.1. Introduction

Understanding the effects of low levels of radiation on the human body is not at all easy. Complications include the fact that, although we talk of "ionising radiation", we are really talking about at least four different phenomena; alpha, beta, gamma and neutron radiation which are very different from each other and the density of their interaction with matter varies over several orders of magnitude. They can also each come in a wide range of energies and intensities, which are important. Further complications arise from the complexity of the human body and its response to radiation damage, which can vary from slight damage to a few cells, damage and death to cells in a particular area or organ, to wide scale damage of cells across the body. Understanding is hampered by having at least two sets of units to measure radiation and its effects (curie, rad, rem against becquerel, gray and sievert), although that is less problematic as time goes by and the becquerel, gray and sievert become increasingly dominant.

Some good books exist to help those seeking more information. "*Living with Radiation*" by the NRPB (forerunner of PHE CRCE) is excellent for the general reader and members of the public (Ref. 132) (I am not sure this is still available), "*An Introduction to Radiation Protection*", by Martin, Harbison, Beach and Cole and now in its 7th edition (Ref. 133) is the standard text for trainee professionals.

4.2. What is radioactivity and radiation?

All chemicals are composed of atoms. At the centre of the atom is the nucleus, containing almost all the mass of the atom in the form of protons and neutrons. Some materials, including most fission products, contain nuclei with the number of neutrons and protons out of balance. These nuclei are "unstable" and they spontaneously "decay" i.e. rearrange themselves into a more stable form spitting out the bits of nuclear matter or energy that they don't want. The bits emitted travel away from the nucleus in all directions – they "radiate" from the nucleus and interact with the materials around them. Materials containing unstable atoms are called **radioactive materials** or **radioactive sources** and the particles or waves of energy emitted are called **radiation**.

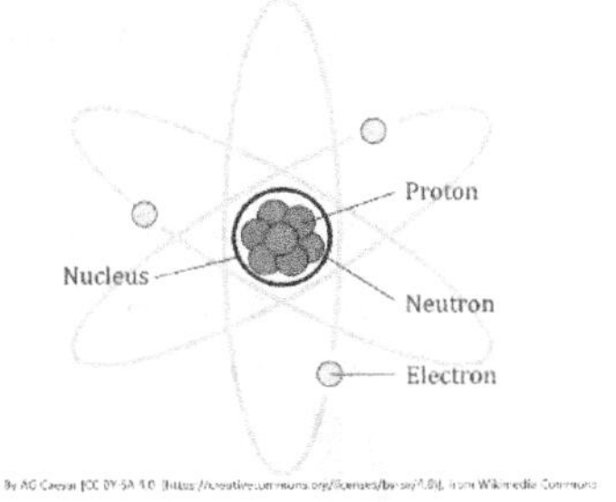

Atom: The smallest particle of a chemical element, consisting of a positively charged nucleus (containing protons and typically also neutrons) surrounded by negatively charged electrons.

Concise Oxford English Dictionary

There are four main ionising radiations associated with the nuclear industry. These are alpha particles, beta particles, gamma radiation and neutrons. These are "ionising radiations" which means that they interact with the material they pass through and can knock electrons out of atoms on the way causing chemical changes.

Each time the radiations interact with the material they are in they are themselves scattered or absorbed (each time conserving mass/energy and momentum). During scattering processes their direction changes and they lose energy by transferring it to the material. The number of interactions before the radiation runs out of energy, the average distance between those interactions, and the change in direction for each interaction all depend on the nature of the radiation and the nature of the material.

The alpha particle is a relatively heavy entity (4 atomic mass units) being composed of two neutrons and two protons. It has a charge of +2e (where e is the electron charge of 1.602×10^{-19} Coulombs). Because of its charge the alpha particle interacts strongly with the medium it is travelling in (mediated by the Coulomb force) causing dense ionisation. Because it loses a lot of energy over a short path length (in the jargon it has a high linear energy transfer (LET)) it runs out of energy quickly and has a short range in matter.

Coulomb force: There is a force between charges that causes attraction of unlike charges and repulsion of alike charges. This force works at a distance significantly greater than the atomic radius and acts to slow down moving charged particles by transferring some of the kinetic energy of the particle to the atoms they interact with.

Since alpha particles stop quickly they cannot generally penetrate a layer of clothing or the dead outer layers of skin. Therefore this radiation is only hazardous if the source enters the body via inhalation, ingestion or injection. If this does occur it is especially dangerous because the density of the energy deposition can cause a lot of local damage.

The beta particle is an electron. It is relatively light (1/1835 atomic mass units) and has a charge of -e. It does not interact as strongly as the alpha particle but when it does interact it can be scattered over a wider angle and lose a larger fraction of its energy in each interaction. It therefore does not penetrate far into materials (but further than alphas). This radiation does not normally penetrate beyond the outer layers of skin but it can cause skin burning at high levels of contamination and, like alpha contamination, is more dangerous if the source is taken into the body. Low atomic number materials (such as water and plastic) are preferred as beta shields because they produce less damaging X-ray radiation as a by-product of absorption than high atomic number (high z) materials.

The neutron, is simply that, a neutron; one of the sub-atomic particles. Because it has no charge the neutron does not experience the Coulomb force and will only interact with atomic nuclei by direct collision. As it does not interact continuously it can travel some distance without a change in direction or loss of energy. Its range in material is very much greater than that of the charged particles and it deposits energy non-uniformly. For fast neutrons, such as those produced in fission, materials with light atoms (such as water or polythene, which both contain hydrogen) are more effective shields as each collision with a light nucleus takes a larger fraction of the energy from the neutron.

Gamma radiation is an electromagnetic radiation like light, x-rays and radio waves but is higher in energy. Gamma radiation has a relatively long range in material and only dense materials such as steel or lead, or a significant path-length in other materials such as water, provide an effective shield. Gamma radiation can deliver a significant dose to internal organs without being taken into the body.

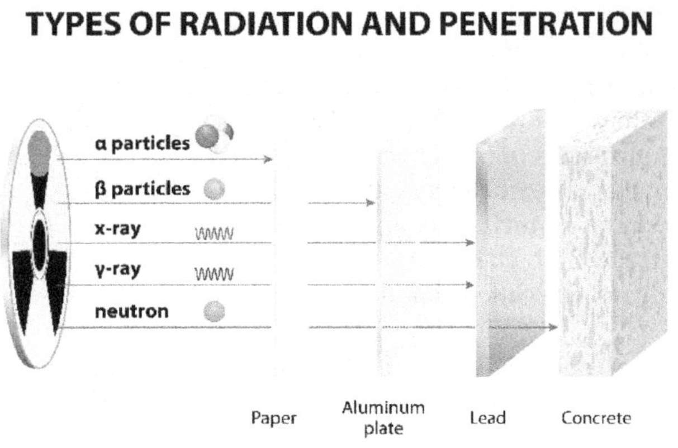

Figure 14 Range of different radiations

Figure 14 illustrates the range of the different radiations. Alpha particles, with their rapid energy loss, can be stopped by a few centimetres of air or a sheet of paper. Beta particles are stopped by thin sheets of plastic. Lead shielding is used for gamma rays although any dense material of suitable thickness will serve. Water or concrete is a good shield for neutrons but the shield needs to be quite thick.

All radiations are produced in a range of energies. This affects their behaviour in matter and the biological effect that they have on human bodies.

4.3. Half-life

Radioactivity involves the decay of unstable atomic nuclei. This process gives out radiation. As the radioactive material decays, it is used up. The time taken for half of the material to be used up is termed "the half-life" or more precisely "the radioactive half-life", often written as $T_{1/2}$. After 2 half-lives there is a quarter of the original radioactivity left, after 3 it is down to 1/8th.

Each isotope has its own characteristic half-life. Figure 15 shows the remaining amount of cobalt-60 (Co-60) and caesium- 137 (Cs-137) in the years after starting with 100% of each. Co-60 has a half-life of 5.26 years and Cs-137 30.17 years.

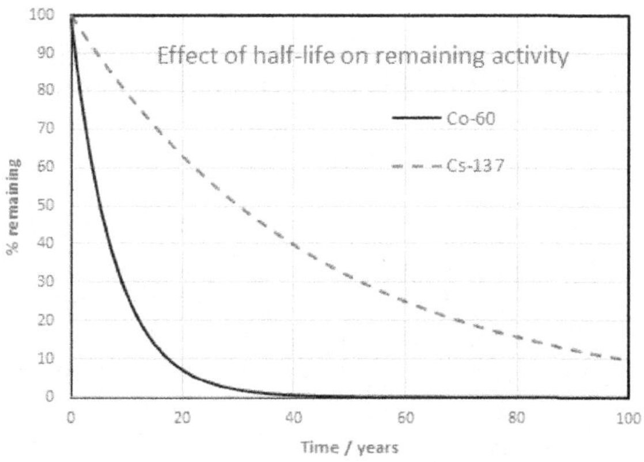

Figure 15 Effect of half-life on populations

If iodine-131 (I-131) ($T_{1/2}$ = 8.02 days) had been plotted on the same axis the line would hardly be noticeable dropping to near zero in the first few months. (As a rule of thumb it is said that a material has effectively gone after ten half-lives where there would be one part in 1024 left).

Plutonium-139 (Pu-139) ($T_{1/2}$ = 24,000 years) would appear as a horizontal line at 100% over this time scale because its activity would not drop appreciably. There would be 99.71% of the original source strength remaining at the end of 100 years.

Half-lives	0	1	2	3	4	5	6	7	8	9	10	20
Remaining activity %	100	50	25	12.5	6.25	3.125	1.563	0.781	0.391	0.195	0.098	0.00095

4.4. Radiation or contamination?

There is quite often confusion between the words "radiation" and "contamination" which does not help in the attempts to protect and reassure the public. The fission products and activation products are radioactive i.e. they give off radiation; alpha, beta, gamma or neutron. If fission products or activation products are released into the environment as dust or gas they "contaminate" it – the same use of the word as for chemical or biological substances.

> **Activation product:** Activation products are produced in and around nuclear reactors when a neutron reacts with nuclei in the material it passes through. They are radioactive materials with a range of half-lives and radiations emitted.

Radiation describes the alpha, beta, gamma and neutrons that stream out of the radioactive source or contamination.

A person can be contaminated with a radioactive material. This means that they have a radioactive material on their clothes or skin or inside their body. This radioactive material will be giving off radiation which can be measured by a radiation detector.

Summary: When a radioactive source gets on or in a person, that person is said to be "contaminated".

4.5. Radiation units

Confusion is caused by the large number of units that can be used to describe radiation. This is in part due to different parts of the world using different units to describe the same thing in the way that metres and feet can be used to measure distance, but it is mainly due to the fact that different things are being measured.

4.5.1. Activity

The activity of a source of radiation is a measure of the rate of disintegrations within it. This is measured in becquerels (Bq).

1 Bq = 1 disintegration (radioactive decay) per second.
1 MBq = 1,000,000 disintegrations per second.
1 TBq = 1×10^{12} disintegrations per second (1,000,000,000,000)

The original unit of activity was the Curie (Ci). This is defined as the activity of 1g of radium-226. The conversion factor is 3.7×10^{10} Bq = 1 Ci and 1 Bq = 2.702×10^{-11} Ci. The Curie is named after Marie and Pierre Curie who discovered radium in 1898.

The Curie is used in some historical records but rarely seen in modern texts.

4.5.2. Radiation dose, equivalent dose and effective dose

Radiation dose is a measure of the impact of radiation on the body. It is a complex concept because of the several different ways radiation can act on your body and the fact that different parts of the body have different levels of sensitivity.

To determine the level of harm caused by a radiation dose it is necessary to understand:

(1) How much energy the radiation deposits in the affected tissue (absorbed dose)
Absorbed dose (Gy) is the energy (in Joules) absorbed in an amount of matter (in kg).
Often seen as mGy = 1/1,000 Gy.

(2) How much harm that deposited energy causes to those affected tissues (equivalent dose)
Equivalent dose (Sv) is a measure of the ability to damage living tissue. It is obtained by multiplying the Absorbed dose (Gy) by a radiation weighting factor (WR) which accounts for the different effectiveness of different radiations. Often used as mSv = 1/1,000 Sv (milli-sievert) or µSv = 1/1,000,000 Sv (micro-sievert).

(3) The level of harm that is caused to the body by the harm done to those affected tissues (effective dose).
Effective dose (Sv) Is a measure of the ability to harm an individual. Obtained as the equivalent dose (Sv) multiplied by a tissue weighting factor WT representing the susceptibility and importance of the organ affected and summed over the whole body.

In the response phase to a nuclear accident the main interest of decision makers is the likely effective dose (a surrogate for harm) to individuals which will be expressed in µSv (micro-sieverts) (hopefully), mSv (milli-sieverts) (possibly), or Sv (sieverts) (we really hope not!).

The limit on effective dose for any person other than an employee or trainee is 1 mSv in any calendar year, a CT scan of the head gives 1.4 mSv, 2.7 mSv/year is the average annual background dose in the UK, 6.9 mSv is the average annual radon dose to people in Cornwall, 20 mSv/yr is the annual dose limit in the UK nuclear industry, 100 mSv is the level at which changes in blood cells can be readily observed, acute radiation effects start at about 1000 mSv and 5000 mSv would be expected to kill half of those exposed.

4.5.3. Projected, residual and averted dose

In a response to a nuclear emergency the authorities take steps to reduce the radiation dose to the public. This leads to yet more jargon which it is useful to be able to understand.

Projected Dose: The dose a particular person would be expected to receive in the course of an emergency without any protective actions being applied.

Residual Dose: The dose a particular person would be expected to receive in the course of an emergency if protective actions were applied.

Averted Dose: The difference between projected and residual dose. That is, the dose saved (averted) by the protective actions.

Projected dose = residual dose + averted dose

The protective actions we put in place are designed to keep radiation doses below the 100 mSv (usually) threshold and to avert radiation doses below that level.

Example - Consider an accidental release that would last for 10 hours and give a dose of 10 mSv/hr at Site Perimeter Cottages. Assume that we can ask the inhabitant to shelter, which would be effective after three hours delay and reduce the subsequent dose rate by 50%. Alternatively we could evacuate them, which would be effective after six hours and reduce the subsequent dose rate to zero. What would the projected, residual and averted doses be?

There are four possible pathways through this problem (Figure 16):

Figure 16 shows the accumulated dose as a function of time for the four options (labelled 1 to 4). The bars within the diagram highlight the projected dose of 100 mSv (the dose received in the absence of protective actions) and the split between the residual dose and averted dose for each scenario.

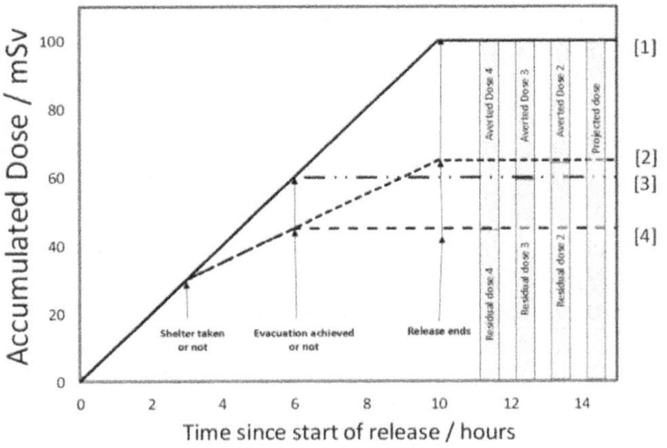

Figure 16 Projected, residual and averted dose

Scenario	Exposure implications	Dose / mSv	
		Residual	Averted
No protective actions:	10 hr at 10 mSv/hr	100	0
Shelter after 3 hr	3 hr at 10 mSv/hr + 7 hr at 5 mSv/hr = 65 mSv	65	35
No protective actions for 6 hr then evacuation	6 hr at 10 mSv/hr + 4 hr at background = 60 mSv	60	40
Shelter at 3 hr and evacuation at 6 hr:	3 hr at 10 mSv, 3 hr at 5 mSv and 4 hr at background = 45 mSv	45	55

For the combination of protective actions (shelter and then evacuation) the dose averted by sheltering is quite low and it may be tempting not to bother advising shelter during this period. However, another advantage of getting the population into shelter, other than just dose reduction, is that you can also ask them to listen to the local radio station which would provide advice. There is also always the possibility that the implementation of evacuation would be delayed by circumstance.

4.6. Internal dose

When a radioactive material is taken into the body by either inhalation, ingestion or injection its fate depends on its chemical and physical form.

Some radioisotopes are excreted quite rapidly, some stay in certain parts of the body or are evenly distributed about the body for a long time.

Models are used to estimate the dose these internal radiations give the body over time. Dose records record the dose over 50 years for adults and up to age 70 for children. Dose is assigned to year 1 which is clearly a pessimism. Various tables of dose per unit intake values can be found in the literature.

Once a nuclide is taken into the body you cannot walk away from it or clean it away. It is also closer to the sensitive tissues. Internal dose is best avoided.

Assessments of internal radiation dose can be performed in a number of ways.

1. Estimation of the amount inhaled by knowledge of the airborne concentration, time in the area and breathing rate and then the application of dosimetric models which provides estimates of dose per unit intake (Sv/Bq);
2. Estimation of the amount ingested by identifying and measuring radionuclides in food then applying dosimetric models;
3. The identification and measurement of radionuclides in faeces and urine and then the application of models of the movement and excretion of these radionuclides;
4. Monitoring - this can be whole body monitoring in a large shielded apparatus or part body monitoring with a hand held devices.
5. Nose blows and nasal swabs, and later urine and faeces, may be collected to confirm or to help disprove a potential intake.

4.7. Background radiation

Radiation is naturally occurring and unavoidable. The average annual radiation exposure in the UK is about 2.7 mSv. This is the radiation dose we get from the materials around us, radiations from space, and from medical procedures. Radiation levels in the UK are kept under review by UKHSA who issue occasional reports summarising the average radiation dose in the UK. (Ref. 134)

A short film on background ionising radiation can be found on YouTube (Ref. 135). A more detailed report can be found on the government website (Ref. 136).

The World Nuclear Association has published an information leaflet on background radiation which states that "*The highest known level of background radiation affecting a substantial population is in Kerala and Madras States in India where some 140,000 people receive doses which average over 15 millisievert per year from gamma radiation in addition to a similar dose from radon*" (Ref. 137).

The average doses from medical diagnosis exposures, the comparison of these with background radiation and an estimate of the increased changes of cancer from PHE can be found at Ref. 138. This shows Myocardial perfusion (Tl-201) at 18 mSv as the highest dose medical treatment.

4.7.1. RIFE - local dose levels

The Food Standards Agency produces an annual report summarising most of the environmental monitoring that takes place around the UK nuclear sites. (Radioactivity in Food and the Environment (RIFE)) (Ref. 139).

A recent publication in this series (Ref. 140) states that for Torness, taken as an example "*In 2015, the total dose from all pathways and sources of radiation was 0.020 mSv or 2 per cent of the dose limit, and unchanged from the previous four years. Direct radiation was the dominant contributor to the dose and the representative person was an adult. [....] The estimated dose to a terrestrial food consumer was 0.006 mSv, which was approximately 0.5 per cent of the dose limit for members of the public of 1 mSv, and unchanged from recent years. The dose to a fish and shellfish consumer was less than 0.005 mSv.*"

To gather this data "*a variety of foods, including milk, crops, fruit, and game as well as grass, soil and freshwater samples, were measured for a range of radionuclides. Air sampling at three locations was undertaken to investigate the inhalation pathway. [...] Samples of seawater, sediment and Fucus vesiculosus, as useful environmental indicators, were collected in addition to seafood. Measurements were also made of gamma dose rates over intertidal areas, supported by analyses of sediment, and beta dose rates on fishing gear.*"

The RIFE reports would be important in a post-accident situation as they provide a measurement of the radiological conditions around the site prior to any accident to use as a baseline.

4.8. Radiation harm

4.8.1. Introduction and history

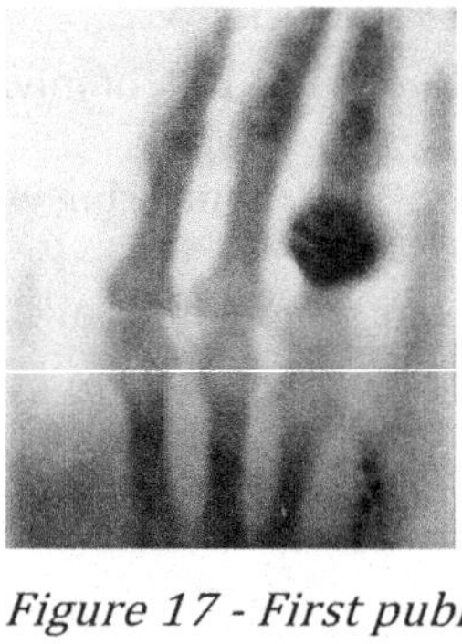

Figure 17 - First published x-ray image

There is no doubt that large doses of radiation over a short period of time are damaging to health. This has been known for a long time. What is less clear is the harm that might be done by small doses of radiation.

Ionising radiation was discovered in 1895 with the discovery of X-rays by Röntgen. Ref. 141 is a translation of his original paper. Figure 17 is an early example of an X-ray image. It shows his wife's hand. Almost immediately their value to the world of medicine was recognized and many people started to experiment with them. Glasgow Royal Infirmary had an X-ray department later that year (Ref. 142).

It was not long before the harmful side effects became apparent. A headline in the New York Morning Journal of November 29th 1896, asked "Does danger lurk in the X-rays?" and recounts the unpleasant experiences of Captain Webster after X-rays were used to locate a bullet in his body. In the same year an American inventor Elihu Thomson deliberately exposed his finger to x-rays over several days and reported "*pain, swelling, stiffness, erythema and blistering*". In 1901 William Rollins reported that two guinea pigs exposed to X-rays for several hours a day died on the 8th and 11th day of the experiment (Ref. 143). It was clear that a balance had to be found in which the benefits of radiation outweighed the health costs.

Two main types of radiation harm are known. The first type, deterministic effects, has a rapid impact while the second, stochastic effects, might take months to decades to be noticeable (see below).

The early limits on radiation exposure were designed to reduce the incidence of deterministic effects so aimed to keep the dose rates below the threshold values. More modern limits are designed to keep the risks of stochastic effects within acceptable bounds.

4.8.2. Radiation effects 1. Deterministic effects

Short term effects (*deterministic effects)* are the effects that happen quickly during and after large exposures. Deterministic effects include sickness, diarrhoea, and anaemia which can be severe, even fatal, at high doses. They worsen with increasing dose (just as sunburn does) and the time before the symptoms show decreases with increasing dose.

ICRP-103 provides a table of estimated dose thresholds for 1% incidence of morbidity and mortality for adult humans. **Error! Reference source not found.** is based on this information. It is unlikely that a member of the public would receive these levels of dose in a UK accident, but they are credible for members of staff, particularly for a criticality accident.

Effect	Organ/tissue	Time to develop	Absorbed dose (Gy) for 1% incidence
Temp Sterility	Testes	3 - 9 weeks	~0.1
Perm Sterility	Testes	3 weeks	~6
Perm Sterility	Ovaries	< 1 week	~3
Depression of blood-forming process	Bone marrow	3 - 7 days	~0.5
Main phase of skin reddening	Skin (large areas)	1 -4 weeks	< 3 - 6
Skin Burns	Skin (Large areas)	2-3 weeks	5 - 10
Temporary hair loss	Skin	2 - 3 weeks	~4
Cataract	Eye	Several years	~1.5
Mortality			
Bone Marrow syndrome			
Without medical care	Bone marrow	30 - 60 days	~1
With medical care	Bone marrow	30 - 60 days	2 - 3
Gastro-intestinal syndrome			
Without medical care	Small intestine	6 - 9 days	~6
With medical care	Small intestine	6 - 9 days	> 6
Pneumonitis	Lung	1 - 7 months	~6

Table 5 Radiation effects and delay

Note: The original table (Table A.3.4 in ICRP-103) has several important footnotes that have been omitted here.

In the UK the NRPB provided a table of emergency planning threshold doses for serious deterministic injuries (Table 6) (Ref. 144). These are for planning purposes and were set cautiously compared to the likely biological thresholds.

Type of radiation	Dose (Gy)	Integration period
Low LET	1 Gy whole body/red marrow 2-3 Gy other radiosensitive organs	Fractions of a second up to a few days
Neutron	0.5 Gy whole body	Fractions of a second up to a few days
Alpha	1 Gy lung	Dose integrated to one year from acute inhalation

Table 6 Emergency planning thresholds for serious deterministic injury (cautious)

4.8.3. Radiation effects 2. Stochastic effects

The second type of radiation effect is the longer term damage leading to effects such as cancer sometime after the exposure. Effects of this type are called *stochastic effects (sometimes called probabilistic effects).* The time delay can be as long as several decades after the exposure. In this case the probability of seeing the effect increases with dose received, but the consequences are not affected by the dose (either you see the effect or you don't). The likelihood of these effects at low doses is very difficult to assess. The assumption currently used internationally is that the chances of an ill-effect such as a cancer decreases with decreasing dose and that there is no dose below which the chances are zero. This is the *linear no-threshold hypothesis* (Section 4.9.1) and is why scientists and government advisors are reluctant to say whether a particular dose is "safe".

Someone exposed to a low level of radiation may, or may not, suffer stochastic effects later in their life. It is impossible to say at the individual level. It is also generally impossible to say whether a cancer discovered later in life was due to the exposure or was fated to happen anyway.

Many people overestimate the danger of radioactivity and most would be surprised to hear that according to the Japanese Ministry of Health, Labour, and Welfare, 80% of the atomic bomb survivors who were younger than 10 years old in 1945 were still alive in 2015. Their mean age was then more than 80 years old (Ref. 145). So, even if you are unfortunate enough to be exposed to a high dose of radiation you still have a very good chance of a long and healthy life.

Deterministic effects: There is a threshold below which the effect is not seen and above which effect gets worse with increasing dose. Thus sunburn is a deterministic effect although not caused by ionising radiation.

Stochastic effect: The probability but not the severity of effect increases with increasing dose. Assumed to be no "safe" level. These are a bit like repeatedly crossing a dangerous road. The more often you cross it the more your chance of getting hit by a car increases. However, the number of times you cross before being hit has no effect on how badly injured you are if you do finally get hit.

4.8.4. Acute Radiation Syndrome

It is considered to be very unlikely that an accident at a nuclear power station will lead to cases of ARS in the general public. It may be seen in a few workers as a result of a criticality accident or an industrial accident involving a powerful source. The symptoms and medical treatment are included here for completeness.

Acute radiation syndrome, or radiation sickness or radiation poisoning, is the name given to the collection of symptoms and effects that the human body suffers if it is subjected to a large dose of ionising radiation to the whole body, or at least the majority of it, in a short period of time. Estimates of the absorbed dose threshold for symptoms to occur range from about 0.3 - 2.0 Gy.

There are three main phases of acute radiation syndrome; prodromal, latent and manifest. The higher the dose received the faster the stages progress and the more severe the symptoms. By following the stages and severity, doctors can estimate the dose that someone received and from that they can plan treatment.

The prodromal (from the Greek for "running before") stage lasts for up to two days after exposure. Symptoms include anorexia, nausea, vomiting, diarrhoea, fever, headache and increased heart rate. This stage is mild or absent for doses below 1 Gy and more rapid about 2 Gy. Above 10 Gy onset is almost immediate.

During the latent phase the patient can seem to make a recovery as the initial symptoms subside. However, cell replacement function is inhibited and as mature cells reach the end of their life organ function is disrupted.

The manifest illness phase lasts from 21 to 60 days after exposure and is often considered as three parallel effects and their associated symptoms; bone marrow syndrome, gastrointestinal syndrome and cardiovascular syndrome.

Bone marrow syndrome, or hematopoietic (associated with blood) syndrome (dose between 0.7 and 10 Gy with mild symptoms from 0.3 Gy). The radiation kills the cells in the bone marrow. This reduces the production of the white blood cells that fight infection and the cells that help blood clot at the scene of an injury. The loss of these cells leads to infection and haemorrhage (internal bleeding). The early symptoms are anorexia, nausea and vomiting. Onset occurs 1 hour to 2 days after exposure and lasts for minutes to days. There is then a latent stage where the patient appears to be well but their bone marrow is dying and their blood being depleted of fresh cells. The bone marrow death later shows itself as anorexia, fever, and malaise and a drop in blood cell counts. The rate of loss of these cells can be used to estimate the received dose and to indicate the severity of the response. Recovery is possible at lower doses but for higher doses death follows, usually within a few months, due to infection and haemorrhage. The LD50/60 (the dose that will kill 50% of those exposed in 60 days) is about 2.5 to 5 Gy (250 to 500 rads).

Gastrointestinal (GI) syndrome (dose from 6 to 10 Gy). The radiation affects the ability of the body to generate new cells to line the gut to replace those worn out. This causes infection, dehydration, and electrolyte imbalance. Survival is unlikely. Short term symptoms are anorexia, severe nausea, vomiting, cramps, and diarrhoea. A latent stage, where the patient may appear and feel well of generally less than a week may occur but the stem cells in the bone marrow and the cells lining the gut are dying. Later symptoms are malaise, anorexia, severe diarrhoea, fever, dehydration, and electrolyte imbalance. Death, due to infection, dehydration, and electrolyte imbalance, usually occurs within 2 weeks. LD100 (the dose that will kill 100% of those exposed) is about 10 Gy (1000 rads).

Cardiovascular (CV)/Central Nervous System (CNS) syndrome (dose from 20 to 50 Gy). This is thought to be due to the collapse of the circulatory system, oedema (swelling), vasculitis (damage to blood vessels), and meningitis (acute inflammation of the protective membranes covering the brain and spinal cord). The initial symptoms are extreme nervousness and confusion; severe nausea, vomiting, and watery diarrhoea; loss of consciousness; and burning sensations of the skin. Onset occurs within minutes of exposure and lasts for minutes to hours. Later symptoms, possibly after a brief period of partial functionality, are return of watery diarrhoea, convulsions, and coma within 5 to 6 hours after exposure. Death occurs within 3 days of exposure.

In addition cutaneous radiation syndrome, that is damage to the skin and related organs, can be important. This can be caused by external radiation, by contamination of the skin or by a combination of the two and can be complicated by conventional injury to the skin, such as thermal burns received, at the same time. The signs of cutaneous radiation syndrome include erythema (redness caused by dilation and irritation of the superficial capillaries) and skin oedema (swelling caused by a build-up of fluid) which are similar to sunburn, blistering, epidermal denudation, dry desquamation (peeling of the skin), moist desquamation (skin thinning and weeping), ulceration and, especially after high dose skin exposure at high dose rate; hair and nail changes, epilation (loss of hair which may be temporary) and Onycholysis (the nail separating from the underlying skin).

The speed of onset of symptoms and their severity can be used to estimate the whole body dose that somebody was exposed to and a number of crib sheets for doctors who find themselves with irradiated patients are available on the internet. For example reference 146 from which Table 7 has been extracted. More comprehensive tables of symptoms and onset times against dose, such as Table 1 in reference 147, should probably be included in the on-site plan for any site that recognises the possibility of a criticality event and high acute exposures and in their receiving hospitals. UK advice on response to CBRN events, including treatment flow charts are given in Ref. 148.

Dose	Time to onset of Vomiting	% of incidence
Mild (1 -2 Gy)	> 2 hr	10 - 50
Moderate (2 -4 Gy)	1 - 2 hr	70 -90
Severe (4 - 6 Gy)	< 1 hr	100
Very Severe (6 -8 Gy)	< 30 minutes	100
Lethal (> 8 Gy)	< 10 minutes	100

Table 7 Time on onset of vomiting with dose

4.8.5. Recognising and treating radiation harm

LD50/60 is defined as the dose necessary to cause death in 50% of an irradiated population in 60 days. For humans this has been estimated to be approximately 3.5-4.0 Gy in persons without supportive care, 4.5-7 Gy when antibiotics and treatment support are provided, and potentially as high as 7-9 Gy in patients with rapid access to intensive care units, reverse isolation, and hematopoietic cell transplantation. There is virtually no chance of survival following a total body exposure in excess of 10-12 Gy (Ref. 147).

Usually someone subjected to a large dose of radiation will be aware of the fact but that is not always necessarily true. People could, in principle, present with radiation harm without understanding the cause of their problems.

In the modern world there are a number of circumstances that can lead to overexposure.

- Nuclear reactors and nuclear fuel cycle (enrichment, fabrication, transport, reprocessing, and storage). Accidents at these facilities could lead to individuals or groups being subject to either or both of contamination and ionising radiation exposure, with or without conventional injury;
- Criticality events can lead to external radiation overexposure and, depending on the circumstances, some contamination issues (See A.8);
- Maloperation of radiation devices such as x-ray generators or particle accelerators can lead to whole body or part body radiation dose. This would normally involve the failure of a number of layers of protection. They are far less likely to lead to contamination events;
- Industrial radiography - this uses sealed sources. Whole body or, more usually part body radiation exposure can happen if the source is handled badly. Contamination is improbable (see A.9);
- Unsealed sources. These can lead to radiation doses and to contamination;
- Nuclear medicine departments. Accidents or maloperation can lead to radiation doses and to contamination;
- Source transport can lead to external radiation exposure if the shielding or distancing provision fail and contamination hazards if the containment is damaged;
- Radioactive waste can lead to external radiation or contamination problems if handled badly.

Outside the mainstream nuclear facilities, the three main scenarios are external exposure of an individual or group from close proximity to a source (radioactive sources lost or stolen or malicious placement of a source in a public area), accidental exposure from an unshielded source (there have been many examples of this from gamma radiography tests going wrong) and exposure following the rupture of a source casing (Goiânia in Brazil being the most famous of several known incidents).

The appropriate treatment for people exposed to large doses of radiation over a short period of time depends on the estimated dose and, to a certain extent, the nature of the exposure. The dose can be estimated from the symptoms observed and the time to onset (Table 8 and Table 9).

Medical personnel should note that they are unlikely to receive a significant radiation dose themselves while treating contaminated casualties. The IAEA claim that the medical personnel in direct contact with the patients from Chernobyl and Goiânia, treating and decontaminating them received no more than a few mSv of radiation exposure (Ref. 149). A possible exception to this is if the patient has a highly active component on them or in them. These would be readily detected using handheld probes and should be removed using long handled tools and placed in a lead box, or at least in a remote location. Use of the universal precautions for the management of patients, coveralls, gloves, and light eye protection will provide adequate protection to avoid contamination of the health care staff caring for the patient.

The treatment of acute radiation dose is a specialist skill, with the treatment being suggested by the initial radiation dose and the patients changing condition. See reference 147 and METREPOL: MEdical TREatment ProtocOLs for Radiation Accident Victims (Ref. 150).

Aside: Comparison of lethal radiation dose with thermal energy from a sip of coffee.

A 5 Gy whole body dose of radiation may kill an 85 kg man. In doing so it deposits 5 J/kg x 85 kg = 425 J.

The specific heat capacity of water is 4.18 J/g/ °C.
Coffee drunk at 60°C compared to body temperature of 37 °C provides 4.18 J/g/ °C x (60-37) °C = 96 J/g.
Assuming a sip is about 20g, a sip of coffee provides about 20 g x 96 J/g = 1923 J

A sip of coffee provides more energy than a lethal dose of radiation. Whereas the radiation energy is localised and causes ionisation leading to chemical and physical changes the coffee's energy is wide spread, low grade and has no significant impact on the body.

Clinical manifestations		Estimated dose		Initial decision
Whole body exposure	Local exposure	Whole body exposure	Local exposure	
No vomiting	No erythema	< 1 Gy	< 3 Gy	Outpatient with five week surveillance (blood, skin)
Vomiting 2-3 hr after exposure	Primary erythema* 12-24 hr after exposure	1 -2 Gy	> 3-8 Gy	Monitoring in a general hospital
Vomiting 1-2 hr after exposure	Primary erythema 8-15 hr after exposure	2 - 4 Gy	> 15 Gy < 25 Gy	Hospitalisation in a haematological or surgical (burn) department) or specialized surgical department.
Vomiting earlier that 1 hr after exposure	Primary erythema within 3-6 hr (or less) associated with itching, oedema and pain	> 4Gy	> 25 Gy	Hospitalisation in a haematological or specialized surgical department.

Table 8 Initial decisions based on vomiting and erythema (IAEA)

Note: Primary erythema is a deterministic manifestation referring to a transient erythema appearing at an early stage after the exposure; also called early erythema.

Procedure	Manifestation	Time of onset*	Minimum exposure (Gy)
Clinical observations	Nausea, vomiting, Erythema, Epilation	With 48 hr With hours to day Within 2-3 weeks	≈1 ≈ 3 ≈ 3
Laboratory examinations: Blood cell counts	Absolution Lymphocyte** count < 1 x 10^9 /l	Within 2 - 72 hours	≈0.5
Cytogenetics* **	Dicentrics/rings, micronuclei, translocations	With hours	≈0.1 (Detection level)

Table 9 Methods for the early diagnosis of radiation injuries (IAEA)

Original notes:
* The latency time is inversely dependent on radiation dose
** The Lymphocyte count may decrease within hours, experts recommend that the first count is obtained as soon as is possible and repeated every four hours for the first day and daily thereafter.
*** Results can be obtainable in three to five days depending on the technique used.

4.9. The risks of radiation

This section looks at the longer term effects of radiation doses. These are the risks to long term health, mainly, cancer induction but also some impacts on fertility and the health of future generations.

Some parts of this section are quite technical. Appendix B is a more concise guide to radiation risk which many may prefer.

4.9.1. The linear no-threshold hypothesis

This section explains the linear no-threshold hypothesis and why radiation protection experts are uncomfortable to say what consitutes a safe level of radiation is. It may be a bit more detailed than most responders need.

We know that radiation in large, quick doses causes measurable harm because of the large number of people that have been exposed to such doses and the symptoms they have shown. With lower doses of radiation, certainly doses below 100 mSv, the evidence of harm is much less clear.

Over the years since the early 20th century a lot of work has been performed comparing groups of people who have been exposed to non-fatal doses of radiation to similar groups who have not been exposed and comparing the health outcomes for the two groups. This work is not without its challenges. A summary of some of the data collected is presented in Figure 18. This shows that when you include error bars to denote the uncertainty it looks as though there is a relationship between dose and probability of effect, but it is not possible to say with confidence that there is not a threshold below which radiation does you no harm. Indeed some have argued that low doses of radiation may be good for you (Radiation hormesis).

Radiation hormesis: The theory of radiation hormesis states that low doses of ionising radiation are not only harmless, but they have beneficial effects by stimulating the immune system and repair mechanisms. The theory of radiation hormesis is in conflict with the LNT (Linear No-Threshold) hypothesis.

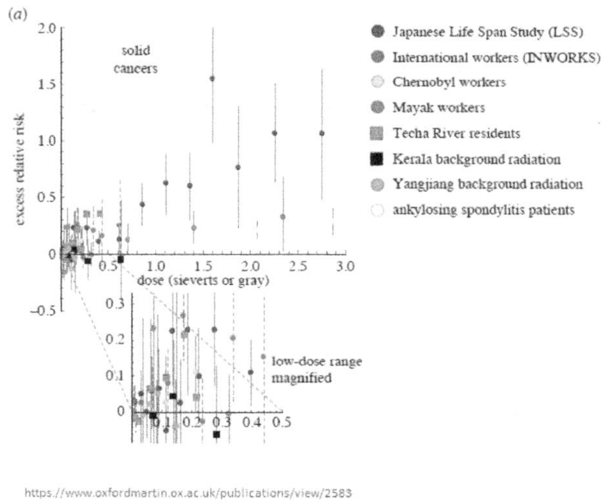

Figure 18 Data associated with low level radiation effects

The linear no-threshold hypothesis (LNT) is "*A dose-response model which is based on the assumption that, in the low dose range, radiation doses greater than zero will increase the risk of excess cancer and/or heritable disease in a simple proportionate manner*" (ICRP).

Because we accept the linear no-threshold hypothesis as a prudent assumption, we also assume that there is no "safe" dose of radiation. Each micro-sievert (μSv) is as likely to induce a cancer as the next. So the next μSv of radiation is as dangerous to you as the first whether or not you've previously had a dose of 10 μSv or 150 mSv.

The upshot of this is that health physicists (radiation protection professionals) try to keep all doses as low as reasonably practical, they do not talk about a "safe" dose and they do not panic when discussing low doses. This presents difficulties when the public ask, "is it safe?" when discussing levels of radiation exposure.

4.9.2. Risk assessment

Because of the lack of clear data of the effects of radiation at low doses the ICRP recommend the adoption of the linear no-threshold hypothesis as the cautious approach. This means that we cannot declare any small dose to be "safe" and makes setting dose limits more difficult. Ref. 151 gives a short history of radiation protection and dose limits.

The ICRP and several other bodies now suggest that, despite the LNT hypothesis, for single doses below 100 mSv for whole body exposures to external radiation no immediate effects will be noticeable and the additional rate of long term effects such as cancer induction in the exposed population would also be too small to be noticeable.

Above about 1 Gy of radiation dose deterministic effects start to be experienced and become more severe and unpleasant with increasing dose. Between 1 and 2 Gy a few people will die within a couple of months without appropriate medical care. For doses above 5 Gy the outlook, even with medical care, is increasingly grim.

This gives us a crude scale. Doses above 1 Gy are very undesirable, doses above 100 mSv should be avoided. For doses below 100 mSv effort should be made to reduce them, particularly if they are at the high end of the scale. Doses below 10 mSv compare to background levels in some parts of the world and doses of a few mSv, if unavoidable, should not be the cause of great alarm.

An important question for the emergency preparedness community is, what are the dose limits you would set for people working in the vicinity of an nuclear accident? These include, for example:

- Responders saving life at the front end;
- Responders preventing major releases;
- Police personnel on road blocks;
- Emergency services operating within affected zones;
- Social Services support to sheltered people;
- Personnel at a nearby chemical works that cannot be safely shutdown.

We then need to consider how we would monitor the doses to these groups and how we could keep those doses as low as reasonably practical (ALARP) while allowing them to complete their missions.

ICRP-103 contains a comprehensive review of the data on radiation harm and presents several tables of sex-averaged nominal risks and detriment based on different models. It concludes that based upon cancer incidence data, detriment-adjusted nominal risk coefficients for cancer are $5.5 \times 10^{-2}\ Sv^{-1}$ for the whole population and $4.1 \times 10^{-2}\ Sv^{-1}$ for adult workers (ICRP-103 Table A.4.4). This can be interpreted to mean that if 100 people each receive a radiation dose of 1 Sv in a short period of time then maybe about 5 or 6 more of them will suffer a future cancer than would have done without the extra radiation dose.

Detriment-adjusted probability coefficients for heritable disease up to the second generation are $0.2 \times 10^{-2}\ Sv^{-1}$ for the whole population and $0.1 \times 10^{-2}\ Sv^{-1}$ for adult workers.

To put that into context WHO statistics show that globally, about 1 in 6 deaths is due to cancer (2018) (Ref. 152). Cancer research UK states that cancer causes more than one in four of all deaths in the UK (2017) (Ref. 153).

IAEA provide the following "plain language" explanation of radiation harm in EPR-Communications 2012 (Ref. 154).

1000 mSv: A dose to the whole body of more than 1000 mSv could result in severe health effects. Therefore doses above 1000 mSv would require immediate medical evaluation by experts in diagnosing and treating the health effects of radiation exposure.

100 mSv: At doses above 100 mSv to a fetus, expert medical evaluation is warranted to determine the possible effects and to provide counselling to allow informed decisions. The effects to the fetus from doses above 100 mSv depend on many factors, such as the stage of development. Furthermore, these health effects can only be assessed fully by experts in diagnosing and treating the health effects of radiation exposure. Others, such as local physicians, probably will not always have the expertise needed to make such assessments. At levels of effective dose towards 100 mSv, there is a small subsequent additional cancer risk of less than 1%.

Below 100 mSv: At doses below 100 mSv there would not be any detectable cancers or other severe health effects even to the foetus. The termination of a pregnancy at foetal doses of less than 100 mSv is not justified based upon the radiation risk. An increase in the cancer rate has not been detected in any group of people who received a whole-body dose from external exposure below about 100 mSv.

Note: This information is for external radiation exposure to the whole body.

4.9.3. Risk comparison

It is commonly held that if you could assess and explain the risk of a low radiation exposure and then compare these risks with risks a person is exposed to either unwittingly, involuntarily or voluntarily then you should be able to convince them that the radiation dose is acceptable. "You are more likely to die from an accident in the home than from an additional radiation dose of x mSv so you should accept x mSv in the course of your job". It doesn't always work.

There are a number of risk comparisons available on the internet. Of these, the tables given in HSE's Reducing Risk, Protecting People (Ref. 119) are probably reasonably trustworthy. These show that the annual risk of death across the entire population of the UK is 1 in 97 per year (10,309 per million people), rising to 1 in 36 (27,777 per million people for men aged 65-74. These are based on 2001 statistics.

The annual risk of death due to cancer is given as 1 in 387 compared to 1 in 3,137 for injury and poisoning and 1 in 4,064 for accidents and all other external causes. Lightning gives you a 1 in 18,700,000 chance of dying per year.

Fatalities to employees was running at 1 in 125,000 per year or 8 per million per year in 2001. For self-employed the rate was 1 in 50,000 or 20 per million. Mining and quarrying of energy producing materials gave an annual risk of 1 in 9,200 for their workforce compared to the service industry's 1 in 333,000.

By comparison the National Radiological Protection Board's (NRPB) publication *Living with Radiation* states that the annual risk of fatal cancer for workers at the average dose of 1.5 mSv is 1 in 17,000; appreciably higher than the general level of fatal accidents in the manufacturing and service sectors and comparable to the risks in the construction industry.

At a continuous exposure of 20 mSv a year the implied risk of fatal cancer is 1 in 1,250. Unlikely to be acceptable in normal circumstances.

It should be noted that an industrial accident can kill immediately or cause life altering injury. Death through radiation-induced cancer can allow many years of healthy life before the symptoms (which admittedly can be very unpleasant) express themselves. Attempts to weigh these different aspects leads to the concept of detriment.

Radiation detriment is a concept used to quantify the overall harm to health from stochastic effects of low-level radiation exposure of different parts of the body. The tissue-specific detriment is determined from the nominal tissue-specific risk coefficient, weighted by the severity of the disease in terms of lethality, impact on quality of life and years of life lost. Total detriment is the sum of the detriments for separate tissues and organs. Detriment values are used to specify tissue weighting factors used in the calculation of effective dose.

ICRP

A very detailed statistical analysis of cancer within the population of the USA is presented by BEIR (Ref. 155). The baseline of solid cancer (i.e. not including leukaemia) rates per 100,000 population, taken from this publication, are shown in Table 10:

Gender	Incidence	Mortality	Years lost
Male	45,500/100,000	22,100/100,000	11
Female	36,900/100,000	17,500/100,000	11

Table 10 Solid cancer incidence rates per 100,000 of the population (BEIR)

For a mixed age group exposed to 0.1 Gy, additional incidence of solid cancer is estimated to be about 480 (ranging from 240 to 980 at the 95% confidence level) per 100,000. So, a single dose of 100 mSv or 0.1 Gy would not make a noticeable increase in a person's chances of experiencing a solid cancer (up from about 45,500 to about 45,980 per 100,000 for a male). Similar statistics are presented for other forms of cancer in the reference.

4.9.4. ICRP Letter to Japan

Apparently the public in Japan were sceptical when they were told that the dose limit of 1 mSv that had been in place before the accident was now changed to a dose constraint of 100 mSv because of the accident and would move to 20 mSv in due course. They naturally assumed that because 1 mSv was the "safe level" when operating then 100 mSv must be considered unsafe.

The ICRP tried to confirm and clarify the situation in an open letter which can be found at (Ref. 156).

The ICRP continues to recommend optimisation and the use of reference levels to ensure an adequate degree of protection with respect to exposure to ionising radiation in emergency and existing exposure situations.

For the protection of the public during emergencies the ICRP continues to recommend that national authorities set reference levels for the highest planned residual dose in the band of 20 to 100 millisieverts (mSv) (ICRP 2007, Table 8).

When the radiation source is under control contaminated areas may remain. Authorities will often implement all necessary protective measures to allow people to continue to live there rather than abandoning these areas. In this case the ICRP continues to recommend choosing reference levels in the band of 1 to 20 mSv per year, with the long-term goal of reducing reference levels to 1 mSv per year (ICRP 2009b, paragraphs 48-50).

The ICRP continues to recommend reference levels of 500 to 1000 mSv to avoid the occurrence of severe deterministic injuries for rescue workers involved in an emergency exposure situation. This means that it will be justified to expend significant resources, both at the planning stage and during the response, if required, in order to reduce expected exposures to below these levels (ICRP 2007, Table 8 and ICRP 2009a, paragraph e).

Furthermore, the ICRP continues to recommend no dose restrictions for lifesaving efforts by informed volunteers if the benefit to others outweighs the rescuer's risk (ICRP 2007, Table 8).

4.10. Radiation Protection today

REPPIR-19 regulations, ACOP and guidance (Ref. 3) contain a number of radiological requirements starting with the principle that protection strategies should do more good than harm in terms of public dose and environmental protection. This is a "golden thread" running from the ICRP principles of justification, optimisation and dose limitation which is repeated in Article 5 of the BSS and in IRR-17 (although justification only seems to sneak into guidance paragraph 88 "*One of the first considerations is to decide whether the risk can be avoided by choosing an alternative technique which does not involve ionising radiation*").

The international standard for radiological protection is Publication 103 (Ref. 157) from the International Commission on Radiological Protection. This provides reviews of the history of the Commission and its recommendations, the current understanding of the biological aspects of radiological protection, quantities for measuring radiation and its effects, systems of protection and a series of recommendations. (If you read one in-depth book on the subject then this would be it. However, it is not a particularly easy read).

There are three key principles of radiological protection. The principles of justification and optimisation apply in all three exposure situations whereas the principle of application of dose limits applies only for doses expected to be incurred with certainty as a result of planned exposure situations.

These principles are defined as follows: (THESE ARE IMPORTANT)

1. The Principle of Justification: Any decision that alters the radiation exposure situation should do more good than harm.
2. The Principle of Optimisation of Protection: The likelihood of incurring exposure, the number of people exposed, and the magnitude of their individual doses should all be kept as low as reasonably achievable, taking into account economic and societal factors.
3. The Principle of Application of Dose Limits: The total dose to any individual from regulated sources in planned exposure situations other than medical exposure of patients should not exceed the appropriate limits specified by the Commission.

KEY DOCUMENT ICRP Publication 103 (Approximately 330 pages)

This publication presents the 2007 Recommendations of the International Commission on Radiological Protection, replacing the 1990 system of radiological protection.

It presents an update of the radiation and tissue weighting factors in equivalent and effective dose, introduces a situation-based approach to the fundamental principle of justification, optimisation and limitation giving planned, emergency and existing exposure situations. It also reviews the biological and epidemiological basis for the system, including support for the linear dose-response model (linear no-threshold hypothesis).

All senior health physicists should be familiar with the key information this document but not necessarily the very technical annexes.

ICRP Publication 103, The 2007 Recommendations of the International Commission on Radiological Protection

The ICRP recognise that a single set of dose limits cannot be used in all situations and have therefore defined three "exposure situations". The three situations are:

1. Planned exposure situations, which are situations involving the planned introduction and operation of sources. (This type of exposure situation includes situations that were previously categorised as practices.)
2. Emergency exposure situations, which are unexpected situations such as those that may occur during the operation of a planned situation, or from a malicious act, requiring urgent attention.
3. Existing exposure situations, which are exposure situations that already exist when a decision on control has to be taken, such as those caused by natural background radiation.

The dose limits and reference levels recommended in ICRP 103 are discussed in Section 5.

The EU sets out standards for radiation protection in the Member States in the revised Basic Safety Standard (EU BSS) which was adopted in 2013 by the European Council (Ref. 158). This document mandates dose limits that are largely in agreement with ICRP 103 although the occupational eye dose is lower.

The UK left the European Union (EU) and the European Atomic Energy Community on 31st January 2020. However, the requirements of the EU BSS are written into UK regulation, notably in the Ionising Radiation Regulations and Radiation (Emergency Preparedness and Public Information) Regulations and will not change anytime soon. Therefore, the EU BSS remains useful as reading it helps understand the intent of the regulation and guidance.

The EU proposes that "*With regard to the management of emergency exposure situations, the current approach based on intervention levels should be replaced by a more comprehensive system comprising an assessment of potential emergency exposure situations, an overall emergency management system, emergency response plans, and pre-planned strategies for the management of each postulated event*".

Emergency management system: "a legal or administrative framework establishing responsibilities for emergency preparedness and response, and arrangements for decision making in the event of an emergency exposure situation;" EU BSS.

The EU BSS engages with the ICRP-103 exposure situations requiring "dose constraints" to be set for the purpose of prospective optimisation of protection (planned exposure situations) and "reference levels" to be set for emergency and existing exposure situations.

Operational radiation protection is embedded in UK law via the Ionising Radiations Regulations 2017 (IRR-17) (Ref. 159). These present a system of notification, registration and consent for work with ionising radiations and require risks assessments, dose limitation, contingency plans, training and procedures to minimise and optimise radiological exposure.

Radiological protection in the event of a radiation emergency is further developed in REPPIR-19 which transposes the emergency planning aspects of the EU BSS.

KEY DOCUMENT REPPIR-19 Regulations, ACOP and Guidance
This document contains the regulation, approved code of practice and guidance for the REPPIR-19 regulations. The emergency planning managers of all local authorities with a DEPZ or OPZ, or a section of a DEPZ or OPZ, in their area must be familiar with this document as should the local emergency services and all operators.
A Guide to the Radiation (Emergency Planning and Public Information) Regulations 2019,

5. Dose limits, reference levels and emergency reference levels

5.1. Dose Limits

The principles of radiological protection tell us that because exposure to radiation can be harmful to health we should not allow exposure without good reason (justification), where we do allow exposure we should work hard to reduce exposure to the lowest levels that are reasonably practical (optimisation) and that we should not allow exposures over a given dose limit (limitation).

Each of these principles has been the subject of debate over the years building up a canon of expectations and good practice guidance.

With regard to dose limitation the government is responsible for the legal and regulatory framework including the setting of dose limits, the employer or licensee have prime responsibility for ensuring safe systems of work that ensure that doses are as low as is reasonably practical and below the set dose limits and workers are responsible for following all safety procedures and working in a manner that keeps themselves and their colleagues safe.

Table 8 of ICRP-103 (Ref. 157) presents the current recommendations for dose limits from ICRP. They include dose limits for planned occupational exposures of 20 mSv/year effective dose averaged over defined periods of 5 years with limits on equivalent dose to certain organs for part body exposures including a 1 mSv limit to the embryo/fetus. Lower dose rates, including a limit of 1 mSv/year effective dose in a year, for the public. This gives the dose limit for those who routinely work in radiation areas. These limits do not apply in emergency exposure situations or existing exposure situations.

For members of the public the dose limit for planned exposure is 1 mSv per year. This defines, for example, the limit for members of the public living near a nuclear power station in normal operation and requires that routine discharges be carefully controlled so that this limit is not breached. It is well below the doses at which effects are apparent (Figure 19).

There are separate limits for people who receive radiation dose as volunteers for biomedical research and for those who receive dose as comforters and carers for people undergoing medical exposure. (This latter limit is 5 mSv per episode. Which may be a useful guide to the appropriate dose limit for local authority workers and others working in areas subject to protective actions).

For emergency exposure situations, occupational exposure is unlimited for informed volunteers saving life and 500 mSv per event for urgent rescue operations.

In the UK the dose limits are set in Schedule 4 of the Ionising Radiations Regulations (2017) (Section 9.5). These regulations make no mention of the three exposure situations (planned, emergency and existing).

The IRR-17 limit on effective dose for any employee of 18 years of age or above is 20 mSv in any calendar year with other limits on part body doses (including "*the abdomen of a women of reproductive capacity who is at work")* and for trainees under the age of 18. IRR-17 regulation 12(2) provides an opportunity for an employer who can show that these dose limits are "*impracticable having regard to the nature of the work undertaken by that employee*" to work to higher dose limits: up to 100 mSv over 5 years or 50 mSv in one year for an employee over 18 years of age.

The limit on effective dose for any person other than an employee or trainee, including any person below the age of 16 (this includes members of the public), shall be 1 mSv in any calendar year.

It should be noted that regulation 19 of REPPIR-19 disapplies the IRR-17 dose limits for radiation workers responding to, or preventing, a radiation emergency. It does not dis-apply the dose limits for the general public which is troubling.

5.2. Reference Level

Reference levels of dose were introduced by ICRP as lines in the sand denoting levels of individual dose which would be considered to be inappropriately high as estimated doses after all attempts to reduce them under different postulated accident circumstances.

ICRP 103 (Ref. 157) has "*In emergency or existing controllable exposure situations, [reference level] represents the level of dose or risk, above which it is judged to be inappropriate to plan to allow exposures to occur, and below which optimisation of protection should be implemented. The chosen value for a reference level will depend upon the prevailing circumstances of the exposure under consideration. Emphasis on optimisation using reference levels in*

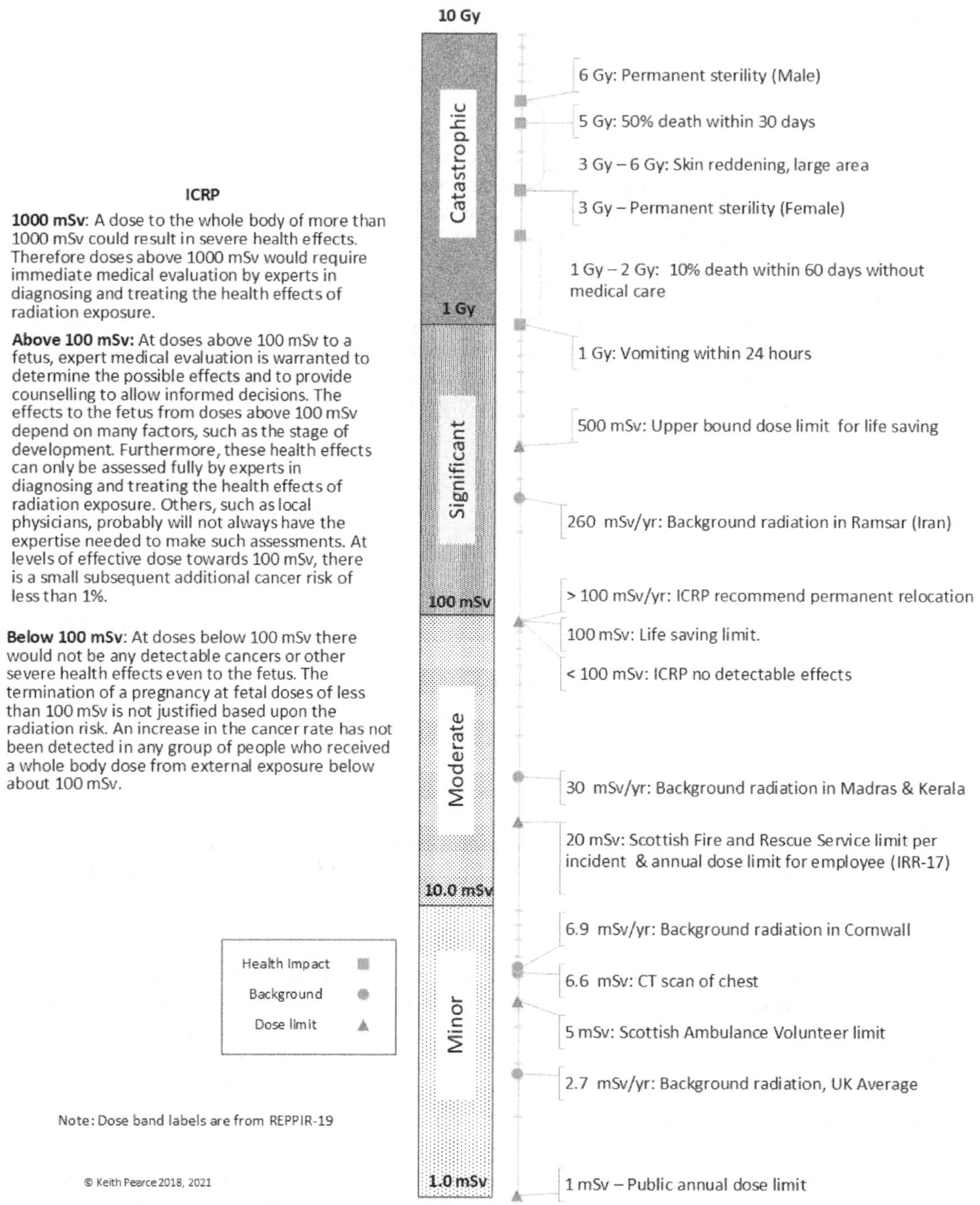

Figure 19 Radiation dose scale

emergency and existing exposure situations focuses attention on the residual level of dose remaining after implementation of protection strategies. This residual dose should be below the reference level, which represents the total residual dose as a result of an emergency, or in an existing situation, that the regulator would plan not to exceed".

The EU BSS (Ref. 158) has *""reference level" means in an emergency exposure situation or in an existing exposure situation, the level of effective dose or equivalent dose or activity concentration above which it is judged inappropriate to allow exposures to occur as a result of that exposure situation, even though it is not a limit that may not be exceeded"*.

The EU BSS requires that "*Member States shall ensure that reference levels are established for emergency and existing exposure situations. Optimisation of protection shall give priority to exposures above the reference level and shall continue to be implemented below the reference level*".

The REPPIR-19 regulations (note: regulation not guidance) defines a reference level as "*the level of effective dose or equivalent dose above which optimisation of radiation protection for members of the public must be prioritised*". This is surprisingly poor compared to the previous two definitions.

REPPIR-19 guidance paragraph 534 states that: reference levels are:
a) an emergency planning tool aimed at achieving an optimised response over all relevant exposure pathways and protective actions;
b) an indicator of the level of exposure considered as tolerable, given the prevailing circumstances; and
c) values to inform decisions on protective action and support the practical implementation of the optimisation principle when the response to an emergency is underway.

The reference levels for emergency exposure situations advised by the ICRP are shown in Table 11.

Aspect	Occupational emergency exposure intervention levels (ICRP)
Life-saving (informed volunteers)	No dose restrictions if benefit to others outweighs rescuer's risk.
Other urgent rescue operations	1000 or 500 mSv
Other rescue operations	< 100 mSv

Table 11 ICRP Emergency Reference Levels (Occupational)

For members of the public the ICRP recommend reference levels "*typically between 20 and 100 mSv according to the situation*".

The EU BSS (Ref. 158) requires reference levels expressed in effective doses to be set in the range of 1 to 20 mSv per year for existing exposure situations and 20 to 100 mSv (acute or annual) for emergency exposure situations although it allows lower levels to be set "*where appropriate protection can be provided without causing a disproportionate detriment from the corresponding protective actions or an excessive cost*".

REPPIR-19 regulation20 (reference levels) requires that the operator's and local authority's plans should prioritise keeping doses during a radiation emergency below 100 mSv, effectively setting a 100 mSv upper bound on the site's reference level. It also allows the local authority to set reference levels of dose as targets (note, not limits) for the maximum dose received where a response to a radiation emergency is underway. The Secretary of State may also set a reference level where applicable locally or nationally in addition to any reference level set by a local authority.

What this means in practice is that:

- the Operator's Consequence Report should not be predicting residual doses to members of the public (the radiation dose they receive over the course of an accident plus two days assuming that the proposed protective action strategy was implemented) above 100 mSv, or other lower reference level if set, without significant effort to introduce additional ways to reduce their dose; and
- after the accident the annual dose to members of the public resulting from contamination (the sum of resuspension dose, ground dose and ingestion dose) should be as low as reasonably practical and every effort should be made to keep it below the post-accident reference level expected to be in the range 20 mSv to 100 mSv.

PHE advice (Ref. 161) states that "*The reference level can be taken as an indicator of the level of exposure considered as tolerable, given the prevailing circumstances. Reference levels are different from dose limits, which are also restrictions on individual doses but only used for planned exposure situations, that is, situations for which authorities permit an operator to introduce a radioactive source. A dose limit is a value not to be exceeded. In contrast, reference levels are values to inform decisions on: urgent protective actions and longer-term protective actions. Reference levels are applicable to areas affected by contamination following the radiation emergency (that is, the DEPZ and outline planning zone)*".

It is important to note that an organisation or area can have different reference levels for different areas, different groups of people or even at different times. For example a site may have different reference levels for those involved in the response and those sheltering on-site until they can be moved to safety or the event ends. For those off-site it may be practical to plan for a lower reference level later in the response where the dose rates are dropping, the situation is better understood and the urgent actions of casualty rescue and evacuation of nearby areas are expected to be complete.

Group	Description	Reference level[1]
On-Site	Trained responders for lifesaving activities	500 mSv (EE)[2]
	Trained responders for urgent and vital activities	100 mSv (EE)[2]
	Responders (those working near the event)	20 mSv
	Non responders and control centre responders	5 mSv
Off-site	Workers – urgent actions near the site fence in the early stages of the event	20 mSv
	Workers – later in the response, working further from site fence, more controlled situation	1 mSv

Table 12 Possible reference levels for site emergency plan

Notes:

1. The health physicists advising on dose control for entries into the affected areas and work in and around the site would strive to keep all potential exposures ALARP and, in particular, below these values (there may be a requirement to inform the Emergency Controller if a planned activity is likely to take personnel over the relevant reference level).
2. EE - using the Emergency Exposure processes.

In an off-site plan reference levels may be set for emergency services and social services personnel working within areas subject to protective action advice. These are likely to be of the order of 5 - 20 mSv with the expectation that the doses will be controlled on the day to lower levels where practical (ALARP).

5.3. Emergency Reference Levels (ERLs)

It has to be recognised that protective actions have costs associated with them. These include the costs of maintaining and exercising the capability to implement them at short notice, the costs of the disruption to those subject to protective action advice and the costs of any worry that the implementation of the protective action may cause. These costs have to be compared to the potential benefits of the protective action, which is generally a reduced radiation exposure.

There are two ways to consider the role of protective actions (1) as a way to ensure that the radiation doses received by individuals do not exceed some pre-agreed level and (2) as an activity that must repay its costs whatever the dose to the individual might be. In the former you are not sensitive to the costs of the protective action (within reason). In the latter you are.

The IAEA provide generic criteria, based on projected dose, for protective actions to reduce the risk of stochastic effects in emergency exposure situations in Table 3 of GSG-2 (Ref. 160). These include:

- Iodine thyroid blocking for $H_{Thyroid} > 50$ mSv in the first seven days,
- Sheltering, evacuation, decontamination, food controls for E or $H_{Fetus} > 100$ mSv in the first seven days,
- Temporary relocation, decontamination, replacement of contaminated foods for $E > 100$ mSv per annum,
- Screening based on equivalent doses to specific radiosensitive organs (as a basis for medical follow-up); counselling > 100 mSv in a month.

Relocation is the displacement of a population from a contaminated area for a period of several weeks or longer. Relocation involves moving people and their belongings from their homes to evacuee centres, or to live in alternate accommodations. This could occur if the incident that caused the initial evacuation led to a moderate deposition of long-lived radionuclides causing the area to be deemed too hazardous for residents to return. It is not considered to be a prompt protective action.

Public Health England have provided advice about the protective action decisions (Ref. 161). They suggest that such decisions be based on a comparison of the dose that might be averted by applying the protective action with the societal cost of the protective action.

Their advice is that protective actions should be applied if the averted dose is greater than the upper Emergency Reference Level (ERL) and not applied if the averted dose is below the Lower ERL. For averted doses between these two values a judgement is needed on the difficulties that imposing the protective action will entail (Table 13). It is recommended that these levels are applied to the dose integrated out to seven days after the start of exposure.

Protective action	Body or organ dose	ERL Averted dose (mSv)	
		Lower	Upper
Sheltering	Whole body	3	30
Evacuation	Whole body	30	300
Stable iodine	Thyroid	30	100

Table 13 - Emergency Reference Levels

In reference 161 PHE stress the three principles of radiological protection for response to radiation emergencies:

- all protection strategies should aim to do more good than harm (justification);
- protection strategies should aim to avoid the occurrence of deterministic effects (avoid deterministic effects);
- protection strategies for exposures below the thresholds for deterministic effects should aim to maximise the benefit achieved (optimisation).

The emergency planning thresholds of dose recommended to protect people from receiving exposures that would lead to deterministic effects are:

- acute doses of 1 Gy whole body (to avoid direct injury to the bone marrow);
- 2-3 Gy to the other most radiosensitive organs for low Linear Energy Transfer (LET) radiation,
- 0.5 Gy whole body for acute exposure to neutrons.
- 1 Gy lung dose, integrated to 1 year, from an acute inhalation of alpha-emitting radionuclides.

These are admitted to be conservative in nature.

For comparison the US, French and Canadian response levels are summarised in Table 14. Note: When comparing these number notice that ERLs are in averted dose US, French and Canadians use projected dose unless stated otherwise. A comprehensive review of the German system for the protection of the public can be found in Ref. 162.

Protective action recommendation	US PAG, or Guideline	French	Canadian
Shelter-in-place or evacuation	PAG: 10 - 50 mSv projected dose over four days.	If public-exposure predictions exceed a whole-body effective dose of 10 mSv	10 mSv in 2 days (averted dose)
Administration of stable iodine	PAG: 50 mSv projected child thyroid dose from exposure to radioactive iodine.	If thyroid-exposure predictions exceed an equivalent dose to the thyroid of 50 mSv	50 mSv thyroid exposure in first 7 days
Evacuation		if public-exposure predictions exceed a whole-body effective dose of 50 mSv	100 mSv in first 7 days (E or H_{fetus})
Temporary relocation			100 mSv in year or for the full period of in utero development (H_{fetus})
Limit emergency worker exposure (total dose incurred over entire response)	Guideline: 50 mSv/year (or greater under exceptional circumstances)		50 mSv over the period of the response for off-site emergency workers

Table 14 Protective Actions for the Early Phase of an Accident - USA, France and Canada

5.4. Emergency exposures

It is recognised that in an emergency it may be necessary for a few people to be exposure to doses in excess of the annual dose limits given in the ionising radiations regulations (Ref. 159) and that exceeding these limits a few times in a working life is likely to cause no great harm. To this end REPPIR-19 provides for emergency exposures. These are defined in regulation 2(1) as "*an exposure of an employee engaged in an activity of or associated with the response to a radiation emergency or potential radiation emergency in order to bring help to endangered persons, prevent exposure of other persons or save a valuable installation or goods, whereby one of the individual dose limits referred to in paragraphs 1 and 2 of Part 1 of Schedule 3 to the 2017 Regulations could be exceeded*".

Such exposures are permitted to exceed statutory dose limits but only for pre-identified authorised employees who have received appropriate information and training and are appropriately equipped.

The regulation controlling emergency exposures (regulation 18) makes a number of conditions on the use of emergency exposures. These include identifying and training those who may be exposed and those authorised to authorise emergency exposures of their own personnel, stating the expected upper bound of dose and mandating the provision of suitable PPE/RPE and dosimetry. Most responding organisations set their emergency exposures limits within the ranges 100 mSv - 500 mSv for life saving and 20 mSv - 100 mSv for other operations.

Role	Possible emergency exposure limit
Emergency responder on-site preventing releases	20 mSv - 100 mSv
Saving of life	100 mSv - 500 mSv

Table 15 Possible emergency exposure limits

6. Protective actions

In this section the purpose of the standard protective actions (countermeasures) of shelter, stable iodine, evacuation and food controls are explained.

6.1. Introduction - Pathways

Protective actions, in this context, are the steps we take to reduce the dose to the public from a release of radioactivity to the atmosphere. The intent is to reduce the exposure to radiation which requires affecting the exposure pathways.

IAEA requires that protective actions are "justified". This is defined as " *The process of determining for an emergency exposure situation or an existing exposure situation whether a proposed protective action or remedial action is likely, overall, to be beneficial; that is, whether the expected benefits to individuals and to society (including the reduction in radiation detriment) from introducing or continuing the protective action or remedial action outweigh the cost of such action and any harm or damage caused by the action.*"

In the event of an atmospheric release people can be exposed to radiation through a number of "pathways" (Figure 20). These include exposure to the radiations emanating from the cloud (cloud dose), exposure to radiations from deposited activity (ground dose) and direct inhalation of the plume.

After the cloud has passed ground dose remains and the deposited activity can migrate into food chains leading to ingestion dose. Inhalation dose following resuspension of deposited activity is also possible.

In the short term inhalation dose tends to dominate (thyroid dose from inhalation of iodine and whole body dose from a range of nuclides) and it is this dose route that the early protective actions concentrates on reducing.

The three most cost-effective protective actions for the period when the plume is airborne are:

- Shelter;
- Evacuation;
- Stable iodine (for operating reactor accidents only).

When the airborne plume has passed but activity remains deposited onto surfaces, food controls are used to reduce ingestion doses and a combination of decontamination and changing habits can be used to reduce ground shine doses and resuspension doses.

The Government view on the application of protective actions in given in (Ref. 163). The formal PHE advice is given in (Ref. 161).

The application of protective actions is subject to the general principles of radiological protection: justification, optimisation and dose limitation.

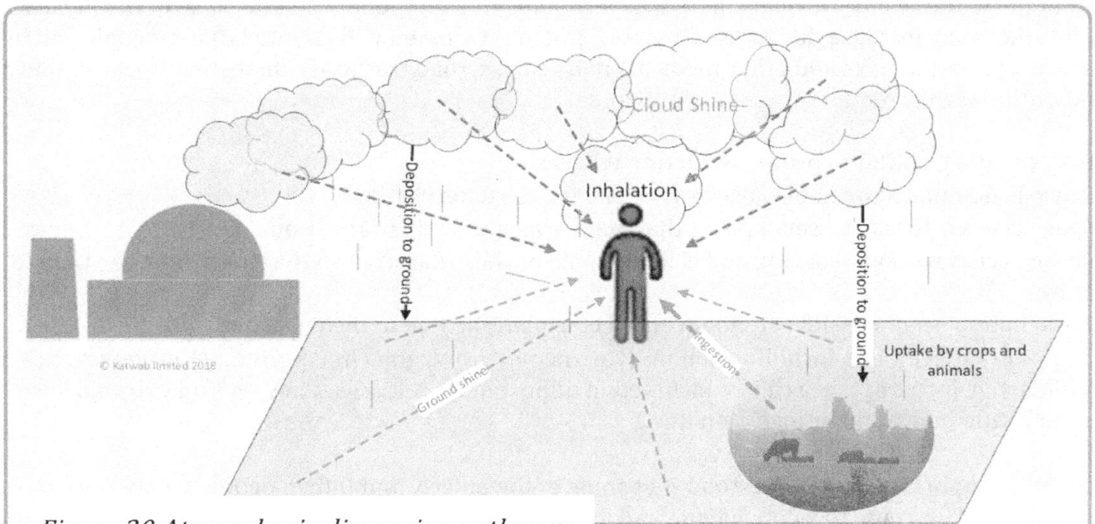

Figure 20 Atmospheric dispersion pathways

6.2. Reducing inhalation doses

To reduce inhalation dose you must reduce the amount of radioactivity retained in the lungs of people exposed and the easiest way to do this is to prevent inhalation. The most effective protective action in terms of fraction of dose averted is to evacuate the person from the area before the radioactivity arrives. Evacuation may still be appropriate after the plume has arrived if the release is expected to last for long enough after the evacuation would be complete for a suitable dose to be averted. Care is needed when implementing an evacuation when a plume is overhead to ensure that people don't get a higher dose than they would have had staying in shelter for the duration of the release.

Sheltering is a commonly used protective action which can reduce inhalation dose in the short term. According to the NRPB "*In this context sheltering refers to staying indoors, with doors and windows closed and ventilation systems turned off. It provides protection from external irradiation from radioactive material in the air and that deposited on the ground, and from inhalation of radioactive material*". This is easier in a built up area than out on moors or hills.

A useful public information film on shelter-in-place "Protective Actions for Radiation Emergencies - Get Inside, Stay Inside, Stay Tuned" from US Centers for Disease Control and Prevention (CDC) can be found at Ref. 164.

It is claimed that a solidly built and reasonably airtight UK house will reduce inhalation of particles by a factor of three. However, it is thought to be ineffective for vapours and the protection only lasts for a few hours. Air exchange between the house and the outside, driven by temperature differences and wind, allows the internal air to become contaminated in time.

PHE (Ref. 161) state that "*the health and wellbeing of sheltered populations may be affected by restricted access to medical care or assistance. In such situations, consideration should be given to supervised entry into the sheltered area by medical professionals and carers, or planned evacuation of these vulnerable groups. Residents of hospitals and nursing homes can face additional challenges, for example, sheltering-in-place without electrical power can be fatal for those dependent on modern technology such as ventilators. Therefore, if electrical power is lost, evacuation of the vulnerable groups is likely to be essential in such situations. Similarly, specific support and advice should be provided to farmers needing to tend livestock and those managing key infrastructure. Sheltering-in-place is not a long-term option and whilst it is potentially straightforward to implement, its use should ideally only be planned to last hours, and, at most, 1 or 2 days. In order to minimise the anxiety and stress created by advice to shelter, it is important to ensure communication with those sheltering is continuously maintained through appropriate communication channels. In particular, those sheltering-in-place for prolonged periods require reassurance that the advice has not changed and that those from whom they may be separated (for example, children at school, partners at work) are being properly cared for."*

As the time in shelter increases, so to might the number and complexity of issues arising. People running out of food and medicine, people requiring their usual home-help, the desire of families to be reunited, particularly at the end of the school day and the working day. Normal emergencies will not stop happening just because of the nuclear emergency (in fact they may increase due to the stresses); pregnant women will go into labour, people will have breathing problems or heart attacks and other medical emergencies, road traffic accidents may occur within and around the zones, utilities may require urgent maintenance.

Sheltering is likely to be the optimum protective action where:

- The release is dominated by noble gases - by reducing the external dose;
- The release is likely to lead to small doses that make evacuation disproportionate;
- The release occurs without warning and is likely to be of short duration so that there isn't the time to effect evacuation;
- Where evacuation is not possible or would entail considerable risk to the evacuees.

Noble gases: Very unreactive gases including helium (He), neon (Ne), argon (Ar), krypton (Kr), xenon (Xe), and radon (Rn). Radioactive forms of these can be included in atmospheric releases. They give an external dose but generally give very little inhalation or ingestion dose.

The last bullet point is important. It is understood that some of the elderly and infirm people who were evacuated from the Fukushima event died en-route to safety having had an uncomfortable and traumatic last few hours. The levels of radiation they would have been subjected to had they stayed put would have been trivial so shelter would have been a more humane and effective response. But that does give us an issue about the perception of the decision to leave infirm people in an area evacuated by the able bodied and also a dilemma about providing for their needs during the event - do you ask able-bodied people to stay behind to provide care when others are leaving? The ICRP suggest a 5 mSv per event limit for comforters and carers but that is in the context of controlled medical exposures in medical facilities.

A useful discussion of some of the factors affecting the success of shelter in the context of a toxic gas cloud can be found in Wanyu et al (Ref. 165). Whitfield et al (Ref. 166) show that greater protection can be obtained by constructing an "expedient shelter" with polythene sheets and duct tape.

The US government recommend that people be ready to improve the airtightness of their home in case they are called upon to shelter in place from an airborne hazard. Part of their advice to the public with regard to shelter in place (found in Ref. 167) is:

- *"Go into an interior room with few windows if possible.*
- *Seal all windows, doors and air vents with thick plastic sheeting and duct tape. Consider measuring and cutting the sheeting in advance to save time.*
- *Cut the plastic sheeting several inches wider than the openings and label each sheet.*
- *Duct tape plastic at corners first and then tape down all edges.*
- *Be prepared to improvise and use what you have on hand to seal gaps so that you create a barrier between yourself and any contamination."*

This idea does not seem to be taken seriously in the UK.

It is important to note that the protection offered by a building against inhalation dose depends on several factors including the airtightness of the building, the weather conditions, the duration of the release and the duration of the exposure. Ref. 166 suggests protection factors between 5.1 and 8.7 (mean 6.3) for a two-hour exposure to a two-hour release and between 2.9 and 4.7 (mean 3.5) if the exposed person stays in the building for the hour after the plume has passed. This shows that any number used in planning is somewhat arbitrary and that it would be important to ask people to ventilate their houses soon after the plume passes, which would not be easy to get right (Figure 21).

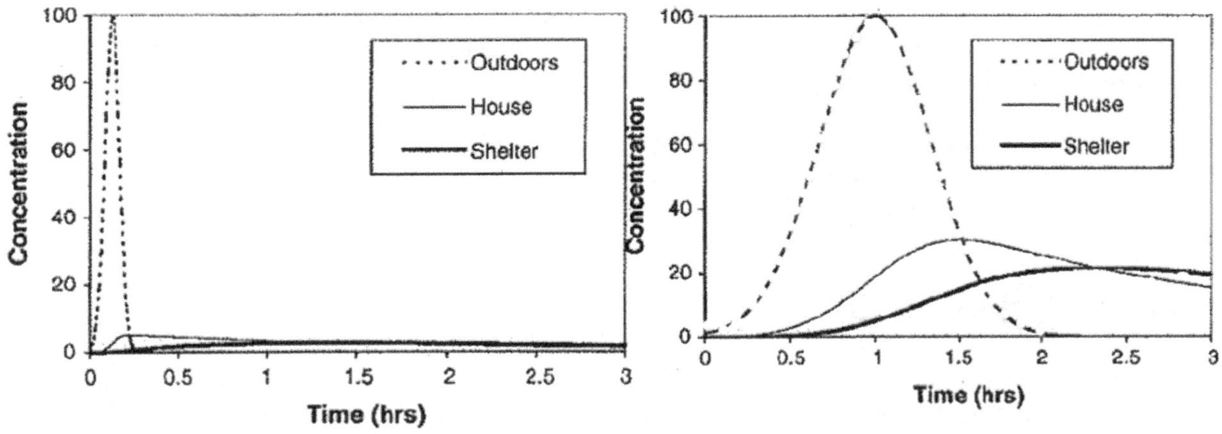

Simulation for 0.25 h outdoor exposure.

Simulation for 2 h outdoor exposure.

Figure 21 Impact of house and improvised shelter on inhalation dose

Inhalation doses can also be reduced by the use of particulate filters in the form of respirators or face masks. UK advice has generally been that the types of face mask that are generally available to the public have limited success, particularly if they are not well fitted but this may be revisited after the Covid-19 episode has made face masks much more acceptable, and reassuring, to the public.

In extreme cases inhalation dose can be reduced after the exposure by a process called lung, or bronchopulmonary, lavage. This is a treatment to reduce the lung burden of inhaled insoluble radionuclides using sequential flushing of the lungs with saline solution. It is a process used in the treatment of various lung conditions but is not a process to be taken lightly (Ref. 168).

6.3. Reducing external radiation dose

External radiation dose is the sum of cloud doses (those external doses received from being in or near a plume) and ground doses (those external doses received from being near radioactively contaminated surfaces).

Cloud doses tend to be much smaller than inhalation doses for people actually inside the plume but can be the dominant dose route (for a smaller total dose) when the plume passes overhead or nearby (within the range of gamma rays) and inhalation doses are zero or small.

Cloud doses are reduced by increasing the distance of people from the cloud (evacuation) or by additional shielding and slightly increasing the distance between them and the cloud (shelter).

It is claimed (Ref. 169) that a solidly built and reasonably airtight UK house will reduce external exposure by a factor of about 10. This protection factor may reduce with time as the inside of the building becomes contaminated, but dose-rates will generally be lower inside than out.

Ground dose, the radiation dose received by being near radioactively contaminated surfaces, is seldom important compared to other dose uptake routes during the release but can be important in the days, weeks and even years after the event so is important when considering recovery and post-event decontamination. Ground dose can be reduced by both time and distance - staying away from contaminated surfaces.

Long term reduction of ground doses after nuclear accidents is usually a combination of decontamination and changing habits. The latter involves reducing the time or the activities in areas of higher contamination.

Decontamination can be used to reduce the dose-rates, and therefore future doses but can be expensive and is often ineffective. It can also lead to a lot of mildly radioactive material being gathered and becoming a disposal problem.

Both approaches have been seen in Fukushima with extensive decontamination taking place and also areas that are totally closed off to access and others where people are recommended to limit their time outdoors and to avoid more highly contaminated areas (change of habits).

Ground doses will naturally reduce as a function of time as the material is subject to radioactive decay and as it migrates into surfaces, down through soil systems or is washed into watercourses and drains. These migration processes are often referred to as "weathering". A potentially unwanted consequence of natural migration is that previously "clean" areas can become contaminated.

6.4. Reducing thyroid dose

Iodine is an essential nutrient. It naturally concentrates in the thyroid gland in the throat where it is used in the production of hormones that regulate metabolism. Any radioiodine taken in will follow this pathway and deposit a radiation dose in the thyroid. This is known to increase the incidence of thyroid cancer.

Stable iodine, taken in sufficient doses, can saturate the thyroid gland and reduce the amount of radioactive iodine it absorbs reducing the dose to the thyroid (any surplus iodine is rapidly excreted in urine). World Health Organisation guidance on the taking of stable iodine can be found in WHO, Guidelines for Iodine Prophylaxis following Nuclear Accidents Update 1999 (Ref. 170), and the Government's (PHE) in (Ref. 161). Importantly stable iodine is in a race against the radioactive iodine to reach the thyroid so it has to be taken before exposure or shortly afterwards to have a real effect (See Figure 22 taken from Ref. 172). This is the main reason why pre-distribution is used for power reactor sites with "no-notice" releases in their safety cases.

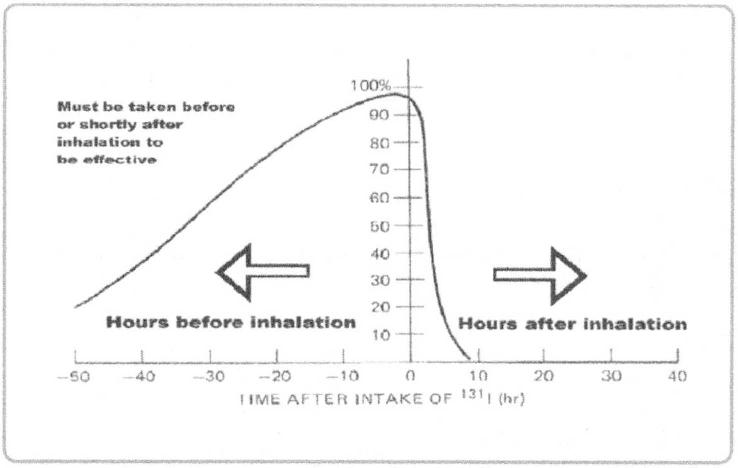

Figure 22 Iodine prophylaxis effectiveness

In the case of a more severe accident there may be a case for giving stable iodine over a wider area. This may cause logistical issues. If it is delivered too late it would have reduced health impact but it may still serve a (dubious) function in providing a level of reassurance to some members of the public.

One dose is usually enough. Evacuation and food restrictions are generally recommended in the event of continued exposure to radioactive iodine. Only in special cases might a new dose be appropriate after one to two days.

In limited circumstances potassium iodate can present a risk factor to those taking it. The only people who should not take the tablets are those who know they are allergic to iodine and those who suffer from the very rare conditions of hypocomplementaemic vasculitis or dermatitis herpetiformis. People who suffer from these conditions will probably have been told by their doctor.

In general the benefits of stable iodine outweigh any risk factor, including the potential need to be outside and collect tablets during a release.

6.5. Reducing ingestion doses

Ingestion doses are mainly the result of consuming food or drinks contaminated with radioactivity but can also be caused by other processes such as putting contaminated fingers etc. in, or near, the mouth or nose. The dose impact of radioactivity inside the body can be very much greater than the same amount of radioactivity outside the body (mainly because the source is closer to the living tissue and it stays there until it is excreted or decays). Radioactivity can be concentrated in food chains long after the release so ingestion dose post-accident must be taken very seriously.

In the event of a significant atmospheric release of radioactive material, ground contamination could be a long term problem. Three isotopes of potential concern following a reactor accident are I-131 (half life 8 days), Cs-137 (half life 30 years) and Sr-90 (half life 29 years). I-131 will decay to low levels within a month or so. The caesium and strontium would be expected to weather with time, by which we mean be washed off surfaces, buried by natural processes or be fixed into the soil structure reducing its bioavailability, it may not be a rapid process.

The risks from external doses and resuspension doses are likely to be small except for areas with high deposition but ingestion dose resulting from eating foodstuffs grown on contaminated land could be a problem beyond the area subject to shelter advice. For this reason, areas which receive a comparatively low level of contamination which did not warrant sheltering protections may nevertheless need food controls to protect consumers against this prolonged exposure from food consumption. This is why some countries define an area around nuclear sites where long-term food pathways must be considered in the planning stage.

Foods can become contaminated directly or indirectly. The former takes place when the radioactivity is deposited on the growing crop, food in open markets or food awaiting preparation. The latter results from activity in soil being taken up by the crop as it grows or soil contamination on the harvested crop. For green vegetables and fruit the first harvest contamination is likely to be dominated by direct contamination of the exposed surfaces. For root vegetables and for crops grown in subsequent years the root uptake becomes more important. Account also has to be taken of radioactivity in any soil that is ingested along with the food although, in some cases, the bioavailability of this may be lower than that of the radioactivity in the food stuff itself.

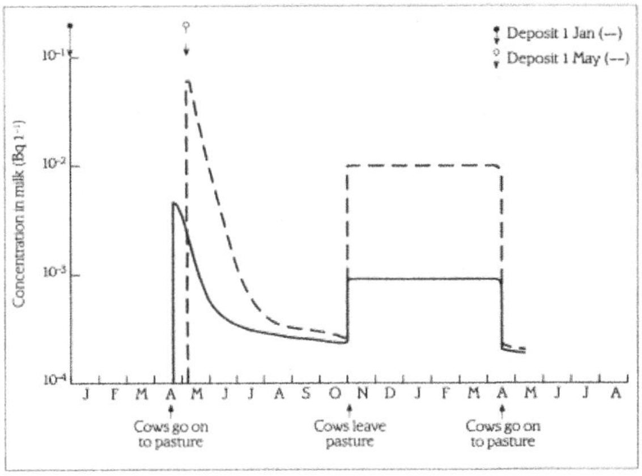

Figure 23 Estimated concentrations in milk for spike deposit at two different times (NRPB W19)

The transfer of radioactivity to the consumed parts of crops and animal products depends on several factors. Important is the time of year that the event takes place. This is well described in the NRPB Emergency Data Handbook (Ref. 173 - although an old publication (2001) this should be in every nuclear emergency planning and response library!). See Figure 23.

In many cases radioactivity in food can be greatly reduced by discarding the outer layers of leaf crops, by washing, by selection of the part of the crop to eat and by cooking. However, this is likely to be unacceptable to the public and there is an expectation within the planning community that contaminated food will be collected and disposed of as contaminated waste.

Changing practices at farms can reduce the amount of activity being transferred from fields to the final foods. This was demonstrated post-Chernobyl by the process of "polishing" sheep from contaminated areas of Wales and Cumbria by moving them off contaminated fields to clean fields for the last few weeks of their lives.

Controlling food in the event of an off-site nuclear emergency can lead to two problems:
1. It is necessary to find alternative sources of food;
2. It is necessary to dispose of food considered not fit for consumption. This can be a particular problem for milk where production cannot be easily reduced and storage quickly becomes a logistical issue.

We are fortunate in the UK that (1) is not considered to be likely to become a problem.

In their document, General Safety Requirements (GSR) Part 7 (Ref. 174), the IAEA recommends the defining of an *Ingestion and Commodities Planning Distance* (ICPD) from the facility for which arrangements shall be made to protect the food chain, water supply and other commodities from contamination following a significant radioactive release and for protecting the public from contaminated foods and drinks.

Restrictions on the sale of food will be put in place by the Food Standards Agency (in Scotland working with Food Standards Scotland (FSS)) based on areas where they believe that the EU Maximum Permitted Levels (MPLs) (Ref. 175) might be exceeded in the food. They are likely to impose a relatively large "precautionary ban" area and reduce it as monitoring provides better estimates of radionuclide uptake by crops and animal products. The regulations give them the powers to:

- prohibit the movement and sale of affected food, animal feed and livestock;
- prohibit the slaughter of animals;
- prohibit the harvesting of food and feed and other agricultural activities;
- issue consents specifying monitoring or other conditions which must be met before food can be released into the market.

Following Chernobyl, a mark and release process was used on some farms in Wales and Scotland. This required that sheep be live monitored before being moved out of the area. Those showing over 1000 Bq/kg were prohibited from slaughter for a minimum of three months. The areas subject to control shrank with time and were removed entirely in June 2012.

Stable iodine can be used as a protective action against radioiodine in foods but, where food is plentiful it is preferable to prevent the consumption of contaminated food rather than reduce the thyroid dose chemically.

6.6. Time, distance and shielding

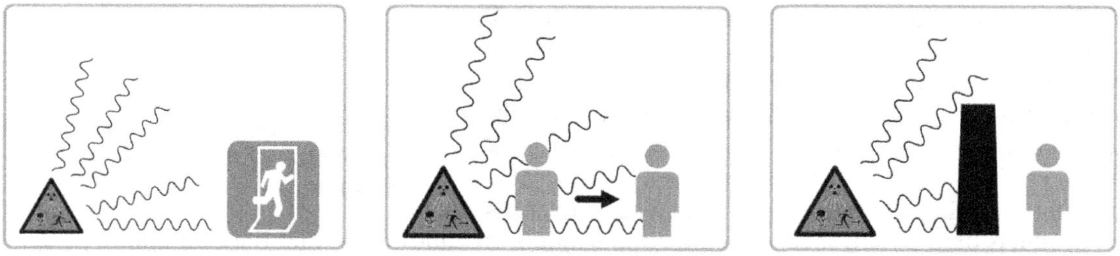

Time Distance Shielding

If you are dealing with an external radiation source, such as a contaminated roadway or a hospital radiation source that has been removed from its shielding you have three factors that you can control that have an impact on the radiation dose.

- The more time you spend within the radiation field of the source the more radiation you will be exposed to. Only stay in the area for the briefest time required to complete the job.
- The nearer you stand to the radiation source the higher the radiation dose-rate is likely to be. Try to stay as far away from the source as possible as much as possible. For physically small sources the inverse square law applies, the dose rate drops as $1/r^2$ so if you double your distance from a source, you reduce the dose rate by a factor of 4 (i.e 2^2), increase the distance by a factor of three reduces the doserate by a factor of nine (i.e. 3^2).
- If you can put a material that absorbs radiation between you and the radioactive source, then it will absorb the radiation dose and you will not (shielding).

It takes practice and experience to make dose-minimisation activities second nature. If a work area has a high dose-rate at one end and a lower dose-rate at the other, then use the lower dose end for unpacking and preparing tools and other equipment and for carrying out briefings (or better still - do these activities outside where there is no additional dose). Someone with nothing to do for a few minutes should retire to the lower dose areas rather than stay and watch others work. Use natural features such as walls and pillars to shield you from the source where practical. Think time, distance and shielding, take advice from experienced radiation protection workers such as your organisation's RPA, get the job done efficiently.

I have seen experienced nuclear submariners performing a radiation survey of the side of a reactor using handheld instruments. They were (a) standing at right angles to a reactor rather than directly facing it (b) holding the detectors at arm's length with arms stretched towards the reactors and (c) not hanging around but getting on with the job. They knew what they were doing, (a) reduces the size of the target their body makes for radiation streaming out of the reactor, (b) takes them further from the source reducing their dose-rate and (c) allows them to get out of the radiation field as soon as possible.

6.7. PPE/RPE

Personal protective equipment (PPE) can protect from internal contamination (the ingestion or inhalation of radioactive material or the entry of radioactivity through a wound) and from external contamination (getting radioactive material onto the skin).

The primary aims of PPE are to prevent contamination of skin and hair to reduce the dose to the wearer and to prevent spreading of contamination beyond its current boundaries. PPE can be as simple as overshoes, gloves and head coverings up to fully impermeable suits with clean air supply. PPE can be reusable, subject to suitable cleaning and monitoring, or disposable.

Personal protective equipment should be selected with due consideration of the hazards involved. The equipment should not only provide adequate protection but also be convenient and comfortable to use. PPE should be selected to fit individual workers.

Personnel wearing PPE may encounter problems such as limited visibility, reduced dexterity, claustrophobia, restricted movement, dehydration, and the effects of heat and cold.

Personnel should be trained in how to don protective equipment, how to work in contaminated areas and, importantly, how to undress and stow the potentially contaminated protective equipment and anything else used in the contaminated area, how to monitor themselves and how to decontaminate themselves should this be necessary. If appropriate, emergency plans should provide facilities for dressing, undressing, monitoring and decontamination and provide trained personnel to assist in the processes. These facilities benefit from posters on the walls reinforcing the correct manner for dressing and undressing and clear demarcation of clean and dirty routes and areas.

A good overview of PPE/RPE in a nuclear response is given in reference 176.

Internal contamination results when radioactive material is taken into the body via inhalation or ingestion or open wounds. Alpha and beta activity will result in intense local exposure where they are in the body. Gamma activity will result in a more widespread exposure. Exposure continues until the radioactive material decays, is flushed from the body by natural processes, or is removed by medical protective actions.
External contamination of a person results when radioactive material is deposited on their skin, hair, eyes, or clothing. This can cause skin damage if left in-situ which can be severe if levels of contamination are high. Irradiation stops when the material is removed by shedding contaminated clothes and/or washing off the contamination.

The environment is contaminated if radioactive material is spread about or uncontained.

6.8. Reducing population contamination doses

It is likely that some of the members of the public who were in areas subject to protective actions might become alarmed at the possibility that they, their children or their pets might be contaminated by radioactive material. In this circumstance it would the local NHS Boards' role to provide reassurance and to provide monitoring and decontamination if required.

On-scene decontamination may follow an event such as a dirty bomb. According to PHE the preferred decontamination methods are (Ref. 177):

1. Disrobing;
2. improvised dry decontamination; or
3. improvised wet decontamination.

A helpful graphic explaining decontamination can be found on the US Centre for Disease Control and Prevention website (Ref. 178). Their 100+ page document on planning for population monitoring is at Reference 179.

Improvised decontamination: Improvised decontamination is the use of an immediately available method of decontamination.
Dry decontamination utilises any available dry, paper tissue (eg 'blue roll'), kitchen towel, toilet roll or paper tissues, towels and clean rags or strips of blanket or sheeting to blot the exposed skin. Care is taken to use sufficient absorbent material to prevent spreading contamination and to bag the waste.
Wet decontamination is the use of water from any available source such as taps, showers, hose-reels, sprinklers, etc. and includes the 'RINSE-WIPE-RINSE' Method of improvised wet decontamination preferably using clean warm water and a sponge or soft brush.

A Radiation Monitoring Unit (RMU) may be set up to monitor contamination on people. A photograph of an RMU set up for an exercise is shown below (Figure 24).

According to the UK government an RMU is "*used to determine levels of radioactive contamination in or on people and any subsequent requirement for decontamination"* (Ref. 180). It is a facility where members of the public and off-site responders will be carefully monitored for radioactive contamination and given advice based on the findings.

The intent of an RMU is to provide reassurance monitoring to people who fear that they may have been contaminated by a release as well as provide measurements and records of any actual contamination. They will be managed by people with suitable medical physics backgrounds and staffed by Public Health England, hospital and operator personnel.

At the RMU the ideal methodology for decontaminating a person is to "*remove all clothing and wash people down using warm water, detergent and a washing aid like a flannel or a sponge*" but if there were going to be a delay before this could be delivered then it is better to react more quickly. Just removing clothing can remove up to 90% of contamination. Gentle, dry or damp, wiping can remove more (Ref. 181).

Fukushima provided experience of using RMUs for real. Their initial action level to trigger decontamination was 13,000 cpm. But they found this impractical as it selected a very large number of people for decontamination and was difficult to measure in areas with elevated dose-rates. The action level was raised to 100,000 cpm to identify those who would most benefit from decontamination. Fukushima prefecture report performing radiation screening for 253,844 residents including those who had evacuated from within 20 km from the NPP from 13 March to 6 June 2012. 102 people were found to have levels in excess of 100,000 cpm. After removing clothes, the counts decreased to less than 100,000 cpm; nobody had contamination levels that would lead to health problems. Changing clothes was one of the recommended protective actions for decontamination however, because of the Tsunami and the manner of the evacuation, there was a shortage of clothing and the reception centres were too cold to remove clothing without replacing it. The reception centres and RMU also suffered from a lack of water, again due to the effects of the Tsunami. (Ref. 182).

Figure 24 Radiation Monitoring Unit

Counts per minute (CPM)
Many radiation monitors count the number of times they detect an ionisation event and display this as counts per minute or CPM. This reading can be related to dose if the radiation type and energy is known but even without this knowledge can be used to distinguish between contaminated and not contaminated. In the context of person monitoring, it is common for an initial rapid survey (screening survey) to be undertaken and the readings compared to a threshold reading (sometimes a multiple of the background count) to select people for more careful monitoring or decontamination.

6.9. Responding to potentially contaminated patients

It is very unlikely that many, if any, of the casualties of radiation accidents will be highly contaminated but much more likely that there will be people with light contamination[9] or concerned that they may be contaminated.

Advice on planning for and responding to a mass casualty event with radioactive contamination can be found from the IAEA (Ref. 182). This gives advice on pre-hospital and hospital arrangements and stresses that "*Triage is initially on the severity of medical conditions and NOT on radioactive contamination. Primary attention should always be aimed at life -threatening conditions*". A "Pocket Guide" on the topic is available as reference 183. If no other training or training material is available then this pocket guide is a very good place to start.

The more generic advice for the UK is found in the NHS's Concept of Operations for managing Mass Casualties (Ref. 184) and their *Guidance for the initial management of self-presenters from incidents involving hazardous materials* (Ref. 185).

In a series of procedure demonstration videos available on the internet (Ref. 186) the US Radiation Emergency Assistance Center/Training Site show how responders should don their own protective clothing, undress a contaminated casualty, decontaminate a wound and skin and remove their own contaminated PPE.

At scene responders should only work in the inner sector if strictly necessary and only with appropriate personal protective equipment (coveralls etc.) and respiratory protective equipment (BA, respirators or face masks) (PPE/RPE), radiological survey and advice - ideally from your Radiation Protection Advisor (RPA).

The UK advice (Ref. 177) to those treating potentially contaminated patients is to wear PPE for standard precautions plus double gloves.

The IAEA advice to first responders (Ref. 187) is that the radiation monitoring should locate and mark areas where ambient dose rates are > 100 mSv/hr in which only lifesaving activities should be performed with a stay time of less than 30 minutes. The inner cordon should be established where the ambient doserate is no greater than 0.1 mSv/hr. This advice applies to all radiological events not just dirty bombs.

The inner cordoned area is the area around a dangerous radioactive source where precautions should be taken to protect the responders and the public from potential external exposure and contamination.
IAEA

Persons contaminated by a radiation emergency or a dirty bomb are unlikely to be a health hazard to those treating them. A possible exception to this could be if the victims have highly radioactive fragments embedded in their wounds. These can be readily detected with hand held radiation detectors (which most hospitals would have access to) and should be removed with long tweezers, avoiding direct contact, and placed in lead lined boxes or at least stored a distance from people if shielded boxes are not available.

The stabilisation of those with life threatening injuries takes priority over decontamination. For other casualties the

[9] We might suspect that many members of the public will not accept the description of any measureable level of radioactive contamination as "light".

bulk of any contamination is likely to be removed by removing their outer clothes. Bare skin can be decontaminated by brushing, showering or washing. Care should be taken to be gentle on the skin, any abrasion can give a route into the body for radioactive material.

6.10. The importance of timely imposition of protective actions

The three main prompt protective actions of shelter, evacuation and stable iodine prophylaxis are designed to protect people from the cloud of radioactivity released by the event. The intent is to avert dose, that is reduce the total radiation dose received by those subject to the protective action in such a way as to provide a nett benefit to that person.

Many of the events listed in the nuclear industry's safety cases give little or no notice before they start releasing radioactivity to the atmosphere and then would be expected to last no more than a few hours. Any Delays in the achievement of the protective action reduce its effectiveness. This is demonstrated by Figure 25 which shows four different simplified release time structures all of which give a projected dose of 20 mSv at the position considered.

- A two hour release with no notice (10 mSv/hr);
- A four hour release with no notice, half the release rate of 1 (5 mSv/hr);
- Like the first (10 mSv/hr for two hours) but with a four hour warning period;
- No notice and a release rate that drops exponentially with a half-life of 2 hours.

Dose rate as a function of time is shown as a solid line and the remaining avertable dose as a dashed line. The 7.5 mSv projected dose line (dotted line) is the boundary between shelter being considered potentially appropriate and not likely to be appropriate (the line marks the lower ERL of 3 mSv avertable dose assuming that shelter reduces doses by about 40%).

These give quite different avertable dose as a function of delay time. Those releases with no notice are particularly unforgiving of delays in implementing protective actions, particularly those with short duration releases.

In this example, if the intention is to impose shelter where it can save 3 mSv (the lower ERL for shelter) then for the two-hour, no notice uniform release (scenario 1) you have no more 1.25 hours to establish shelter at this point, after this it is not worth it. Implement it one hour sooner and you avert 7 mSv rather than 3 mSv.

The situation with half the release rate for twice as long (scenario 2) (a dose-rate of 5 mSv/hr) is that sheltering would avert a dose of 2 mSv an hour when implemented. If you cannot get people into shelter more than 1.5 hours before the release end, then it is not indicated as a protective action based on the PHE guidance.

With the delayed release (solid straight lines) you can, in principle, avert 8 mSv (20 mSv x 0.4) by achieving shelter any time within the warning period. Thereafter the avertable dose drops with every delay.

The exponentially declining dose-rate gives shelter as an indicated protective action if it can be implemented in about 3.5 hours.

This example highlights a difficulty with the ERL system being based on averted dose. To be able to estimate averted dose you need a crystal ball rather than a radiation meter. You need to be able to predict how the off-site dose rate will vary with time in the future and you need to be able to predict how quickly you can implement protective actions which, in turn, depends on your ability to communicate advice and the recipient's ability and willingness to comply with the advice.

The teams of dispersion experts working on this issue in real time tend to convert measured or estimated off-site dose rates into a measurement called "time to ERL". For shelter this found by dividing 7.5 mSv by the dose rate. A measured dose-rate of 15 mSv/hr therefore translates into a time to ERL (shelter) of 0.5 hours. This means that, if the dose-rate is expected to remain as it is for half an hour or more after you can get people into shelter then it is worth doing and shelter should be advised for that point. Dispersion models would then be used to attempt to determine the full area over which that protective action advice should apply.

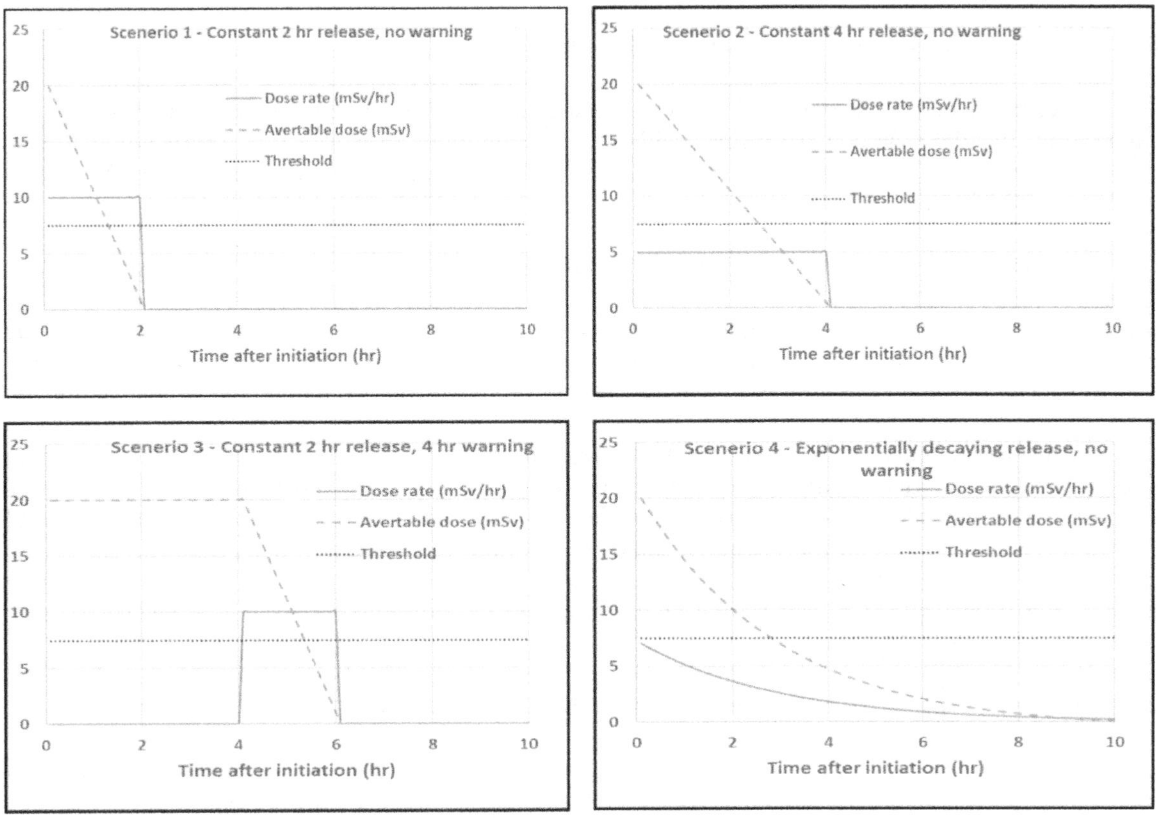

Figure 25 Avertable dose as function of implementation time

The amount of dose averted by the protective action strategy will depend on a number of things including;

- the actual projected dose - protective action advice can be improved by an accurate and clearly communicated assessment of the projected dose to the public;
- The speed and clarity with which the protective action advice can be formulated and promulgated;
- The extent to which the advice is acted upon in a timely manner;
- The ability of the population, supported by the local authority and the police, to implement and maintain the protective action strategy in a timely fashion.

The recognition of the importance of speedy implementation of protective actions led to the development of automatic protective actions whereby, at reactor sites with the potential for no notice events, stable iodine tablets are pre-distributed, a telephone warning system is in place for those in the area to register with, prior information is provided to ready those in the protective action and permissions are obtained from the directors of public health to initiate shelter and the taking on stable iodine on declaration of an off-site nuclear emergency without any debate.

These automatic actions tally with IAEA expectations for a "precautionary action zone" (PAZ) which they define in the box below.

> Precautionary action zone (PAZ)
> *"An area around a facility for which emergency arrangements have been made to take urgent protective actions in the event of a nuclear or radiological emergency to avoid or to minimize severe deterministic effects off the site. Protective actions within this area are to be taken before or shortly after a release of radioactive material or an exposure, on the basis of prevailing conditions at the facility".*
>
> IAEA GSR Part 7

6.11. Severe accident protective actions

The US NRC have published a relatively short and readable analysis of the effectiveness of protective action recommendations for two rapidly progressing accidents in a high population area 200,000 residents in the EPZ and a medium population area 50,000 residents in the EPZ (Ref. 189). The faults had a warning time of 0.5 hr and 0.7 hr and a plume duration of 2.8 hr and 1.6 hr.

Important for this analysis is the determination of the Evacuation Time Estimate (ETE) for the sites.

The implications of different strategies and delay times were investigated by mathematical modelling of the dose to members of the public for a number of different scenarios including immediate evacuation (starting within 30 minutes of the initiation) compared to shelter in place within 30 minutes of the initiation followed by evacuation after a delay.

The results of the consequence analysis show that, unsurprisingly, shelter-in-place is more protective when evacuation cannot be accomplished within suitable timeframes. The conclusion was that licensees developing site-specific protective action strategies for rapidly progressing accidents should consider that evacuation protects public health and safety more than sheltering-in-place when it can be completed within the following timeframes.

For the 0 to 2 mile area around a nuclear power plant: evacuation is more protective when the evacuation time estimate (ETE) is less than 2 hours.

For the 2 to 5-mile area: Evacuation is more protective when the ETE is less than 3 hours.

For the 5 to 10-mile area: Shelter-in-place would likely be the initial protective action to allow a staged evacuation to proceed. If evacuation is the initial protective action for this area, it is more protective than shelter-in-place when the ETE is less than 3.2 hours.

These conclusions depend on the assumed time structure of the release and the estimated time to achieve the protective action.

The current expectation for outline planning around UK nuclear licensed sites falls short of this level of analysis but uses the experience of the local planners and responders to consider what might and what probably will not work in their area if faced by a severe accident.

6.12. Withdrawal of protective actions

There are issues with planning the withdrawal of protective actions. No predetermined radiological criteria for initiating withdrawal of sheltering and evacuation advice is given in the UK *Public Health Protection in Radiation Emergencies document* (Ref. 161) but it does state that in general terms, this advice should only be issued when these urgent protective actions have achieved their desired effect by averting doses, or when their continued application will cause more harm than good in the broadest sense.

There are two views that are frequently expressed. The first is that shelter in place should not be withdrawn until sufficient environmental monitoring and dose estimation has been completed to declare it “safe”. The second is that if the remaining avertable dose is expected to be well below the lower ERL then the protective action has no great value going forward. Unfortunately, level 2 exercises tend to finish before this decision is required so it does not get discussed to conclusion.

Ref. 161 shows how reference levels of dose can be used as a planning tool in the optimisation of doses to the public following the end of the acute response phase of an emergency. The recommendation is to choose a reference level of dose which most people are expected to be under but some exceed. The priority is then to identify the most exposed people (those over the reference level) and to reduce their exposure. If this is successful then over time, the number of people receiving doses above the reference level will decrease and only a few people with atypical behaviours are likely to receive exposures exceeding it. At this stage most people would be below the reference level, but their doses should still be considered to see if they can be reduced in any reasonably practical way. After progress has been made in this manner it may be appropriate to reassess the reference level and reduce it. In this way exposure to radioactivity is reduced in the general population with emphasis on identifying those with higher exposures due to their location or activities and prioritising reducing their exposure.

6.13. Emergency planning areas

Part of the risk assessment is the process of estimating the areas that might be affected by the event. Different size areas will be affected by different aspects of the event and, for wind borne contaminants the area downwind will be affected further from the site.

The IAEA defines several planning zones around major nuclear facilities:

- The on-site area is the area under the control of the operator or first responders,
- The off-site area is the area beyond that area under the control of the facility, operator or first responders which is divided into:
    - A Precautionary Action Zone (PAZ), for facilities in threat category I, for which arrangements shall be made with the goal of taking precautionary urgent protective action, before a release of radioactive material occurs or shortly after a release of radioactive material begins, on the basis of conditions at the facility (such as the emergency classification) in order to reduce substantially the risk of severe deterministic health effects,
    - An urgent protective action planning zone (UPZ), for facilities in threat category I or II, for which arrangements shall be made for urgent protective action to be taken promptly, in order to avert dose off the site in accordance with international standards.

The IAEA suggests sizes for the PAZ and UPZ depending on the nature of a site including PAZs of 3 - 5 km and UPZs of 5 - 30 km for reactors with power ratings of 1000 MWth or more but it is important to note that the IAEA state that "*a different distance should be used when this is substantiated by a detailed safety analysis*" (Ref. 190).

In the USA, the NRC defines two emergency planning zones NRC, 2018, Emergency Planning Zones (Ref. 191):

- the plume exposure EPZ, about 10 miles in radius for which predetermined protective action plans are in place designed to avoid or reduce dose from potential exposure of radioactive materials. These actions include sheltering, evacuation, and the use of potassium iodide where appropriate; and
- the ingestion exposure pathway EPZ, about 50 miles in radius for which predetermined protective action plans are in place designed to avoid or reduce dose from potential ingestion of radioactive materials. These actions include a ban on contaminated food and water.

This results in a keyhole shaped protective action zone which extends further in the downwind direction. A detailed historical justification for these distances along with a discussion about significantly reducing them for future advanced reactors including Small Modular Reactors can be found in reference 192.

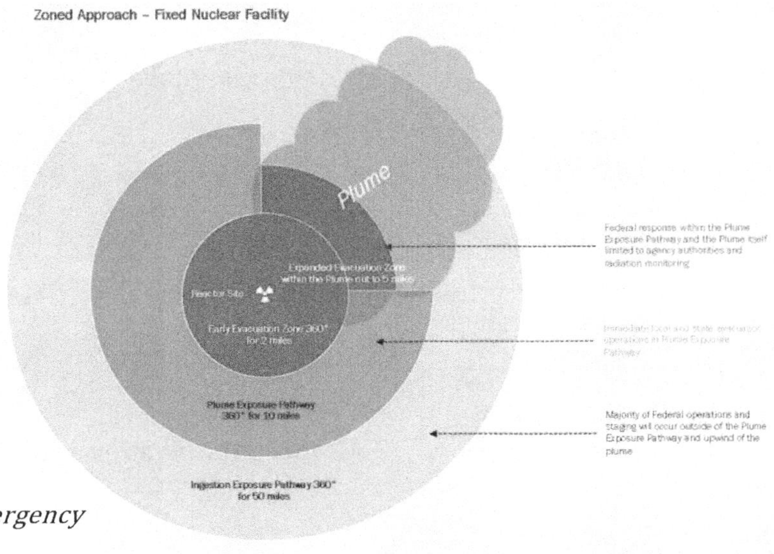

Figure 26 US Emergency Planning Zones

In the UK REPPIR-19 defines the Detailed Emergency Planning Zone (DEPZ) and the Outline Planning Zone (OPZ). The DEPZ is the area nearest the site where it is proportionate to pre-define protective actions which would be implemented within a few hours (Guidance paragraph 192) but protective actions do not need to be applied equally across the detailed emergency planning zone.

> Detailed Emergency Planning Zone
> ""*Detailed emergency planning zone" means a zone determined in accordance with regulation 8 and covered by the local authority's off-site emergency plan*".
> ACOP 190: "*The detailed emergency planning zone must be based on the minimum geographical extent proposed by the operator in the consequences report and should:*
> *(a) be of sufficient extent to enable an adequate response to a range of emergencies; and*
> *(b) reflect the benefits and detriments of protective action by considering an appropriate balance between;*
> *i. dose averted; and*
> *ii. the impact of implementing protective actions in a radiation emergency across too wide an area."*
>
> Guidance 192: "*The definition of a detailed emergency planning zone is a defined zone around premises where it is proportionate to pre-define protective actions which would be implemented without delay (e.g. within a few hours) to mitigate the most likely consequences of a radiation emergency. The protective action taken should provide prompt protection to those who may be affected maximising effectiveness which would be reduced if time was taken to consider and implement appropriate action*".

Planning zones as defined in GSR-7 and REPPIR-19 are more useful in national concept of operations type documents rather than when applied to single site.

6.14. Summary

The dose to the public can be reduced by the timely imposition of protective actions (Figure 27). The most important of these during the accident are shelter, evacuation, the taking of stable iodine and food/water interventions. After the accident the food interventions are likely to remain important, decontamination maybe attempted and restrictions on the usage of areas is possible.

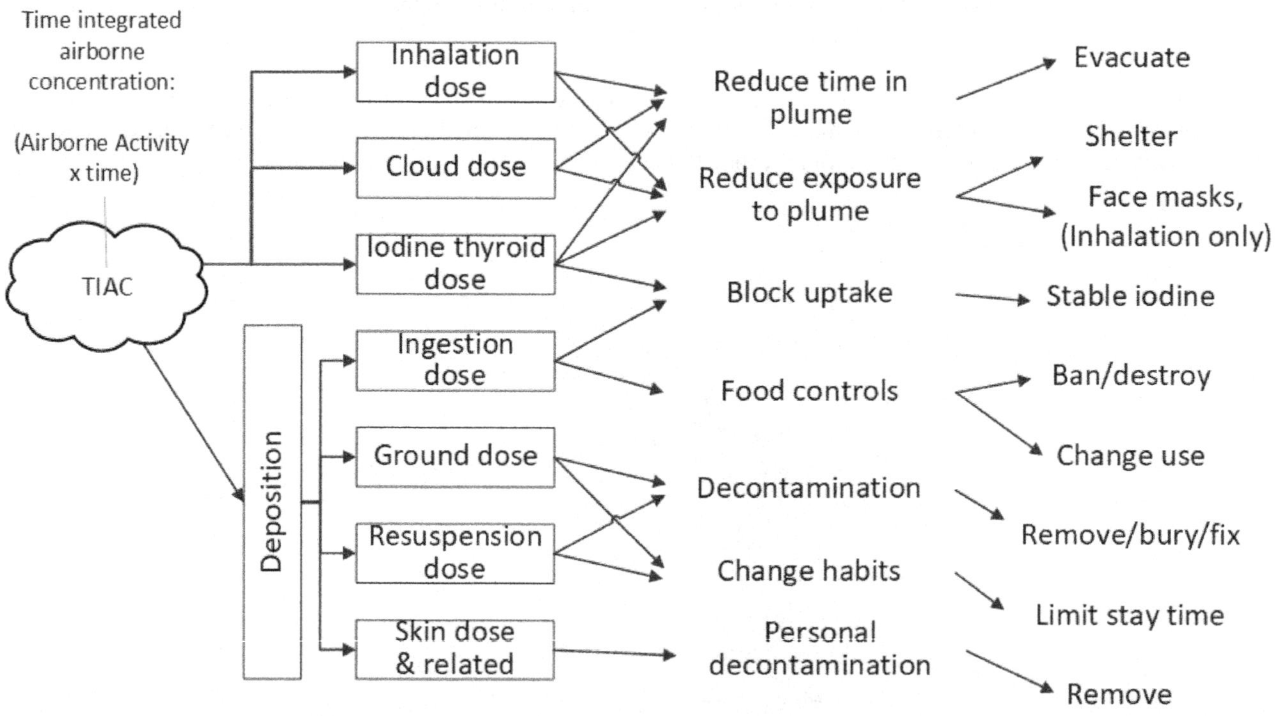

Figure 27 Summary of protective actions

7. Detecting and measuring radiation

The detection and measuring of radiation is a highly specialised subject and this section is intended to be the lightest of overviews.

All radiation detectors have a sensitive region where the incident ionising radiation makes a change that can be detected and turned into a signal which is processed and displayed in some way or sounds an alarm.

It is more important for emergency responders to understand the types and purposes of detectors rather than how they actually work.

Radiation detectors can be divided into:

- **Counters:** Which measure the number of interactions occurring in the sensitive volume but give you know information on the type of radiation or its energy other than that that you can imply from the fact that it is capable of triggering the detector.

  The best known example is the Geiger-Muller counter (or Geiger counter to those less sensitive to Muller's feelings) which can be connected to a meter to give "counts per second" or to a speaker which gives clicks whenever a radiation is detected allowing the user to be warned of varying intensities while using their eyes for other tasks. A detailed description of how a GM detector is used to survey a contaminated patient can be found in reference 193.
- **Spectrometers:** Which give more information about the radiation, specifically the energy it deposits in the sensitive volume which is often related to its total energy. In some instruments this allows the identity of the sources of radiation to be determined as well as the quantities.
- **Dosimeter:** These are devices that are designed to measure the energy deposited in the sensitive volume, or the quantity of ionisation. There are often used to measure dose rate and record dose over a period of time. Personal dosimeters are designed to be worn close to the body and give a reading of equivalent dose (in Gy) which is related to the long-term harm the radiation might be doing the person exposed and is the basis of legal dose records and limits.

All detectors should be calibrated which is a process of seeing how they respond to well defined radiation fields to confirm that they give an adequate level of accuracy when measuring the radiations they are designed to measure. "Adequate" in this context can vary greatly depending on the circumstances.

It should be notes that instruments that work well in the laboratory may not work so well in the field where they may be subjected to physical challenges, which prevent them working as designed, as well as mixed radiation fields or stray electric fields which may confuse the reading.

When presented with a detector or dosimeter it is important to know what it can and cannot do and to know how to use it. Batteries should be checked before use and radiation detectors switched on in low radiation areas before entering areas that might have elevated radiation levels to check that they are not giving anomalously high readings.

The NAIR handbook (Ref. 194) has a very helpful section on the selection of instruments to detect the radiations from a wide range of radionuclides and measure the resulting dose rate or contamination levels.

Part 2. Emergency preparedness

This part of the document looks at the function of emergency preparedness; the process of scoping emergency plans, consulting, writing testing and reviewing plans, training personnel and preparing facilities.

"Plans Are Worthless, but Planning Is Everything"
(used by Dwight D. Eisenhower, Winston Churchill, Richard M. Nixon and others)

In almost all crisis in the past it has been found that the emergency plan was wanting in some, quite often significant, way. However, the planning process and the training has been found to be invaluable. The trick is to gather the right people with the right information and resources and give them the understanding of each other and each other's role and the flexibility to work their way through the problem at hand, which is probably not exactly the one you had in mind when writing the plan.

8. Emergency Response: Principles and purposes

The IAEA state that the goals of emergency response are (Ref. 174):

- *(a) to regain control of the situation and to mitigate consequences;*
- *(b) to save lives;*
- *(c) to avoid or to minimize severe deterministic effects;*
- *(d) to render first aid, to provide critical medical treatment and to manage the treatment of radiation injuries;*
- *(e) to reduce the risk of stochastic effects;*
- *(f) to keep the public informed and to maintain public trust;*
- *(g) to mitigate, to the extent practicable, non-radiological consequences;*
- *(h) to protect, to the extent practicable, property and the environment; and*
- *(i) to prepare, to the extent practicable, for the resumption of normal social and economic activity.*

The operator will be focussed on (a) to (e) on site. For a local authority the focus will be on items (f) to (i) but the two plans should dovetail together and provide the appropriate support for each other.

Similarly, according to REPPIR-19, the purpose of emergency plans are:

- *(a) to reduce or stop the effects of the radiation emergency;*
- *(b) to reduce the exposure to individuals and to the environment resulting from the release of ionising radiation;*
- *(c) if necessary, to ensure that provision is made for the medical treatment of those affected by the radiation emergency; and*
- (d) *to prioritise the implementation of the plan in relation to a person exposed to a dose consequence in excess of the reference levels set out in regulation.*

According to REPPIR-19 Schedule 7 the principles to which emergency planning must have regard are:

- *(a) the necessity for the plan to respond to the particular characteristics of a given radiation emergency as those characteristics emerge;*
- *(b) the necessity to optimise protection strategies to ensure that the proposed response, as a whole, is predicted to do more to mitigate the radiation emergency and facilitate transition from that emergency than to increase its duration or consequence, taking into account—*
  - *(i) the health risks arising from exposure to ionising radiation as a result of the radiation emergency, in both the long and the short term;*
  - *(ii) the economic consequences of the radiation emergency;*
  - *(iii) the effects of the disruption, both on the premises and the area immediately surrounding it, and on the public perception of the effects of the radiation emergency;*
- *(c) the necessity of avoiding, so far as possible, the occurrence of serious physical injury to any person or persons;*
- *(d) the necessity of ensuring that an appropriate balance is struck between the expected harms and benefits of any particular protective action so as to maximise the benefit of that action.*

Principle 1a would seem to be requiring that any emergency response is based on situational awareness so that the appropriate response for the situation is enacted. In fact, so far as protective actions to protect the public are concerned, the initial actions tend to be pre-planned and automatically applied on declaration as a prudent response to lack of decision time and expected lack of detailed situational awareness. The response then begins to understand the situation and responds to the "particular characteristics" of the event by considering if changes to the protective action advice are required.

Principle 1b requires optimisation of the protective actions to ensure that dose is averted in the short term and the long term while not causing undue disruption to the people living in the area and beyond. This is not an easy balancing act to achieve, and it is quite likely that any post-event reconstruction will be unfairly critical of the decision makers who will have had to have made important decisions rapidly and with a lack of clear information on the current situation let alone on the prognosis. At the planning stage thought should be given to range of accidents credible, how the extent of the accident could be rapidly measured on the day and how to maximise dose reduction and reduce uncertainty in the minds of those affected while preventing unnecessary disruption.

Of the purposes (a) is largely about the operator's on-site plan bring the situation under control in a manner that reduces the amount of radioactivity released to the environment or otherwise modifies the release in some way to reduce the consequences.

Purpose (b) is served by the steps taken under (a) and also by using the principles of time, distance and shielding, as appropriate to reduce external dose and using shelter, evacuation, stable iodine and PPE/RPE, as appropriate, to reduce inhalation, cloud and contamination doses to members of the public and responders.

Purpose (c) requires planning for receiving hospitals for people who may potentially have some or all of; conventional injuries, radionuclide contamination and large external radiation doses.

Purpose (d) is based on the principle that the plan should prioritise action to reduce the radiation doses of those likely to get the highest doses. We would expect the protective actions to start near to the site and work outwards. Thus if the area within 3 km of a site is to subject to a controlled evacuation we would expect those nearest to be evacuated first.

Missing from this list of purposes, maybe because it was considered that it is so obvious that it need not be said, is the protection and reassurance of the population around the site and beyond. If people are evacuated from their homes and businesses then they need somewhere to go to in the short term and, possibly, in the long term. Their needs, for shelter, sustenance, hygiene and privacy should be met taking into account different needs as a result of age, gender, religion and culture so far as is practical. This is one area where planning can make a big difference to outcome.

The question about how severe an accident to prepare for is a potentially contentious issue. If you prepare for outrageously unlikely accidents with extreme consequences you are likely to devote significant resources to maintaining capabilities that you are very unlikely to use. This is arguably prudent but you need to justify the expenditure year after year and convince people that this is best way to spend money rather than spending it in any of the competing fields including environmental, social, health, educational and defence. REPPIR-19 provides guidance on risk assessment and detailed and outline planning but it will take a few years for the impact to be seen clearly.

UK National Nuclear Emergency Planning and Response Guidance is found in Ref. 195.

9. Regulation

While it is not necessary to have a full knowledge of the regulations to be a competent emergency responder, or even planner, it is useful to have at least an overview of the regulations and guidance that set the context in which responders work.

This section gives a brief overview regulation in the UK. Please read the regulations or seek expert advice when attempting to be compliant with regulation.

The regulation of the nuclear industry is based on the Nuclear Installation Act 1965 (NIA) (Ref. 196), the Health and Safety at Work Act (1974) (H&SAWA) (Ref. 197) and the Energy Act (2013) (Ref. 198). The civilian industry is regulated by the Office of Nuclear Regulation (ONR) whose mission statement is: "*to provide efficient and effective regulation of the nuclear industry, holding it to account on behalf of the public*". The Defence Nuclear Safety Regulator (DNSR) (Ref. 199) performs the same function for military facilities that are exempt from regulation. Ref. 201 is a comprehensive overview of the regulation of radiological and civil nuclear safety in the UK.

The regulations discussed in Sections 9.1- 9.3, 9.5 & 9.10 are mainly of concern to the operator. 9.6 concerns the local authority and emergency services and 9.7(COMAH) & 9.8 (REPPIR) concern mainly the operator and the local authority but also the emergency services.

9.1. The Convention on Nuclear Safety

The UK has ratified the IAEA convention on nuclear safety (Ref. 200 & 202). In doing so the UK government accepted obligations to implement a certain number of safety rules and standards at all civil facilities related to nuclear energy. These include issues of site selection; design and construction; operation and safety verification; and emergency preparedness.

The objectives of this Convention are:

- to achieve and maintain a high level of nuclear safety worldwide through the enhancement of national measures and international co-operation including, where appropriate, safety-related technical co-operation;
- to establish and maintain effective defences in nuclear installations against potential radiological hazards in order to protect individuals, society and the environment from harmful effects of ionizing radiation from such installations;
- to prevent accidents with radiological consequences and to mitigate such consequences should they occur.

Article 16 of the convention (Emergency preparedness) requires that appropriate steps are taken to ensure that there are on-site and off-site emergency plans that are routinely tested and tested before the reactor is first operated. It also requires that each contracting party ensure that its own population and the competent authorities of the States in the vicinity of the nuclear installation are provided with appropriate information for emergency planning and response. This is achieved by REPPIR-19 in the UK.

9.2. The Nuclear Installations Act 1965 and site license conditions

This is an Act to consolidate the Nuclear Installations Acts 1959 and 1965 and has been modified by newer regulation (Ref. 117). Regulation 1 states that "*No person may use a site for the purpose of installing or operating (a) any nuclear reactor (other than a nuclear reactor comprised in a means of transport, whether by land, water or air), or (b) any other installation of a prescribed kind, unless a licence to do so has been granted in respect of the site by the appropriate national authority and is in force*".

It then sets rules about ownership of a site license (notably that they cannot be transferred) and allows "the appropriate national authority" which grants site licenses to attach conditions to those licenses. This is currently the ONR.

It places duties on the licensee including a duty to secure that no occurrence involving nuclear matter or involving the emission of ionising radiations causes an injury to any person, damage to any property of any person other than the licensee or significant impairment of the environment. It is made clear that this includes any events that happen to other people's property whilst on the licensed site (site license holders can't blame such occurrences on sub-contractors).

Subject to a few caveats the regulations allow that, where as a result of a breach of a duty imposed by the regulations there is significant impairment of the environment, a qualifying public authority may make a claim under this Act for compensation in respect of the reasonable cost of relevant measures of reinstatement relating to that impairment.

The nuclear site licence conditions are listed in the ONR's Licence condition handbook (February 2017) (Ref. 203) which has been produced as an aide-memoire for nuclear inspectors. It reproduces the licence conditions from

Schedule 2 of the standard nuclear site licence and highlights the regulatory powers within the conditions.

The Naval equivalent authorisation conditions are listed in JSP 518, Regulation of the Naval Nuclear Propulsion Programme Part 1: Directive (Ref. 204).

The key condition for emergency preparedness is licence Condition 11.

> Licence Condition 11: Emergency arrangements (Required under NIA)
>
> *Without prejudice to any other requirements of the conditions attached to this licence the licensee shall make and implement adequate arrangement for dealing with any accident or emergency arising on the site and their effects.*
>
> *The licensee shall submit to ONR for approval such part or parts of the aforesaid arrangements as ONR may specify.*
>
> *The licensee shall ensure that once approved no alteration or amendment is made to the approved arrangements unless ONR has approved such alteration or amendment.*
>
> *Where any such arrangements require the assistance or co-operation of, or render it necessary or expedient to make use of the services of any person, local authority or other body the licensee shall ensure that each person, local authority or other body is consulted in the making of such arrangements.*
>
> *The licensee shall ensure that such arrangements are rehearsed at such intervals and at such times and to such extent as ONR may specify or, where ONR has not so specified, as the licensee considers necessary.*
>
> *The licensee shall ensure that such arrangements include procedures to ensure that all persons in his employ who have duties in connection with such arrangements are properly instructed in the performance of the same, in the use of the equipment required and the precautions to be observed in connection therewith.*

The military Authorisation Condition 11 uses nearly identical wording but adds a very useful footnote "*Even though nuclear installations are designed and operated to high safety standards it is recognised that it is prudent to plan for accidents. The purpose of this Condition is to ensure that the Authorisee has adequate arrangements in place to respond effectively to any incident or accident. The Authorisee is required to have arrangements in place to cover a wide range of events from minor incidents which are restricted to on-site locations to large incidents or emergencies which can result in a significant release of radioactive material to the environment. The Condition gives DNSR the powers to ensure that the Authorisee's emergency arrangements are exercised. DNSR uses its powers to ensure the Authorisee's exercises demonstrate adequate performance to protect both workers and the public*".

The words "adequate" and "any" in the first sentence have been the subject of long debate. Does "any" mean that the operator has to produce an emergency plan that would put up a credible response even to the most unlikely chain of events? How do you confirm that you pass on that criterion?

When passing judgement on an emergency exercise the ONR often use the key phrase "*that was an adequate demonstration of the plan*". Some find this a disappointingly underwhelming endorsement after a hard exercise well executed.

9.3. The Nuclear Installations (Liability for Damage) Order 2016

The UK is party to the Paris and Brussels conventions (Ref. 205 and 206). These Conventions establish an international regime governing liability to pay compensation for damage following a nuclear incident. Recent changes to the Conventions upgrade the existing liability regime so that, in the unlikely event of a nuclear incident, an increased amount of compensation will be available to a wider category of claimants in respect of a broader range of damage than is currently the case.

The Paris and Brussels conventions are enacted in UK regulation by The Nuclear Installations (Liability for Damage) Order (2016) (Ref. 207) and the changes it makes to the Nuclear Installations Act.

The key principles are (Ref. 208):

1. *Strict (no fault) liability of the operator, i.e. liability without having to prove fault;*
2. *Exclusive liability of the operator i.e. no other party (such as a supplier or contractor) is liable;*
3. *the liability of the operator is limited in amount, time and the types of damage that are compensable;*
4. *an obligation on the operator to cover its liability by insurance or other financial security;*
5. *claims may only be made in respect of damage suffered within a specified geographical area; and*
6. *rules on jurisdiction to ensure that the courts of one Paris country deal with all claims relating to a particular incident. Judgements of the court with jurisdiction must be recognised by the courts of the other Paris countries.*

The operator is required to pay compensation to victims in respect of specified heads of damage up to a capped amount, and operators must have in place financial security to cover their potential liability. Once the operator's capped amount is reached the host government provide compensation from public funds up to a maximum amount and then, if necessary, the other parties to the convention provide compensation from their public funds, again up to a maximum.

The maximum liability for operators will increase from €163 million to €1,200 million per incident for operators of nuclear power plants; from €11.6 million to €70 million per incident for low hazard installations, such as research reactors; and from €11.6 million to €80 million per incident for low-risk transport of nuclear substances.

In addition to personal injury, death and damage to property, the following four new categories of qualifying nuclear damage were added:

1. economic loss arising from property damage or personal injury;
2. loss of income from an economic interest in any use or enjoyment of the environment (for example, fishermen could seek compensation for loss of income caused by a nuclear incident);
3. costs of environmental reinstatement; and
4. costs of preventative measures and further loss or damage caused by such measures.

See References 209 & 210 for more explanation. Reference 211 provides a handy 172 page explanation of the 1997 conventions.

Very few emergency planners need trouble themselves with these conventions. Operator's emergency preparedness teams should be aware of them and should consider how their company may manage appeals for support and claims for damages from members of the public and local businesses and how they access and administer the funds made available by this process.

9.4. The Health and Safety at Work Act and Regulations 2014

The Health and Safety at Work Act 1974 (Ref. 212), the Management of Health and Safety and Work Regulations (1999) (Ref. 213) and the Ionising Radiations Regulations (2017) (Ref. 159) all require work to be risk assessed and appropriate actions to be taken to protect the public, the environment and the workforce from risks inherent in the work undertaken.

There are requirements placed on employers and requirements placed on employees to ensure that people work in a safe manner and are suitably trained and equipped for their jobs, including being able to respond to accidents and emergencies.

These regulations are built into the very fabric of industrial work in the United Kingdom. The emergency preparedness team should be aware of the contingency plans prepared under these regulations and consider the implications of any statement such as "*if the situation worsens the site emergency plan will be called upon*". What resources are required in these situations and are they available without a declaration? They might not be; without a formal declaration of a site incident or off-site nuclear emergency the emergency plan's responding resources will be carrying out their day job or may be off-site. The equipment provided for the emergency scheme might be incompatible and the training inappropriate for the situation. Are the expectations defined or is this sentence used as a lazy way to truncate contingency planning at the work level?

9.5. Ionising Radiations Regulations 2017

The Ionising Radiations Regulations (IRR-17) (Ref. 159) come with an approved code of practice and guidance (Ref. 214). They require that every employer whose employees are engaged in work with ionising radiation provide training on the risks inherent with radiation and the steps taken to minimize those risks.

The principle that those engaged in potentially hazardous occupations be suitably trained and equipped and that the work be risk assessed is also inherent in the Management of Health and Safety at Work Regulations (Ref. 215).

There are requirements placed on employers and requirements placed on employees to ensure that people work in a safe manner and are suitably trained and equipped for their jobs, including being able to respond to accidents and emergencies.

> The status of an Approved Code of Practice in UK law.
> The IRR-17 regulations are provided with an Approved Code of Practice. This is introduced thus:
>
> *"This Code has been approved by the Health and Safety Executive, with the consent of the Secretary of State. It gives practical advice on how to comply with the law. If you follow the advice you will be doing enough to comply with the law in respect of those specific matters on which the Code gives advice. You may use alternative methods to those set out in the Code in order to comply with the law.*
>
> *However, the Code has a special legal status. If you are prosecuted for breach of health and safety law, and it is proved that you did not follow the relevant provisions of the Code, you will need to show that you have complied with the law in some other way or a Court will find you at fault."*

The regulations apply to "any practice" (see box below) and any work performed in atmospheres with radon-222 gas above a given threshold (regulation 3) and place duties on employers, mine and quarry operators and holders of nuclear site licenses (regulation 4). There is a hierarchy of control with some work being subject to notification (regulation 5), registration (regulation 6) and consent (regulation 8). All employees must "*make a suitable and sufficient assessment of any risk to any employee and other person for the purpose of identifying the measures the employer needs to restrict the exposure of that employee or other person to ionising radiation*". Having done this, any risk identified must be minimised by prevention, limitation of consequences and training and equipping employees (regulation 8). Doses must be restricted "*so far as is reasonably practical*" (regulation 9) and kept below dose limits (regulation 12). Contingency plans are mandatory where a radiation accident is reasonably foreseeable (regulation 13).

Also of relevance to this book is regulation 14, the need to consult a suitable RPA, and regulation 15, the requirement to provide information, instruction and training. The remainder of the regulations are details about managing the sites subject to the regulations.

> *"practice" means work involving –*
>
> *(a) the production, processing, handling, disposal, use, storage, holding or transport of radioactive substances; or*
> *(b) the operation of any electrical equipment emitting ionising radiation and containing components operating at a potential difference of more than 5kV, which can increase the exposure of individuals to ionising radiation;*

Planning should be proportionate to the potential impact of a radiation accident which requires that thought be given to:

- *the correct courses of action to take in an accident situation;*
- *recording those actions in a plan;*
- *training those identified to implement the plan;*
- *providing appropriate equipment for dealing with the events or accidents and making sure this is available and maintained.* (Guidance 239)

The plan should identify:

- *who is responsible for putting the plan into effect;*
- *what immediate actions for assessing the seriousness of the situation are necessary, for example the use of suitable radiation and contamination monitors;*
- *what immediate mitigating actions are needed, for instance in clearing the accident area and establishing temporary means of preventing access to that area;*
- *what emergency equipment is required to deal with identified accidents and where this can be found;*
- *other sources of information and guidance, such as equipment manufacturers and contact details;*
- *what PPE is needed and where it can be found;*
- *what personal dosimetry requirements there are for people involved in controlling the accident;*
- *what training is required for employees;*
- *how to obtain radiation protection expertise so that proper judgements can be made about the seriousness of the situation and the measures necessary to recover from it;*
- *under what circumstances to contact the emergency services and who is responsible for doing this;*
- *what dosimetry follow-up is needed so that the people affected by the accident are identified and provision is made for their dose assessment.* (Guidance 224)

This regulation is more relevant to operators than to local authorities. As for the health and safety at work situation, a site's emergency preparedness team needs to be aware of any radiological situations for which the contingency planning passes the buck for unexpected events to the emergency scheme.

9.6. Civil Contingencies Act 2004

The Civil Contingencies Act (CCA) (Ref. 216) and accompanying non-legislative measures, delivers a single framework for civil protection in the UK. A short guide is provided (Ref. 217). These regulations are more relevant to the off-site responders such as the local authority and the emergency services rather than operators.

See also The Civil Contingencies Act 2004 (Contingency Planning) Regulations 2005 (Ref. 218).

Those organisations that are at the core of the response to most emergencies (the emergency services, local authorities, NHS bodies) are labelled "Category 1 responders" and are subject to the full set of civil protection duties outlined in the regulations and supporting documents. They are be required to:

- assess the risk of emergencies occurring and use this to inform contingency planning;
- put in place emergency plans;
- put in place business continuity management arrangements;
- put in place arrangements to make information available to the public about civil protection matters and maintain arrangements to warn, inform and advise the public in the event of an emergency;
- share information with other local responders to enhance co-ordination;
- co-operate with other local responders to enhance co-ordination and efficiency;
- provide advice and assistance to businesses and voluntary organisations about business continuity management (local authorities only);

Category 2 organisations (the Health and Safety Executive, transport and utility companies) are "co-operating bodies". They are less likely to be involved in the heart of planning work, but will be heavily involved in incidents that affect their own sector. Category 2 responders have a lesser set of duties - co-operating and sharing relevant information with other Category 1 and 2 responders.

These organisations come together in resilience fora such as the Local Resilience Forum (LRF) (see Section 11.12.3), which are based on police areas.

9.7. Control of Major Accident Hazards Regulations 2015

Some nuclear licensed sites are also subject to the Control of Major Accident Hazards Regulations 2015 (COMAH) (Refs. 219 & 220). This is triggered by the site holding above set levels of certain chemicals in the same way that REPPIR is triggered by having above set levels of radionuclides on site.

COMAH requires the operator to risk assess chemical holdings and do all that is reasonably practical to reduce the probability of a major accident and the potential impact of any accident (they have to prepare a major accident prevention policy (reg. 7)). They also require a plan to respond to such an event. Upper tier sites must "*prepare and test an internal emergency plan*" (schedule 4), and supply information to local authorities for external emergency planning purposes (reg. 13).

The local authority hosting an upper tier site is required to produce external emergency plans (reg. 13) to make sure that arrangements are in place to deal with the off-site effects of an incident. Unless "*the competent authority is of the opinion that the establishment is incapable of creating a major accident hazard outside the establishment*" (reg. 15). These plans must be reviewed and tested regularly (reg. 14).

COMAH sites have public information zones (PIZs) set by the competent authority. Like the REPPIR prior information requirements the operator is required to ensure that every person likely to be in the area and every school, hospital and public place in the area in the PIZ "*is sent in the most appropriate form, without having to request it, clear and intelligible information on safety measures and requisite behaviour in the event of a major accident at the establishment*" (reg. 18).

9.8. Radiation (Emergency Preparedness and Public Information) Regulations 2019

9.8.1. Introduction

The principle nuclear emergency planning regulations are the Radiation (Emergency Preparedness and Public Information) Regulations 2019 which come with an Approved Code of Practice (ACOP) and with Guidance (Ref. 3) (see above for the status of the ACOP). These regulations are based on an EU Basic Safety Standard (Ref. 158) which, in turn is based on IAEA advice. These regulations updated and replaced REPPIR-01, which had been guiding nuclear emergency planning in the UK since 2001.

The REPPIR-19 regulations establish a framework of preparedness measures to ensure that arrangements are in place to effectively respond to an emergency both on the site and, where members of the public might be affected, off-site. The Regulations seek to ensure that members of the public are provided with information, both before and during an emergency, so that they are properly informed and prepared, in advance, about what they need to do in the unlikely event of a radiation emergency occurring.

9.8.2. Regulation 4 Hazard Evaluation

Risk assessment is the first stage of any emergency preparedness project. It is the elucidation of the hazards and risks posed by a site, the response capabilities that might be needed if an identified risk materialises and the resources that would be needed to meet the needs of the moment.

In the UK the government produces an assessment of the risks facing the UK that could cause a civil emergency. These risks are written in the form of scenarios or events, such as a severe storm or a disease outbreak. The seriousness of any risk depends on two things: (a) how likely it is that the risk will occur; and (b) the expected impacts were it to happen. Government considers both of these factors when assessing a risk scenario, see the 2017 and 2020 national risk registers of civil emergencies (Ref. 97 and Ref. 98) and Figure 28 (the risk matrix which was extracted from the 2020 version). The risk matrix presents the impact and likelihood of a plausible worst case scenario manifestation of each risk. The "impact" scale takes into account the economic impacts, fatalities in the UK, the extent of evacuation and shelter, public perception, environmental damage or contamination, loss of essential services, electricity supply and impact on international relations. These are based on "reasonable worst-case scenarios". This means that "*they represent the worst plausible manifestation of that particular risk (once highly unlikely variations have been discounted) to enable relevant bodies to undertake proportionate planning*". The National Risk Register is "*the public-facing version of the National Security Risk Assessment (NSRA) - a classified cross-government and scientifically rigorous assessment of the most serious risks facing the UK or its interests overseas*".

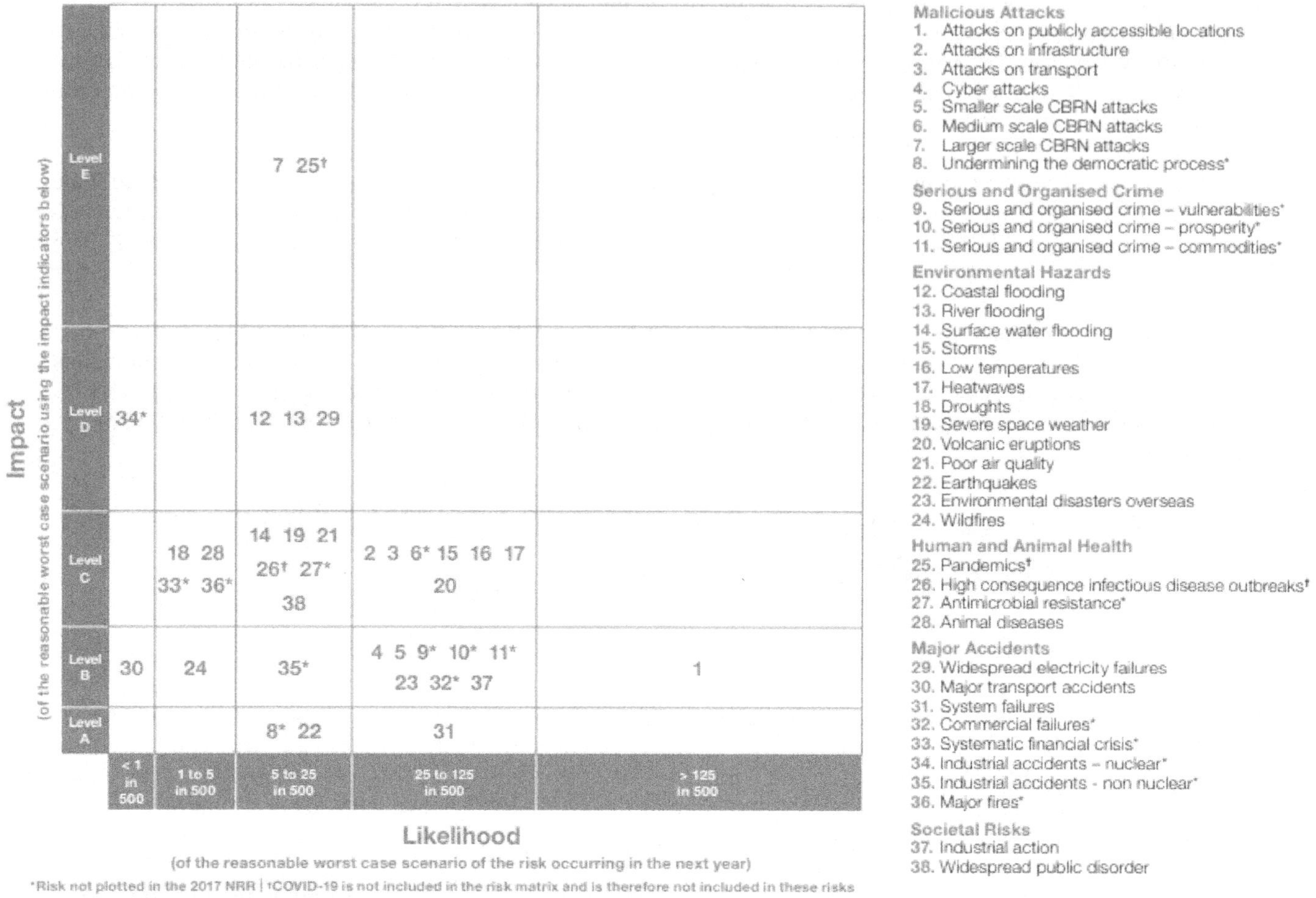

Figure 28 UK National Risk Matrix (summary)

The 2020 version recognises "radiological contamination" and states that "*nuclear sites are designed, built and operated so that the likelihood of releases of radiological material in the UK is extremely low. Historical accidents include Windscale (UK) in 1957, Three Mile Island (US) in 1979, Chernobyl (Ukraine) in 1986 and Fukushima (Japan) in 2011. Of these, Chernobyl and Fukushima were the most severe.*" This is largely unchanged from the previous version.

Based on those impacts, what capabilities should our community and nation have?

What are the gaps between the required capabilities and the current capabilities and how do we close those gaps?

By putting nuclear accidents in impact band D the government is suggesting that such an accident may cause some or all of:

- *Economic impacts: £1 billion to £10 billion;*
- *Fatalities in the UK: circa 201 to 1000;*
- *Evacuation and shelter: 20 thousand people evacuated over 3 days;*
- *Public perception: high impact owing to millions of UK citizens feeling more vulnerable;*
- *Environmental damage or contamination: of a county OR city(ies) for approximately 1 year;*
- *Essential services: lack of health and care services affecting 20% of the population for 7 days;*
- *Electricity supply: major disruption to electricity supply to 1 million people for longer than 18 hours;*
- *International relations: moderate damage to UK relationship with key allies.*

It is hard, and probably fruitless, to relate this to the REPPIR Risk Framework as they use different criteria and the National Risk Matrix compresses a nuclear accident to a single point rather than a point for each determined scenario.

The less than 1 in 500 per year puts the likelihood in the Medium (1 in 20 - 1 in 2 over 5 years) and lower REPPIR likelihood descriptors. Interestingly CBRN attacks of significantly greater impact (level E) are considered to be at least 5 times more likely than an industrial accident.

The US system is summarised in their "National Threat and Hazard Identification and Risk Assessment (THIRA)" (Ref. 221). This is described as a risk assessment process that helps communities understand their risks and what they need to do to address those risks by answering the following questions:

- What threats and hazards can affect our community?
- If they occurred, what impacts would those threats and hazards have on our community?

The US system appears to be a bit more comprehensive than the UK system in that it follows the threat and risk identification exercise with the development of standard scenarios with detailed impacts and then "*specific, quantifiable capability targets, representing the most critical and measurable elements of selected core capabilities*". This gives the required scope of the response capability.

An example of the US scenario play is:

"In early October, the Centers for Disease Control and Prevention (CDC) reports a new (novel) strain of influenza virus in the National Capital Region. Less than two weeks after the first confirmed case is identified at a local hospital, the illness causes hundreds of fatalities and thousands of people seeking medical attention. As the virus spreads, approximately 30 percent of the population across the United States and other countries becomes severely ill. Conventional flu vaccines are ineffective against the current strain, and the CDC estimates that a new vaccine could be months away from mass production.

Because of the pandemic, social distancing is in widespread effect. Utilities, police, fire, government, and other essential services are disrupted due to social distancing and employee absenteeism. Businesses close, resulting in a large-scale loss of services across the region (e.g. banking, food stores, gas stations).

There is a shortage of medical supplies, equipment, beds, and healthcare workers as hospitals are quickly overwhelmed, with up to millions of individuals seeking outpatient medical care and millions more requiring hospitalization. Civil disorder contributes to the high rate of absenteeism and the overcrowding of hospitals and medical centers".

The concept of "*Plausible Concurrent Operations*" is also introduced to consider the potential for multiple challenges to occur at the same time due to common mode causes or bad luck providing further challenge to the national capabilities.

An example of a "Response Mission Area Standardized Target" is "*Within (#) (time) notice of an impending incident, complete the evacuation of (#) people requiring evacuation, including (#) people with access and functional needs (requiring evacuation)".* See Tables 5, 6 and 7 of reference 221 for others.

Regulation 4 of REPPIR-19 requires the operator to "*identify all hazards arising from the work undertaken which have the potential to cause a radiation emergency*" in a "*written evaluation*" named a hazard evaluation after the regulation. Nuclear operators put significant effort into the preparation of site safety cases and ONR scrutinise these very carefully. See *The Safety Assessment Principles for Nuclear Facilities* (Ref. 118) and the ONR technical guide (Ref. 222) for an impression of the depth of work associated with safety cases. Ref. 223 gives a very readable summary of a presentation made in 2015 entitled "The Introduction of Risk Assessment to the Regulation of Nuclear Reactors".

Appendix 2 of the REPPIR-19 Approved Code of Practice and Guidance document shows how risk assessment applies to potential nuclear emergencies introducing a risk matrix. The five impact classes are defined in terms of their impact on human health (deterministic effects), health and Safety (cancer induction), quality of life, property and environment (See Figure 31).

Reading this list it seems possible that the direct effects of radiation i.e. deterministic effects and potential cancer induction may not be the worst outcome of an accident. The post-accident decisions about the habitability of the area, restrictions on activities within the affected areas and economic impacts such as asset depreciation may have longer lasting and, ultimately, more damaging effects at some scales of event. Great care would be needed in the early recovery phase of a major nuclear accident to prevent decisions being made that have a disproportionate effect on the lives of those in the areas affected.

The contribution of this risk analysis process to the defining of an off-site plan is explained in more detail in Section 11.4.

The guidance (paragraph 28) identifies 1 mSv annual effective dose to one or more persons as the threshold for a radiation emergency. This seems a ridiculously low value as it is less than the variation of background radiation across the UK.

Where an operator identifies a potential radiation emergency, they must take steps to prevent it happening and steps to reduce the consequences. This is the role of the site's design and safety case process. In Reference 224 the IAEA give examples of how reactor designs can be modified to reduce the probability and severity of reactor accidents. It includes techniques such as nitrogen inerting, igniter and catalytic devices to prevent the build-up of explosive gases in containment structures and enhanced makeup systems, enhanced or fast acting boron systems and automated dilution control to manage reactivity accidents.

When designing a new nuclear reactor, or when assessing the safety of an existing reactor, UK nuclear utilities are governed by the Safety Assessment Principles last revised in 2020 (Ref. 118) and the Tolerability of Risk document (Ref. 1).

From the Tolerability of Risk, the boundary between the 'tolerable' and 'unacceptable' regions for risk entailing fatality is 1 in 1000 per year for a worker and 1 in 10 000 per year for a member of the public. The boundary between the 'broadly acceptable' and 'tolerable' regions for risk entailing fatality is 1 in 1 000 000 per year for both workers and members of the public. The designers and operators of nuclear power stations estimate the consequences of a large number of potential fault sequences and require that the summed frequency in each decade of off-site dose be below a maximum tolerated value and, ideally below, a broadly acceptable value (See Table 16, taken from the EPR pre-construction safety report (Ref. 225)).

The doses are estimated using the following assumptions:

- the person remains at the point of greatest dose for the maximum duration, although for extended faults a more realistic occupancy may be assumed after a suitable interval;
- the conditions under which the fault is analysed has characteristics which produce the highest dose to that person; and
- no emergency protective actions are implemented, other than those whose implementation is shown to be highly likely.

Effective dose (mSv)	Total predicted frequency per year	
	Maximum Tolerable Limit Basic Safety Limit (BSL)	Broadly Acceptable Level Basic Safety Objective (BSO)
0.1 - 1	1	1×10^{-2}
1 - 10	0.1	1×10^{-3}
10 - 100	0.01	1×10^{-4}
100 - 1000	1×10^{-3}	1×10^{-5}
> 1000	1×10^{-4}	1×10^{-6}

Table 16 EPR design targets

The expectation is thus that the nuclear reactor will be designed, sited, built and operated to minimise both the probability and consequences of accidental releases but accidental releases cannot be totally ruled out. The understanding is that accidental releases could happen, these releases will have consequences somewhere on the range between trivial and catastrophic and, generally speaking, the higher the potential consequences of an accident the lower the probability. Determining the appropriate scale of emergency preparedness is then a question of risk appetite against costs, which is largely a political discussion which we will come to later.

Arguably an off-site plan is only rationally needed in those situations where the existence of the plan may allow useful additional protection of the welfare of the public or environment (largely by reducing radiation dose in line with the ERL documents) compared to the situation where no plan exists. It may also be expected that the potential gains are worth the costs of maintaining the plan.

In order to produce a sensible off-site plan the planner should be guided by a comprehensive review of the site's safety case, outlining a range of faults with different characteristics in terms of warning time, release duration, total release, projected doses and avertable doses and discussing those factors that would affect the dose and avertable dose on the day.

There is a significant amount of guidance associated with hazard evaluation. This would be of more interest to the operator's technical team than to emergency planners.

9.8.3. Regulation 5 Consequence Assessment

The operator is then required to produce another document the *consequence assessment*, the contents of which are mandated by Schedule 3 of the regulations. These require an assessment "*based on a suitable and sufficient range of source terms representing the range of potential emergencies ...*". For each source term the operator is expected to estimate the likely consequences, and consequences which are less likely, but with greater impact based on the range of weather conditions possible. From this information the operator is expected to recommend protective actions based on the distance at which relevant Emergency Reference Levels (ERLs) would be exceeded and hence the dimensions of the Detailed Emergency Planning Zone (DEPZ).

The guidance suggests that to determine a full range of radiation emergencies the operator should include a number of the most bounding cases deserving detailed planning (i.e. with a frequency of at least 1 in 20,000 over five years), a sufficient number of radiation emergencies to demonstrate in a transparent way how it has determined the recommended extent of the DEPZ and include as sensitivity studies any faults that lie just outside the criteria for detailed planning. Paragraphs 147 - 150 of the guidance discuss the types of faults that might merit inclusion for complex sites, gas cooled reactors, light water reactors and non-nuclear facilities, respectively.

The requirement is then to estimate the distances in which urgent protective actions may be warranted for each source terms by comparison to the Emergency Reference Levels (ERLs) (see section 5.3 for a discussion of ERLs) for a range of weather conditions, persons exposed, effective and equivalent doses and all dose pathways of importance (see section 0 for a discussion of pathways).

This document is sent to the ONR but they are under no regulatory obligation to comment on it. It does not need their blessing.

9.8.4. Regulation 6 Review of Hazard Evaluation and Consequence Assessment

Regulation 6 requires that the operator review, and where necessary, revise their hazard evaluation and consequence assessment whenever a material change to their work with ionising radiation is planned, or is recognised to have consequentially taken place, and within three years of the previous assessment.

9.8.5. Regulation 7 Consequence Report

The information from the consequence assessment is summarised in a *consequence report* (CR) (regulation 7) whose contents are mandated by Schedule 4 of the regulations. The operator must send the CR the local authority and, should the local authority wish it, attend a meeting to explain the report. The operator is also required by the regulations to "*comply with any reasonable request for information made by a local authority*".

This consequence report gives information about the site (basically its address) and recommendations about the minimum distances out to which the local authority should be able to implement urgent protective actions and the time scales for their implementation which a supporting rational.

The planners can then put in place automatic protective actions where there is a reasonable probability that, if an off-site nuclear emergency is declared, the avertable dose in those areas will exceed the appropriate ERL and put in intelligence gathering and decision-making processes to determine if the default strategy should be modified on the day.

The consequence report presents a very brief summary of the preceding risk assessment. The operator need not give any information on the range of faults conceivable and their particular characteristics, how they would be detected, distinguished from each other and reported. The local authority will usually require a lot more information in order to draw up their functional specifications for their off-site plan.

Another quirk of REPPIR-19 is that the operator's advice on the downwind distances that prompt protective actions may be appropriate and their advice on the minimum range of the DEPZ are not subject to regulatory approval. Since the regulations and guidance are unclear on whether the distances should be based on best estimate or pessimistic dispersion assumptions the operator has considerable leeway to affect the size of the DEPZ without the need for justification. This compares with REPPIR-01 which involved both clearer guidance on the methodology and detailed challenge from the regulators leading to the regulator justifying the outcome. See, for example, Reference 226 in which the ONR rejected the case that no DEPZ was needed under REPPIR-01 for AWE Burghfield and present their arguments and conclusions over 24 pages of careful argument.

9.8.6. Regulation 8 the Detailed Emergency Planning Zone

Regulation 8 sets a duty on the local authority to determine the detailed emergency planning zone on the basis of the operator's recommendation.

The approved code of practice provides more detail of what is expected.

> ACOP 190 *The detailed emergency planning zone must be based on the minimum geographical extent proposed by the operator in the consequences report and should:*
>
> *(a) be of sufficient extent to enable an adequate response to a range of emergencies; and*
>
> *(b) reflect the benefits and detriments of protective action by considering an appropriate balance between;*
>
> *i. dose averted; and*
>
> *ii. the impact of implementing protective actions in a radiation emergency across too wide an area.*
>
> 191 *In defining the boundary of a detailed emergency planning zone, geographic features should be used for ease of implementing the local authority's off-site emergency plan. Physical features such as roads, rivers, railways or footpaths should be considered as well as political or postcode boundaries, particularly where these features and concepts correspond with other local authority emergency planning arrangements.*

In the first year of application of these regulations the steps that have been taken seem to be:

1. The operators engage in a significant programme of work to satisfy regulations 4, 5, and 7. This includes careful reviews of their safety cases, reports written and formerly verified by subject matter expects, reports taken to their nuclear safety committee, or equivalent, for independent scrutiny and a brief consequence report submitted to the local authority.
2. The consequence report is reviewed by PHE and a short report written. This takes the choice of the accident sequence and source term on trust but looks carefully at the dispersion and dose estimates. PHE favour the use of dispersion under weather conditions that minimise horizontal and vertical dispersion and thus maximise the distance downwind that the doses to members of the public falls below the REPPIR threshold. PHE also insist on using projected dose rather than averted dose as the comparator against the ERLs even when there is no warning of a spike release of activity and therefore no realistic hope of averting dose.
3. ONR does not systematically review and report on the documents it receives.
4. An officer within the local authority, in discussion with others including the operator, prepares a briefing paper for their authority which outlines the regulations and proves a precis of the consequence report. This makes recommendations to the authority which may be a single option for adoption or may include a number of options for discussion.
5. The local authority take the briefing paper to a committee which has a number of other decisions to make within the meeting. A discussion is held and a decision reached.

The result of this process is a bottleneck in the information flow. The consequence assessment is a significant project built on a large body of knowledge and analysis, the consequence report contains very little of the useful information and in reality, the only information the council decision makers have to go is "*the operators tell us that we need a DEPZ of at least x km radius and we've looked at the maps and have made one or two suggestions about the boundary of the DEPZ*".

Figure 29 illustrates the ONR view of the setting of a DEPZ from early 2020 which demonstrates the process the local authority are expected to adopt. It shows the technical suggestion from the operator (in REPPIR-01 agreed by the ONR after extensive challenge) as the dotted line and then the DEPZ as the red line, encompassing the operator's suggested zone but moving outwards to find markers on the ground that mean more than "so many metres from the site fence" such as roads, field boundaries, paths etc. It also shows the DEPZ bulging to encompass settlements that are bisected by the operator's range in order to try to minimise the neighbour issue discussed above.

The local authority are then required to revise their off-site plan to suit the new regulations and the new planning zones.

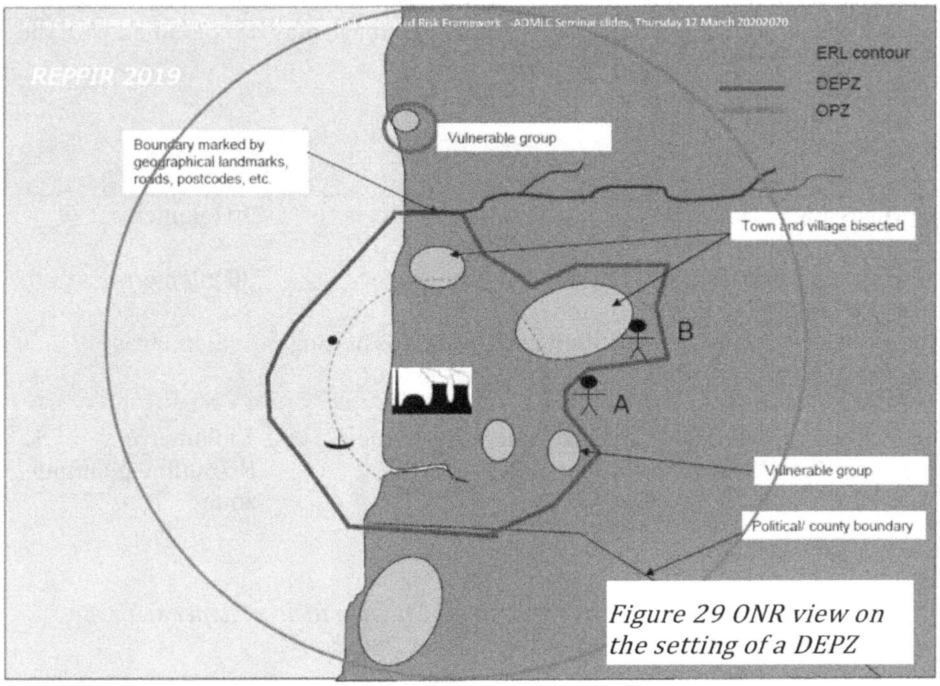

Figure 29 ONR view on the setting of a DEPZ

A potential problem with setting a DEPZ or an area to receive automatic protective actions is that when a line is drawn there are likely to be people just inside the line and other people just outside the line. Those inside the line will get the full prior information and, if near an operating commercial reactor site, will get pre-distributed stable iodine tables. Their neighbours the other side of the line get less prior information and no stable iodine tablets. On the day of an event one set will be asked to stay indoors and take stable iodine whereas their near neighbour will not and will be able to work in the garden or walk to the shops as usual. This is unavoidable but that does not mean that people will like it or accept it. See "*Leicester lockdown: Boundary line branded 'stupid' by residents*" (Ref. 227) for comments on the Leicester local lockdown in response to Covid in 2020 with several quotes such as "*And how are they going to police it? If the pub round the corner opens there will be an open pub within walking distance where you can get a pint. What's stopping people from going? It's ridiculous.*"

9.8.7. Regulation 9 the Outline Planning Zone and outline planning

It has to be recognised that there is a remote possibility that an event may have a greater off-site effect than that determined by the REPPIR risk analysis process. Arguably the Windscale fire, the Chernobyl accident and the Fukushima meltdowns were all in this class which makes it worrying frequent.

Outline planning is not a particularly new feature of nuclear emergency planning in the UK It was already an expectation when the HSE wrote "Arrangements for Responding to Nuclear Emergencies" (affectionately known as the Grey Book in the industry) in 1994 (Ref. 124). Regulation 9 of REPPIR-19 schedule 5 gives the default ranges shown in Table 17. Sites regulated by the HSE have a DEPZ as agreed between the local authority and the operator. For other sites, including defence sites the distance is set by the Secretary of State.

The intention of outline planning is for the agencies involved in the off-site plan to give a little thought to what they would do if faced with an event that was larger than that used to scale the emergency arrangements. In general this involves thinking about how shelter advice could be extended further out, how more stable iodine tablets could be sourced and distributed to areas just beyond the pre-distribution zone in a timely manner and how the area near to the site could be evacuated. Local authorities have collated data about the populations beyond the DEPZ; looked at roads and other features that might be used to delineate wider areas for protective actions; looked at obtaining stores of stable iodine tables and how they could be distributed on the day and looked at the traffic control issues that might arise.

9.8.8. Regulation 10 the Operator's (on-site) emergency plan

Regulation 10 and the associated ACOP and guidance put a duty on the operator of a site where it has been shown that a radiation emergency may occur to "*make an adequate emergency plan designed to secure, so far as is reasonably practical, the restriction of exposure to ionising radiation and the health and safety of persons who may be affected by radiation emergencies identified by the evaluation*".

In fact the operator probably has further risk assessments, they are required by other regulations, looking at the conventional hazards on site, the risks associated with the work done on site (working at height, confined space working, working with chemicals, fire hazards etc.) and the planned response to these may be combined with the nuclear hazards in one set of plans.

Category	Nature of site	Outline planning zone
1	Sites involved in the processing of High Level Waste or storing in excess of 100 tonnes of Plutonium.	50 kilometres
2	Operating nuclear power plants and decommissioning nuclear power plants with a presence of irradiated fuels.	30 kilometres
3	Sites with a significant presence of enriched uranium and decommissioning nuclear sites (other than power plants) with a significant presence of irradiated fuels.	5 kilometres
4	Decommissioned sites without a significant presence of irradiated fuels.	1 kilometre
5	Sites involved in the production of radiopharmaceuticals.	No outline planning zone

Table 17 Default sites of Outline Planning Zones

The ACPO for regulation 10(1) states that 248 "*the operator should prepare the plan in accordance with the requirements of regulation 10 and the associated Schedules and should ensure the plan:*

(a) is a written document, or set of documents; and
(b) can be put into effect without delay when required by:
(i) providing the necessary information, instruction and training and making the necessary equipment available, in accordance with regulations 10(6)-10(8); and
(ii) ensuring any other underpinning capabilities required to implement the plan are in place and readily available".

This requires that the operator has a sufficient number of trained employees with the facilities and equipment required continuously ready (which generally means back up facilities are available in case the primary is useable for some reason), good communications systems in place and adequate written instructions which are complete, up to date and readily available to all who needs access. The plan should be expected to cope with variations in the event such as occurring on a Bank Holiday or when conditions affect several facilities at once or the availability of the emergency services. Emergency preparedness, at this level, is a continuous process rather than something revisited every now and again.

Regulation 10(3) and Part 1 of Schedule 6 sets out the minimum contents of the operator's plan and the principles and purposes to consider to the plan development.

Regulation 10(5) requires the operator to consult with:

- *the operator's employees;*
- *any persons carrying out work on behalf of the operator and who the operator considers might be affected by a radiation emergency;*
- *the lead local authority;*
- *the health authority in whose area the premises to which the emergency plan relates is situated;*
- *Public Health England;*
- *in addition to Public Health England, if the premises to which the emergency plan relates is in—*
  - *Wales, Public Health Wales, and*
  - *Scotland, Health Protection Scotland;*
- *the Category 1 responders in whose area in which the premises to which the emergency plan relates is situated; and*
- *such other persons, bodies or authorities as the operator considers appropriate.*

A lot of this consultation takes place in the EPCC (Section 11.12.2) and in the exercise preparation, execution and debriefing process but it is also common for emergency services to attend shift exercises on site and take part in other site familiarisation exercises.

To ensure that the plan can be put into effect without delay when required the site requires alarms, suitable shift patterns or call out rotas of suitably qualified and experienced persons, trained and equipped in their role.

9.8.9. Regulation 11 the local authority off-site plan

Regulation 11(1&2) states that:

"(1) Where premises require a planning zone under either or both of regulations 8 or 9, the local authority must make an adequate off-site emergency plan covering that zone or zones.

(2) The plan required by paragraph (1) must be designed to mitigate, so far as is reasonably practicable, the consequences of a radiation emergency outside the operator's premises".

The associated ACOP states that the plan must be a "*written document or set of documents*" and can be put into effect without delay. There is more to this regulation detailing the minimum contents (reg 11(3) and schedule 6), consultation (reg 11(5)), training and equipping of responders (reg 11(6)) and reporting completion of the plan (reg 11(7)).

Schedule 7 gives the principles that must be considered when an emergency plan is being prepared. These are:

(a) *the necessity for the plan to respond to the particular characteristics of a given radiation emergency as those characteristics emerge;*

(b) *the necessity to optimise protection strategies to ensure that the proposed response, as a whole, is predicted to do more to mitigate the radiation emergency and facilitate transition from that emergency than to increase its duration or consequence, taking into account—*

(i) *the health risks arising from exposure to ionising radiation as a result of the radiation emergency, in both the long and the short term;*

(ii) *the economic consequences of the radiation emergency;*

(iii) *the effects of the disruption, both on the premises and the area immediately surrounding it, and on the public perception of the effects of the radiation emergency;*

(c) *the necessity of avoiding, so far as possible, the occurrence of serious physical injury to any person or persons;*

(d) *the necessity of ensuring that an appropriate balance is struck between the expected harms and benefits of any particular protective action so as to maximise the benefit of that action.*

The requirement to "*respond to the particular characteristics of a given radiation emergency as those characteristic emerge*" essentially states that the plan cannot simply say something along the lines of "*if the emergency plan is initiated we will impose the following automatic protective actions until the event ceases*". Instead the plan has to allow for the collection of information about the event (situational awareness), must have some form of decision making process to determine the appropriate actions and must have the co-ordinated resources to deliver the protective actions decided upon by the system. 1(b - d) requires people in the decision making process with some understanding of the balance between the steps that can be taken to avoid radiation exposure and the full range of costs that they entail.

9.8.10. Regulation 12 Reviewing and testing emergency plans

Regulation 12 requires that local authorities and operators with a duty to write plans also have a duty to review and, where necessary revise, their plans and the test the plans "*taking reasonable steps to arrange for all those with a role in the plan to participate in the test to the extent necessary to ensure that the plan is effective*" at suitable intervals not exceeding three years.

The tests of the on-site plan, known as Level 1 exercises, concentrate primarily on the operator's actions on site with the support of the emergency services but with minimal input from the off-site centres. The frequency of these exercises are determined by ONR on a case by case basis but are usually carried out annually.

These three yearly tests of the off-site plan (known as a Level 2 exercise) concentrate on the functioning of the Strategic Co-ordination Centre (SCC), responsibilities of individual local agencies within the SCC and the support provided to the SCC from other organisations such as the operator, national agencies, and ONR.

Each year one civil site and one defence site is chosen as a Level 3 exercise with a wider scope. These exercises include the exercising of the various government departments and agencies, including the various government departments' crisis rooms. The decision on which exercise should be selected as the national exercise is made jointly between the licensees, the lead government departments (DECC, MOD or Scottish Government), the relevant local authority, and in consultation with ONR.

The local authority usually takes the lead in organising the off-site exercises and generally select an exercise co-ordinator to lead the effort. This generally involves meetings to agree the objectives of the exercise and then a scenario that allows all of those objectives to be achieved, developing the scenario and driving script, agreeing who will play in the exercise and who will umpire it and arranging for all of the facilities to be available.

Although the level 2 is supposed to a test of a ready plan it is usual for a programme of training and plan review to precede the exercise. The ONR expectation is that "*All participants in an exercise should arrive at the exercise with a good understanding of their role, the roles of other organisations both inside and outside the SCC, and a good understanding of the off-site emergency plan commensurate with their roles and responsibilities*" (Ref. 228).

Annex B2 of the Nuclear Emergency Planning and Response Guidance (Ref. 195) provides gives a useful timeline for the preparation and execution of a level 2 exercise and Appendix E provides a list of possible objectives. Key objectives are:

- Notification;
- Setting up of facilities;
- Supply of information;
- Interpretation of information;
- Interfaces and exchanges/team-working;
- Decision making;
- Communications & public information;
- Facilities and equipment; and
- Competence of participants.

There is also a considerable amount of approved code of practice and guidance related to this regulation which the exercise co-ordinator should be familiar with (Ref. 3).

9.8.11. Regulations 13 - 16 Cooperation and charging

Regulations 13, 14 and 15 are about cooperation and consultation and regulation 16 about what costs the local authority can recover from the operator.

9.8.12. Regulation 17 Implementation of emergency plans

This regulation requires that operators and local authorities with emergency plans take reasonable steps to ensure that they can be put into effect without delay. They also require the operator to inform the local authority and the regulator that the plan is in play.

The operator and the local authority are required to make a provisional assessment of the situation after consultation with a number of parties including the emergency services, health bodies and the relevant environment agency. This must be made "*as soon as reasonably practical in order to respond effectively to the particular characteristics of the radiation emergency*".

There is also a requirement on the operator to "*make a full assessment of the consequences of the radiation emergency or other event and the effectiveness of the emergency plan put into effect*" within 12 months.

9.8.13. Regulation 18 Emergency Exposures

It is recognised that in an emergency, employees of the operator and members of the emergency services may be asked to undertake tasks that entail radiation doses which may exceed dose limits, some by a large margin without great harm to their health. These potentially high doses are regulated by the "emergency exposure" systems within REPPIR-19 regulation 18.

REPPIR regulation 2 has the following definition: *""emergency exposure" means an exposure of an employee engaged in an activity of or associated with the response to a radiation emergency or potential radiation emergency in order to bring help to endangered persons, prevent exposure of other persons or save a valuable installation or goods, whereby one of the individual dose limits referred to in paragraphs 1 and 2 of Part 1 of Schedule 3 to the 2017 Regulations could be exceeded".* Regulation 18 then sets up a system to control such exposures.

Important words in the definition include: "Employee" - the emergency exposure permission is not available to members of the public, nor can an organisation's process be applied to someone not in their employ and "Radiation emergency or potential radiation emergency" -the process cannot be used for high dose activities outside emergencies.

At the planning stages each employer has to consider the roles of their employees in an emergency response and estimate the additional doses that these may entail. They then need to consider if it is practical to reduce those doses in any way. If the estimated doses are above the IRR-17 dose limits then the employer must follow the processes in regulations 18(1) - 18(5) in the planning stage (identify potential exposure routes, identify those potentially exposed, provide equipment and training, ready medical surveillance, dosimetry and dosimetry record keeping and identify and train those authorised to permit employees to be subject to emergency exposures, notify the regulator of proposed emergency exposure limits and adjust them if required to do so).

In the response stage the employer must follow the process in regulation 18(6) - 18(8) which prevent certain classes of employees being subject to emergency exposures and after the event must record and report given details as required by regulations 18(9) - 18(11).

Regulation 18(12) makes it clear that regular volunteer members of an organisation such as the British Red Cross, Police Special Constabulary etc. are to be treated as employees with regard to these regulations.

Anyone seeking to set up an emergency exposure process for their organisation would be well advised to carefully read the several pages of guidance provided and to seek advice. This is a difficult regulation to get right. A response might require attendance and use of emergency dose procedures by some of the site's employees and by officers from the fire, ambulance and police services. It may be relatively easy to identify, train, authorise and brief the relevant site employees but it is harder to identify the individual emergency services officers who may be called upon remembering that these situations do not allow enough time to pull specialist teams from the other side of the county. Deciding who to train and how to train them is a major consideration. The processes also require close cooperation of the two emergency planning structures before the event and of the event strategic managements and radiological protection managements during the event.

9.8.14. Regulation 20 Reference Levels

Regulation 20(1) requires anyone writing an emergency plan to prioritise keeping doses to any individual below 100 mSv. This is the threshold below which organisations such as ICRP are claiming that there is no real harm, notwithstanding the linear no threshold hypothesis, which suggests a slightly increased rate of cancer in later years. This means that if dose estimates associated with the plan show a potential dose above 100 mSv the planners need to do more work to reduce the dose. This can be by limiting the potential for dose (smaller batches of radioactive material, more shielding, better processes) or by improving the speed of response (better alarms, clear evacuation routes that are exercised regulatory).

Regulation 20(2) appears to confuse reference levels (a line in the sand which you strive not to exceed in planning) with emergency exposures (the limits you place on doses for responding to particular situations).

Regulation 20(3) allows reference levels to be revised in real time which is unlikely to happen in the response stage of an emergency as breaching them is not a misdemeanour and all health physicists work to the principle of as low as reasonably practical (ALARP) anyway.

Regulation 20(4) allows reference levels to go up to 500 mSv in certain circumstances, again confusing them with emergency doses.

The remaining regulations provide the ability to change reference levels, giving the power to both the local authority and to the Secretary of State.

The role of reference levels in the recovery stage has been outlined in Section 6.12.

9.8.15. Regulation 21 Prior information to the public

Regulation 21 places a duty on the local authority to provide prior information to members of the public within the detailed and outline planning zones if a site in their area has either or both.

For members of the public within the DEPZ regulation 21(1) requires that the local authority "*must, in cooperation with the operator, ensure that members of the public are made aware of the relevant information, and, where appropriate, are provided with it*". This has not yet been tested with time and experience (this written in June 2020) so we can only speculate about the eventual understanding of the level of evidence required to "*ensure that members of the public are made aware of the relevant information*" which seems much more onerous than just posting them a leaflet and putting information on your website.

The guidance suggests that the local authority and operator should cooperate with one another when preparing a public communication strategy and that information could be provided in the form of leaflets, calendars and "*some combination of advertising in the local media, forming local forums, organising exhibitions, holding public meetings and arranging visits to, and open days at the premises*". When preparing a public communications strategy, the local authority may wish to refer to Reference 233 which is guidance from the cabinet office for communicating with the public.

The minimum information required is specified in schedule 8 of the regulations; part 1 for the DEPZ and part 2 for the OPZ.

Interestingly the Carriage of Dangerous Goods (Amendment) Regulations 2019 (Ref. 234) place a duty on the ONR to provide prior information to the public about the nature and effect of a radiation emergency. They discharge this duty by publishing information on their website (Ref. 235), barely more than two pages of text.

9.8.16. Regulation 22 Duty of local authority to supply information to the public in the event of an emergency

Regulation 22 places a duty on all local authorities, not just ones with licensed sites within their area, to have a mechanism in place to provide information to affected members of the public while admitting that it "*will be more difficult for events of unpredictable location and nature, such as transport radiation emergencies, a fallen nuclear-powered satellite, or radioactive contamination being present in a public area*".

The information required is detailed in schedule 9 of the regulations. The duty holder is required to have in place mechanisms to inform the public about the nature of the event, give advice on what the public should do (protective actions such as shelter, evacuate, take "protective substances", self-decontaminate) and on where further advice will be coming from. If time permits and it is relevant the local authority should also provide information about the "*basic facts about radioactivity and its effects on persons and on the environment*".

This is a big ask for local authorities without nuclear facilities in their patch. These authorities should undertake a risk assessment and consider the types of events they want to protect against. This could be:

- Transport accidents where radioactivity material has been, or may have been, spilled onto or besides roads or railway lines;
- Radioactive satellite crash, or warning that one is imminent, where radioactive dust and fragments might to deposited over a wide area;
- Dirty bomb, terrorist dispersal of contamination, terrorist positioning of radioactive source, repeat of Po-210 murder, where large areas might need to be monitored and large numbers of people traced, monitored and counselled.
- Radioactive source lost or stolen and then dispersed, again potentially involving an investigation over a large area, tracing large numbers of people who need monitoring and counselling.

The local authorities should consider who might be available to explain the situation to the STAC and to the SCG, if they were set up (and if not, how would information be collated and considered?). They also need to consider what mechanisms would be used to communicate with the public.

It is especially important to realise that the failure to promulgate the alarm in an efficient and timely manner and the failure to provide consistent, clear and concise information to the public are both opportunities for reputational harm from which responders will find it hard to recover.

9.9. The Ionising Radiation (Basic Safety Standards) (Misc. Provisions) Regulations 2018

The Ionising Radiation (Basic Safety Standards) (Miscellaneous Provisions) Regulations 2018 (Ref. 88) came into force on the 8th May 2018.

Regulation 4 is concerned with "*land [that] is contaminated as a result of the after-effects of an emergency, past practice or past work activity and the level of exposure of members of the public to ionising radiation cannot be disregarded from a radiation protection point of view*".

This is a very important aspect of post-emergency recovery as learning how to live with the increased levels of environmental contamination that would follow from a significant accidental release of radioactivity is vital to the recovery and well-being of the local community. The regulation requires that an appropriate minister sets a reference level for the land. Paragraph 4 requires the appropriate minister to ensure that "*appropriate arrangements are established for the on-going control of exposure of members of the public to ionising radiation, with the aim of establishing living conditions that can be considered as normal, including–*

a) the establishment of an infrastructure to support continuing self-help protective measures in the affected area, which may include the provision of information, advice and monitoring;
b) remediation measures; and
c) *the delineation of the area.*"

This must be in place "*before the resumption of habitation, or economic or social activities, on the land*". We need to be careful that this does not lead to people being kept away from their homes and offices for longer than is necessary as this is known to add to the social and economic stresses of the situation. We also need to avoid hasty decisions about reference levels and area delineation and, in fact, these are likely to be very fluid in the days and weeks following a serious accidental release of radioactivity.

The establishment of support infrastructure is likely to be a moving target as things become clearer and the affected members of the public become better informed of the situation and its ramifications and more capable of playing their part in the decision making. To say it must be in place before the public move back seems entirely wrong.

Another major concern is that "*this regulation does not apply while any part of an emergency plan is in effect in relation to the land in accordance with [REPPIR 2001]*". It is, of course very difficult to define when an emergency plan closes. Most off-site plans make provision for Recovery Working Groups which morph into the leadership of the recovery phase of the response which, it is recognised, could last for several years. Arguably, since this is a "*part of an emergency plan*" it stalls the application of this regulation pretty much indefinitely.

In summary this regulation touches on an important and difficult aspect of emergency planning; that of how do we keep the public safe, informed, healthy and at ease in areas that have been contaminated with radioactive material? The regulations require that a government minister establishes areas that have been significantly affected and defines reference levels, infrastructure, information streams and remediation plans to support these aims in those areas. There can be no argument that this is the right level for these decisions to be made. In the event efforts should be made to prevent the minister saying that they intend "to build back better" and that the environment will be decontaminated to pre-accident levels. It can't be done, and it isn't sensible to try.

This is a process that is very hard to exercise. It is likely that attention will only be given to this regulation should the need to wield it occur.

9.10. Nuclear Industries Security Regulations 2003

The Nuclear Industries Security Regulations 2003 (Ref. 236) set conditions for on-site security arrangements.

Civil licensed nuclear sites and approved carriers of nuclear material are required to protect sites and material against acts of theft and sabotage in accordance with Nuclear Site Security Plans (NSSPs) and Transport Security Statements (TSSs) approved by ONR (CNS).

According to regulation 2 "*A security plan must describe in writing the standards, procedures and arrangements adopted or to be adopted by the responsible person to ensure the security of–*

- *a) the nuclear premises in relation to which he is the responsible person,*
- *b) any Category I/II nuclear material and Category III nuclear material used or stored on the premises,*
- *c) any equipment used or stored on the premises in connection with activities involving nuclear material,*
- *d) any sensitive nuclear information kept on the premises, and*
- *e) in the case of nuclear premises which are or form part of a nuclear site—*
    - *i. any nuclear material (so far as not already mentioned in sub-paragraph (b)) and other radioactive material used or stored on the premises, and*
    - ii. *any equipment used or stored on the premises in connection with activities involving other radioactive material*".

Regulation 3 requires the plan to "*describe the standards, procedures and arrangements relating to ..... the steps to be taken by the responsible person or any person acting on his behalf if any event of a kind specified in regulation 10(5)(a), (b), (e) or (h) that requires immediate action occurs, and the regular practice of the activities required in connection with those steps.*"

Regulation 10(5) "*the events and matters are–*

- *(a) any unauthorised incursion on to the premises or any attempted or suspected such incursion;*
- *(b) any incident occurring on the premises involving an explosive or incendiary device or suspected such device, or a firearm or replica firearm;*

(c) any damage to any building or equipment on the premises which might affect the security of the premises or any material or equipment mentioned in regulation 4(2);
(d) any malicious damage to any building or equipment on the premises, other than any trivial damage that does not affect the security of the premises or any material or equipment mentioned in regulation 4(2);
(e) any theft or attempted theft, or any loss or suspected loss, or any unauthorised movement—
(i) of any nuclear material used or stored on the premises or in transit to or from them, or
(ii)in the case of premises which are or form part of a nuclear site, of any other radioactive material used or stored on them;
(f) any theft or attempted theft, or any loss or unauthorised disclosure, of sensitive nuclear information kept on the premises, or any suspected such theft, loss or disclosure;
(g) any unauthorised access to any sensitive nuclear information kept on the premises, or any attempt to gain such access;
(h) any threat to do anything which would fall within any of sub-paragraphs (a) to (g);
(i) any failure to comply with any of the standards, procedures and arrangements described in the approved security plan for the premises or in any approved temporary security plan to which for the time being they are subject;
(j) *any other event or matter which might affect the security of the premises or the material, equipment or information mentioned in regulation 4(2)".*

There is an issue between security plans and emergency plans in that the aspiration of the former is usually to close and lock doors to deny access and egress to intruders, while the aspiration of the latter is to open doors to allow people to seek a place of safety. Security people are concerned that when fire alarms, smoke alarms or equipment alarms sound that the site may be under attack. Under the same circumstances the emergency planners are concerned that an industrial accident may be in progress. The IAEA recommend an "Integrated Management System" to help overcome these issues (Ref. 237).

This regulation is a matter for the site operator, the security regulator and the police service (civil and nuclear).

9.11. Food and Environment Protection Act (1985)(FEPA)

FEPA is "*an Act to authorise the making in an emergency of orders specifying activities which are to be prohibited as a precaution against the consumption of food rendered unsuitable for human consumption in consequence of an escape of substances ....*" (Ref. 238).

It allows the Secretary of State (in England and Wales) and Scottish Ministers (in Scotland) the power to make emergency orders by statutory instrument (see box) when there exist or may exist circumstances which are likely to create a hazard to human health through human consumption of food.

> Statutory instruments are the most common form of secondary (or delegated) legislation.
>
> The power to make a statutory instrument is set out in an Act of Parliament and nearly always conferred on a Minister of the Crown. The Minister is then able to make law on the matters identified in the Act, and using the parliamentary procedure set out in the Act. SIs may follow affirmative or negative procedure, or have no procedure at all, but which to use is fixed by the Act.
>
> UK Parliament Glossary

These emergency orders "*shall be laid before Parliament*" and shall expire after 28 days (not counting days when "*Parliament is dissolved or prorogued or during which both Houses are adjourned*"!).

An emergency order can prohibit the following in a designated area:

- agricultural activities;
- the gathering or picking of wild plants;
- the slaughter of creatures;
- fishing for and taking fish; and
- the preparation and processing for supply to purchasers or others of food and anything from which food could be derived.

It may also prohibit

- the movement of food or anything from which food could be derived into or out of the designated area or from one place to another within that area; and
- some activities across the entire UK if those activities may have been affected by contamination in the affected area.

The expected response to an emergency with a significant off-site release is a precautionary food ban over a wide area followed by a smaller, more considered ban when the event is over and more information collected.

9.12. The Regulators

9.12.1. The Office for Nuclear Regulation (ONR)

The Office for Nuclear Regulation (ONR) regulates the safety of nuclear installations (including conventional safety) and the transport of radioactive materials in Great Britain. It also regulates nuclear security and safeguards in the United Kingdom. They express their high level expectations for nuclear safety in the Safety Assessment Principles (SAPs) (Ref. 118) and for nuclear security in the Security Assessment Principles for the Civil Nuclear Industry (SyAPs) (Ref. 239).

The fundamental principle for emergency preparedness and response (FP.7) is that "*arrangements must be made for emergency preparedness and response in case of nuclear or radiation incidents*".

The ONR has an emergency planning and response team as well as site inspectors assigned to each of the major licensed sites (Ref. 240). Their strategic objectives are:

- World-class UK EP&R arrangements, driven by continuous improvement culture.
- Unified EP&R arrangements, 'to provide efficient and effective regulation of the nuclear industry, holding it to account on behalf of the public.
- EP&R arrangements and regulatory regime proportionate to the hazard profile of the nuclear site.
- Competent, capable and sustainable UK EP&R response capability.

9.12.2. Defence Nuclear Safety Regulator

The Defence Nuclear Safety Regulator (DNSR) is part of the Defence Safety Authority (Ref. 241) and is responsible for the regulation of nuclear safety across the Defence Nuclear Enterprise. This includes the Naval Nuclear Propulsion Programme and the Nuclear Weapon Programme, where the Ministry of Defence has specific exemptions from statute.

Key documents are: Joint Service Publication 518 Regulation of the Naval Nuclear Propulsion Programme (Ref. 242) and, Joint Service Publication 538 Regulation of the Nuclear Weapons Programme (Ref. 243).

9.12.3. The (Environment) agency

The REPPIR regulators refer to "the agency" which is the Environment Agency (EA) in England, Natural Resources in Wales and the Scottish Environmental Protection Agency (SEPA) in Scotland.

These agencies are responsible for the regulation of the storage and disposal of radioactive wastes and they are a statutory consultee in the preparation of an off-site plan.

In normal operation the agency provides permits for receiving or disposing of radioactive waste and for keeping or using mobile radioactive apparatus. See reference 244 for a review of the regulation of radioactive substances activities on nuclear licensed sites.

9.13. Tolerability of risk

One of the recommendations of the Sizewell B public enquiry (1982 - 1985) was that the Health and Safety Executive should "*formulate and publish guidelines on the tolerable levels of individual and social risk to workers and the public from nuclear power stations.*" The intent was that public, expert and parliamentary opinion on risks and their tolerability should be available to inform discussions on the acceptability of nuclear power stations.

The resulting document *The Tolerability of Risk from Nuclear Power Stations* was issued in 1988 and subject to challenge at the Hinkley Point C Enquiry and by public consultation and revised and reissued in 1992 (Ref. 1).

The paper explains that "*To tolerate a risk means that we do not regard it as negligible or something we might ignore, but rather as something we need to keep under review and reduce still further if and as we can. For a risk to be 'acceptable' on the other hand means that for purposes of life or work, we are prepared to take it pretty well as it is*".

The paper discusses different types of risk such as those we choose to accept in return for some obvious gain, those that affect us personally as opposed to those that affect our families, particularly our children, those that are imposed by nature and those imposed by industrial activity and it provides tabulations comparing the risks of different activities.

Three bands of risk are defined. The "broadly acceptable" region is where the risks are so low as to be generally accepted. The "Tolerable" region is where the risks are accepted but there is an expectation that the operator will do all that is reasonably achievable to reduce the risks further (This is sometimes called the "ALARP region" - As Low as is Reasonably Practical). The "Unacceptable Region" where the risks are not tolerated, and the operation would be prohibited.

The paper discusses at some length how the risk to an individual and to society can be estimated and how radiation can damage organs and health. It also discusses how the probability of accidents of different sizes can be estimated for nuclear power plant. It then grapples with the problem that the safety analysis may find a number of potential fault sequences with a wide range of probabilities and an equally wide range of consequences. Plant design will be such that faults with high, indeed unacceptable, consequence will be unlikely but will never have zero probability of occurrence.

In 2001 the tolerability of risk document was generalised by the Health and Safety Executive in *Reducing Risks, Protecting People* (Ref. 119). This claims that we "*all recognise that, as an inescapable fact of life, we are surrounded by hazards – all with a potential to give rise to unwanted consequences*" and, furthermore "*consciously or unconsciously, we all view hazards and evaluate their risks to determine which ones we choose to notice, ignore or perhaps do something about*".

On the acceptability of risk, the HSE states that "*HSE believes that an individual risk of death of one in a million per annum for both workers and the public corresponds to a very low level of risk and should be used as a guideline for the boundary between the broadly acceptable and tolerable regions*". They hesitate to give values for the boundary of unacceptable but state that "*Nevertheless, in our document on the tolerability of risks in nuclear power stations, we suggested that an individual risk of death of one in a thousand per annum should on its own represent the dividing line between what could be just tolerable for any substantial category of workers for any large part of a working life, and what is unacceptable for any but fairly exceptional groups. For members of the public who have a risk imposed on them 'in the wider interest of society' this limit is judged to be an order of magnitude lower – at 1 in 10 000 per annum*".

The guidance to REPPIR-19 (paragraph 152) states that "*In the REPPIR risk framework under Regulation 4, the value of 1 in 20,000 in a five-year period is the lowest likelihood considered in the national risk assessment (paragraph 131). It is considered to be appropriate to use this as the lowest likelihood for which detailed emergency planning should be required and the point at which outline planning (or even no emergency planning in the case of low consequence events) is sufficient*". You should note that 1 x 10^{-5} per year is the same value as 1 in 20,000 in a five-year period.

The March 2012 national risk assessment policy sponsored by the Cabinet Office (Ref. 245) agreed that 1 x 10^{-5} per year is the boundary between events that should be prepared for and those that are too unlikely for detailed planning. This is obviously a somewhat arbitrary number but seems to have widespread, but not universal approval. The latest national risk register (Ref. 98), which defines its lowest category as "less than once in 500 years", is not directly comparable as it has no lower boundary.

10. Guidance

10.1. IAEA guidance

It is perfectly possible to be an excellent UK based emergency planner and responder without reading a single IAEA document. This subsection is for those who are particularly interested in how international advice, mainly from the IAEA, informs the EU Basic Safety Standard upon which our REPPIR-19 regulations are largely based. People usually only dig this deep when having an erudite argument about the intent of a regulation.

The International Atomic Energy Agency (IAEA) is the international centre for cooperation in the nuclear field. The Agency works with its Member States and multiple partners worldwide to promote the safe, secure and peaceful use of nuclear technologies. The IAEA has documented fundamental safety principles, general and specify safety guides, a collection of more focussed safety guides and a number of other documents.

The lessor IAEA documents, such as the TECDOC series, which is used to develop ideas and of which there are now over 1000 (Ref. 246) tend to be written by ad-hoc committees pulled together from interested parties from around the world, led and supported by full time employees. Higher documents are drafted by standing standards committees. A document goes through several stages of drafting and review. The top documents are endorsed by a standards committee before publication. The results are documents that have been carefully considered and carry authority.

In terms of emergency planning the IAEA documents of particular relevance are:

- Preparedness and Response for a Nuclear or Radiological Emergency (Series No. GSR Part 7) (Ref. 174);
- Arrangements for Preparedness for a Nuclear or Radiological Emergency (Series No. GS-G-2.1) (Ref. 190);
- Criteria for Use in Preparedness and Response for a Nuclear or Radiological Emergency (Series No. GSG-2) (Ref. 160);
- Planning and Preparing for Emergency Response to Transport Accidents Involving Radioactive Material, Safety Guide No. TS-G-1.2 (ST-3) (Ref. 247).

There are also an increasing number of useful documents in the Emergency Preparedness and Response series (Ref. 248). Of particular interest to off-site organisations might be:

- Manual for first responders to a radiological emergency (Ref. 187);
- Actions to protect the public in an emergency due to severe conditions at a light water reactor (Ref. 249);
- Communication with the public in a nuclear or radiological emergency (Ref. 250).

The IAEA advice does not always agree with UK regulation and current practice, so it is important to note that IAEA GSR Part 3 and GSR Part 7 both state that.

"Resolution of conflicts: The requirements of this standard are established in addition to and not in place of other applicable requirements, such as those of relevant binding conventions and national laws and regulations. In cases of conflict between the requirements of this standard and other applicable requirements, the government or the regulatory body, as appropriate, shall determine which requirements are to be enforced."

Among the IAEA's key publications are its Safety Standards, which provide the fundamental principles, requirements and recommendations to ensure nuclear safety. They serve as a global reference for protecting people and the environment and contribute to a harmonized high level of safety worldwide. The IAEA documents generally talk to governments rather than operators or local authorities but they are expert documents worthy of consideration. Many IAEA documents are embedded in an online database (Ref. 251) which has hyperlinking providing easy navigation.

In its advice to countries embarking on a nuclear power programme (Ref. 252) the IAEA envisages a four-phase process involving:

- Safety infrastructure considerations before a decision to launch a nuclear power programme is taken;
- Safety infrastructure preparatory work for the construction of a nuclear power plant after a policy decision has been taken;
- Safety infrastructure activities to implement the first nuclear power plant;

- Safety infrastructure during the operational phase of the nuclear power plant.

In phase 1 it expects that the national strategy will recognise the possibility of a nuclear emergency including one resulting from a severe accident and will consider what capabilities the state will need to respond adequately. Thereafter the requirements of IAEA Safety Standards Series No. GSR Part 7, *Preparedness and Response for a Nuclear or Radiological Emergency* are introduced in the appropriate phases. **Emergency planning is not expected to be an afterthought or an optional bolt-on but embedded in the national capabilities and in those of the operator and the local agencies.**

The Fundamental Safety Principles are articulated in document SF-1 (Ref. 253). This is effectively the top document in the IAEA library. The stated objective "*is to establish the fundamental safety objective, safety principles and concepts that provide the bases for the IAEA's safety standards and its safety related programme. Related requirements are established in the Safety Requirements publications. Guidance on meeting these requirements is provided in the related Safety Guides." It states that "The fundamental safety objective is to protect people and the environment from harmful effects of ionizing radiation*".

A number of principles are proposed including: **Principle 9: Emergency preparedness and response** "*Arrangements must be made for emergency preparedness and response for nuclear or radiation incidents.*"

According to SF-1 the primary goals of emergency preparedness include:

- *To ensure that arrangements are in place for an effective response at the scene and, as appropriate, at the local, regional, national and international levels, to a nuclear or radiation emergency;*
- *For any incidents that do occur, to take practical measures to mitigate any consequences for human life and health and the environment.*

Basically, this is requiring nations to prepare well to respond well. The document goes on to discuss the need for local, regional and national arrangements that reflect the likelihood and possible consequences of a nuclear or radiation emergency and the scope and extent of arrangements.

GSR-Part 1 Governmental, *Legal and Regulatory Framework for Safety* (Ref. 254) has Requirement 8: Emergency preparedness and response "*The government shall make provision for emergency preparedness to enable a timely and effective response in a nuclear or radiological emergency.*" This puts a requirement on government to ensure that it assigns responsibilities to the correct organization for components of emergency plans and clearly defines how any response will be co-ordinated and how decisions on protective actions to protect the public are made. Further details of the government's responsibilities can be found in GSR Part 3 *Radiation Protection and Safety of Radiation Sources: International Basic Safety Standards* (Ref. 255) of which Requirement 43 states that: "*The government shall ensure that an integrated and co-ordinated emergency management system is established and maintained.*"

GSR Part 7 (Ref. 174) is a key emergency preparedness document. It "*establishes the requirements for an adequate level of preparedness and response for a nuclear or radiological emergency. The application of these requirements is also intended to mitigate the consequences of a nuclear or radiological emergency if such an emergency arises despite all efforts made to prevent it*". It is largely talking to national governments and setting the standards that they should achieve.

It states that: "*In a nuclear or radiological emergency, the goals of emergency response are:*

(a) To regain control of the situation and to mitigate consequences;
(b) To save lives;
(c) To avoid or to minimize severe deterministic effects;
(d) To render first aid, to provide critical medical treatment and to manage the treatment of radiation injuries;
(e) To reduce the risk of stochastic effects;
(f) To keep the public informed and to maintain public trust;
(g) To mitigate, to the extent practicable, non-radiological consequences;
(h) To protect, to the extent practicable, property and the environment;
(i) To prepare, to the extent practicable, for the resumption of normal social and economic activity."

The document states 26 "*requirements*" all of which start "*the government shall*". The first five of these require the government to ensure that an "*integrated and co-ordinated emergency management system for preparedness and response for a nuclear or radiological emergency is established and maintained*". This includes establishing roles and responsibilities for responders and ensuring that "*protection strategies are developed, justified and optimized at the preparedness stage for taking protective actions and other response actions effectively in a nuclear or radiological emergency*".

In a crude hazard assessment the document recognises five threat categories of which the most severe is Threat Category I which is described as "*Facilities, such as nuclear power plants, for which on-site events (including those not considered in the design) are postulated that could give rise to severe deterministic effects off the site that would warrant precautionary urgent protective actions, urgent protective actions or early protective actions, and other response actions to achieve the goals of emergency response in accordance with international standards, or for which such events have occurred in similar facilities*". The other categories are (ii) smaller research reactors, (iii) facilities such as irradiation facilities with very strong sources, (iv) processes such as the transport of radioactive material and radiography and (v) areas within the planning zones of sites in neighbouring countries.

There are several footnotes to the definitions of the threat categories which should be understood when working with this document. For example, "*those not considered in the design*" is clarified thus "*This includes events that are beyond the design basis accidents and, as appropriate, conditions that are beyond design extension conditions*". This is a difficult footnote as it calls into question the meaning of requirement 4 hazard assessment which requires the government to ensure that there is a hazard assessment which provides a basis for arrangements which "*shall be commensurate with the hazards identified and the potential consequences of an emergency*".

KEY DOCUMENT	GSR Part 7 (Approximately 100 pages)

This publication addresses Principle 9 from IAEA Safety Standards publication Fundamental Safety Principles (SF-1), which is concerned with the arrangements that must be made for preparedness and response for a nuclear or radiological emergency. It is also consistent with Essential Element No. 11 of the IAEA Nuclear Security Fundamentals. It replaces GS-R-2.

It provides a high-level review of the needs for off-site emergency preparedness and response including the roles of the government, regulator and operator, dose limitation, protective actions with criteria for their application, emergency planning areas with suggested sizes, and shows how a proportionate level of preparedness is based on functional requirements that depend on risk assessment.

This document should be read by anyone with a place on national nuclear emergency co-ordinating bodies who needs to understand the justification for their country's emergency plans.

IAEA, General Safety Requirements No. GSR Part 7, Preparedness and Response for a Nuclear or Radiological Emergency

The purpose of the hazard assessment (which in this case is determining which of five categories your site belongs in) is stated to be to determine if the facility warrants any of the following:

(a) *Precautionary urgent protective actions to avoid or to minimize severe deterministic effects;*
(b) *Urgent protective actions and other response actions to avoid or to minimize severe deterministic effects and to reduce the risk of stochastic effects;*
(c) *Early protective actions and other response actions;*
(d) *Other emergency response actions such as longer term medical actions and emergency response actions aimed at enabling the termination of the emergency; or*
(e) *Protection of emergency workers.*

In the UK this duty is discharged by the REPPIR-19 risk assessment and reporting process (Regulations 4 - 7) but an important, and sensible, difference is that the emergency preparations in the UK are expected to be proportionate to the risks and hazards identified in the site's safety case, rather than the site's hazard category.

Jargon Buster - Protective actions (called "countermeasures" in some older documents)

A **protective action** is an action for the purposes of avoiding or reducing doses that might otherwise be received in an emergency exposure situation or an existing exposure situation.

An **urgent protective action** is a protective action in the event of a nuclear or radiological emergency which must be taken promptly (usually within hours to a day) in order to be effective, and the effectiveness of which will be markedly reduced if it is delayed. These include shelter, evacuation, stable iodine and food restrictions.

A **precautionary urgent protective action** is an urgent protective action taken before or shortly after a release of radioactive material, or an exposure, on the basis of the prevailing conditions to avoid or to minimize severe deterministic effects. These include shelter, evacuation and stable iodine.

An **early protective action** is a protective action in the event of a nuclear or radiological emergency that can be implemented within days to weeks and still be effective. These include relocation, access restrictions and food restrictions.

Requirements 6 - 19 are concerned with the ability to recognise an emergency, raise the alarm and alert responders and affected members of the public, co-ordinate a response and inform and protect the public.

Requirements 20 - 26 are concerned with the infrastructure, personnel and training required to respond to an emergency and the government's role in ensuring that it is in place.

Requirements 6 - 26 are covered in REPPIR regulations 10 & 11 (operator's and local authority's emergency plans) (ACOP "*can be put into effect when required*"), regulation 17 (implementation of emergency plans), 21 and 22 (provision of prior and real-time information).

The appendices to the document provide dose limits for workers and the public. In the UK these are set by regulation (IRR-17 and REPPIR-19) and are broadly comparable with the IAEA recommendations.

KEY DOCUMENT IAEA GS-G-2.1

This document was written to support the superseded GS-R-2 so could be thought of as out of date. However, it still seems to be part of the IAEA canon and it contains a wealth of useful advice that still rewards the reader. It provides more detail on the appropriate responses to a range of emergencies, discussing the concepts and then how to meet the requirements leading to a general concept of operations.

It provides a sufficient overview of radiological protection in nuclear and radiological emergencies for most emergency planners (assuming that they have health physics advice when appropriate). It discusses proportionate emergency preparedness in terms of a crude risk assessment and shows how this informs the nature and size of off-site emergency planning zones, strategies for the protection of the public and for facilities and capabilities (such as medical facilities able to cope with radiation injury and contamination) beyond the site.

This document should be read by anyone responsible for writing and justifying the functional specification of an off-site emergency plan and the concept of operations.

IAEA GS-G-2.1 Arrangements for Preparedness for a Nuclear or Radiological Emergency (2007)

The Safety Guide GS-G-2.1 *Arrangements for Preparedness for a Nuclear or Radiological Emergency* document (Ref. 190) recommends the assignment of responsibilities between the operators and off-site officials for threat category I and II sites.

The Operator is responsible for arrangements to promptly classify an emergency; protect those on the site and on-site emergency workers; mitigate the consequences of the emergency; notify and recommend protective actions for the public to off-site officials; obtain off-site assistance; conduct environmental monitoring near the facility; and assist off-site officials in keeping the public informed.

There are off-site officials (local authorities, emergency services, health services etc.) responsible for response within the emergency zones including arrangements to promptly: implement urgent protective action within the emergency zones; conduct environmental monitoring; control consumption of contaminated food; provide emergency services to the facility; provide medical treatment for individuals who have been contaminated or overexposed and register them for long term medical follow-up on the basis of predetermined criteria; tell the public and media in plain language of the risks and of the actions they should take; respond to inappropriate public reactions; report transnational emergencies to the IAEA; respond to IAEA notifications; and request IAEA assistance when needed.

Section 4 of the report discusses the functional requirements. Which are:

- *Identifying, notifying and activating:* this has a requirement *"to make arrangements for the prompt identification of an actual or potential nuclear or radiological emergency and determination of the appropriate level of response" and "that declaration of a particular class of emergency shall promptly initiate the appropriate level of co-ordinated and pre-planned emergency response on and off the site. The responsibilities and initial response actions of all response organizations shall be defined for each class of emergency"*

Taking urgent protective actions and assessing the initial phase. In this document an urgent protective action is defined as *"A protective action in the event of an emergency which must be taken promptly (normally within hours) in order to be effective, and the effectiveness of which will be markedly reduced if it is delayed."* These include:

- *For radiological emergencies, isolation of a contaminated area or radioactive source;*
- *Prevention of inadvertent ingestion;*
- *Evacuation;*
- *Sheltering;*
- *Respiratory protection and protection of skin and eyes;*
- *Decontamination of individuals;*
- *Prophylaxis with stable iodine;*
- *Protection of the food supply and prevention of the consumption of significantly contaminated foodstuffs and water;*
- *Management of the medical response;*
- *Protection of international trade.*

Further, it is required that *"arrangements shall be made for promptly assessing the results of environmental monitoring and monitoring for contamination on people in order to decide on or to adapt urgent protective actions to protect workers and the public, including the application of operational intervention levels (OILs) with arrangements to revise the OILs as appropriate to take into account the conditions prevailing during the emergency."*

Jargon Buster - Operational intervention level (OIL).

A calculated level, measured by instruments or determined by laboratory analysis that corresponds to an intervention level or action level. OILs are typically expressed in terms of dose rates or of activity of radioactive material released, time integrated air concentrations, ground or surface concentrations, or activity concentrations of radionuclides in environmental, food or water samples. An OIL is a type of action level that is used immediately and directly (without further assessment) to determine the appropriate protective actions on the basis of an environmental measurement.

Using simple measurements to support rapid decisions - based on a number of assumptions.

Other requirements include:

- Keeping the public informed - "*Arrangements shall be made for: providing useful, timely, truthful, consistent and appropriate information to the public in the event of a nuclear or radiological emergency....*";
- Managing the medical response - this clause refers to situations where patients present with symptoms (orphan sources or malicious actions);
- Taking agricultural protective actions, protective actions against ingestion and longer-term protective actions. "*Agricultural protective actions and longer-term protective actions in accordance with international standards shall be taken to avert doses.*"

- Mitigating the non-radiological consequences of the emergency and response "*Arrangements shall be made for responding to public concern in an actual or potential nuclear or radiological emergency*" and for "*monitoring for and responding to any related health effects and preventing inappropriate actions on the part of workers and the public.*"

The document goes on to recommend emergency planning areas around sites based on their threat category in Appendix II. This suggests a precautionary action zone (PAZ) of radius 3 - 5 km for a >1000 MW(th) reactor. 0.5 - 3 km for a 100 - 1000 MW(th) reactor and urgent protective action planning zones (UPZ) of 5 - 30 km in both cases. These distances are based on very severe accidents with extremely low probability. The PAZ is based on two-day doses to bone marrow or lung that could prove fatal i.e. the induction of acute radiation syndrome. The UPZ is based on 48 hour doses exceeding 1 - 10 times the Generic Intervention Level (level of avertable dose that the action should be taken) for evacuation.

No UK safety case recognises a reasonably foreseeable fault that could lead to acute radiation effects in the general public, it would take an extremely unlikely severe accident at a large nuclear power reactor or a catastrophic release from a very high-level waste storage facility to get into that dose regime. But precautionary urgent protective actions, such as recommending shelter and the readying of stable iodine tablets would be taken if a nuclear reactor was damaged in such a way as to make possible a large release as the event progresses.

The safety guide "*Criteria for Use in Preparedness and Response for a Nuclear or Radiological Emergency for protecting people and the environment*" No. GSG-2 (Ref. 160) provides more discussion of the threat categories, expectations for the size of precautionary action zone (PAZ) and urgent protective action zones (UPZ) around major nuclear facilities, declaration states, urgent protective actions, response time objectives and facilities. It is a useful read for emergency preparedness personnel, particular health physics planners.

It also discusses operational criteria based on "dose received" for actions such as medical care and medical follow-up, counselling and reassurance which is largely missing in UK provisions.

KEY DOCUMENT GSG-2

This document refers to the superseded (by GSR Part 7) GS-R-2 but still contains a wealth of useful information. Its objective was to provide a coherent set of generic criteria to support decision making concerning protective actions and other response actions necessary to meet the emergency response objectives. It addresses health consequences due to external exposure and internal exposure of specific target organs, for which the generic criteria were developed.

It develops the concept of different classes of protective actions (precautionary urgent, urgent and early) with different roles and provides criteria for their selection. It also discusses operational criteria based on "dose received" for actions such as medical care and medical follow-up, counselling and reassurance which is largely missing in UK provisions. It also has a useful discussion on operational intervention levels (OILs).

This document should be read by people responsible for planning the radiological protection aspects of nuclear emergency preparedness and response and those teaching emergency responder health physicists.

GSG-2, Criteria for Use in Preparedness and Response for a Nuclear or Radiological Emergency for protecting people and the environment (2011)

It points out that projected dose and dose that has been received are not measurable quantities and cannot be used as a basis for quick actions in an emergency. There is a need to establish – in advance – operational criteria (values of measurable default quantities or observables) as a surrogate for the generic criteria for undertaking different protective actions and other response actions. This advice is largely absent from UK advice including the recent PHE paper *Public Health Protection in Radiation Emergencies* (Ref. 161).

IAEA advice on the transport of radioactive material is signposted at Ref. 256. Transport emergency planning is discussed in "*Planning and Preparing for Emergency Response to Transport Accidents Involving Radioactive Material*", TS-G-1.2 (ST-3) (Ref. 247). The objective of this Safety Guide is to provide guidance to the public authorities and others (including consignors, carriers and emergency response authorities) who are responsible for developing and establishing emergency arrangements for dealing effectively and safely with transport accidents involving radioactive material.

This safety guide states that it is the government's responsibility to define the roles and responsibilities of local and national government, to ensure that suitable radiation protection services exist, to ensure that there is a notification system and that the response to a transport accident is tested. In the UK transport of radioactive material is covered in the Carriage of Dangerous Goods (Amendment) Regulations 2019 (Ref. 234) which outline the duties of drivers, escort commanders, masters, carriers and consignors in the event of a radiation emergency and also defines the notification system in place. The roles and responsibilities of local and national government and the other potential responders are found in the Civil Contingencies Act (Ref. 216).

In a transport accident involving radioactive material the normal priorities of lifesaving, first aid, firefighting, traffic control and investigation may all be required and may all be complicated by the presence of radioactive material. Steps will be needed to confirm that the radioactivity remains contained or, if not, to determine the steps necessary to protect people and the environment from any released or uncontained radioactivity, to recover radioactive material and to decontaminate the environment to an acceptable level. This may need specialist input.

At the local level planning should at least encompass the development of the capability to recognize a radioactive material package, being familiar with basic precautions and knowing who should be called on to provide further assistance. It should cover the operation of their own organizations and the deployment of their own resources. (Part of para 3.9).

The minimum requirements of the national plan are given as (Para 4.5):

- *(a)* *The planning basis;*
- *(b)* *The responsibilities, capabilities and duties of the organizations involved;*
- *(c)* *The procedures for alerting and notifying key organizations and persons;*
- *(d)* *The methods for warning and advising the public;*
- *(e)* *The intervention and action levels for exposure and contamination;*
- *(f)* *The protective measures;*
- *(g)* *The procedures for response actions;*
- *(h)* *The resources for medical and public health support;*
- *(i)* *The procedures for training, exercises and updating plans;*
- *(j)* *Public information.*

Control at the scene is likely to take the form of an inner-cordon around the accident, into which no one goes without a specific mission (lifesaving, firefighting, first aid or remedial work and ideally only then with appropriate PPE/RPE and briefing), a control point and access control point on the cordon upwind of the accident, and a wider outer cordon with access limited to those that have a good reason to be within it (see Figure 43 on page 181).

Appendix 1 in this document is a comprehensive review of transport packaging and labelling which makes good training material for those who may have to identify such a package if called to an accident scene as a first responder.

In TECDOC-1162 (Ref. 257) the IAEA aim to provide practical guidance for emergency response that, if implemented, will provide a basic assessment and response capability needed to protect the public and the workers in the event of different types of radiological emergencies (excluding reactor accidents) consistent with international guidance.

10.2. Specialist nuclear emergency planning guidance (UK)

In addition to the REPPIR-19 regulation, ACOP and guidance, the UK government has published nuclear emergency planning guidance to help local planners, cross-Whitehall departments, Devolved Administrations and agencies carry out nuclear emergency planning. The Nuclear Emergency Planning and Response Guidance (NEPRG) (Ref. 195) is the primary source of guidance for local planners to enable them to write effective plans.

The guidance is published in five parts:

- Concept of Operations
- Part 1 - Preparedness
- Part 2 - Response
- Part 3 - Recovery
- Annexes

KEY DOCUMENT	Nuclear Emergency Planning and Response Guidance (NEPRG) (Ref. 195)
This provides the UK government non-statutory advice to nuclear emergency planners.	

10.3. General emergency planning guidance to local authorities

This section looks at guidance that is intended for the non-specialist who has a role in the oversight of a local authority's emergency preparedness.

The Ministry of Housing, Communities & Local Government has issued advice to local authorities on preparedness for civil emergencies (*A Good Practice Guide for Chief Executives*, Ref. 258). This document, which is not intended to be prescriptive, lists 10 aspects it would expect to see in a well-prepared authority.

In the introduction Jake Berry MP states that the nature of emergencies facing us continue to increase in variety and complexity "*terror attacks in London and Manchester, the use of nerve agent in Salisbury and the devastating fire at Grenfell Tower have tested the resolve of our communities and reminded us all of the importance of local authority leadership in times of crisis*".

KEY DOCUMENT	A Good Practice Guide for Chief Executives
This is a document that should be read by the local authority manager responsible for emergency planning and by the Chief Executive of the authority fairly regularly. It might be a useful one to have on the desk when performing periodic readiness or accountability reviews. At a mere 27 pages it makes a welcome change from some of the door stop guidance documents that sap the will to live. Local authorities' Preparedness for Civil Emergencies: A Good Practice Guide for Chief Executives, November 2018	

The section entitled "are you ready" asks nine questions about:

- the ability of residents to contact the authority including out of hours and on bank holidays;
- the authority's ability to communicate risks and to warn and inform people in the area, including tourists and other visitors, before, during and after an emergency;
- the ability of senior staff and elected members to manage in an emergency, including managing the authority's role and as a senior officer within the strategic response and recovery mechanisms;
- the testing of plans to ensure that they address local risk and resilience standards and describe how to maintain essential services, ensure business continuity and contain agreements with other local authorities for support in a crisis;
- the resilience of supply chains;
- the role of the authority within the LRF and the accuracy of the role description in multi-agency plans and the ability to deliver;
- resource plans for no notice and sustained emergencies, including over holiday periods;
- a knowledge of the authority's strengths and weaknesses and where support and further resource can be obtained;
- the ability to assess the full impact of an emergency on community needs.

The report discusses the role of the local authority in civil resilience in terms of the CCA, in which the local authority is a category 1 responder.

Importantly the regulations or guidance require that the local authority "*to ensure that they can continue to exercise their functions in the event of an emergency. The duty relates to all functions, not just emergency response functions*". This appears to include the ability of an authority to provide support to any population within protective action zones during a nuclear emergency. "*Plans should be clear about what operational support the local authority will put in place for different emergencies, and how this can be activated in and out of business hours...*". "*Plans must be clear about how this support will be activated and managed. This support could include on the ground community alerting, for example, door knocking, checking on vulnerable residents, operating rest centres and providing on-going welfare support for people directly affected by emergencies*".

To achieve this, all business-critical functions should have robust business contingency plans for the services that the authority delivers and those that are contracted out. An example that is pertinent to nuclear authorities is "*consider building 'all-risk resilience' into contracts (for example, how to ensure domiciliary care is delivered during petrol shortages or severe disruption to transport networks)*". Reinforcing this is the statement that "*Local authorities are expected to manage the humanitarian aspects during emergencies. This requires staff at all levels to be effectively trained to deal sensitively with victims and survivors, including their friends and family. Training and exercising reserves and volunteers builds a further level of resilience in the event of concurrent or long duration incidents*".

An example of communicating with the public given is that of Calderdale Council which utilised social media during floods in December 2015. It is claimed that the council reached over 420,000 people on Facebook and received over 1 million impressions on Twitter.

On the matter of community leadership, the document states that "*The public, media and politicians will also look to the council to provide information and clarity on what has happened, what is still happening and what will happen next*".

The importance of training and exercising is stressed. It builds "*confidence and competence to enable robust delivery of the local authority role whether it is delivered in response or recovery*".

A related document, "A councillor's guide to civil emergencies" (Ref. 259) has also been updated recently (November 2018). The forward mentions that the last version was issued in 2016 but reports that a lot has happened since then with terrorist attacks, tower fires and nerve agent attacks within the UK. This document introduces some core terms including the definition of a civil emergency, the difference between a rising tide and a no-notice event, categories of responders and levels of response. Several case studies are discussed.

The overview section outlines the responsibilities of councils and individual councillors and, later, the role of leaders, portfolio holders and Ward councillors. These headings are then revisited in sections devoted to "preparedness and resilience", "response" and "recovery". Appendices give suggested questions for leaders/portfolio holders (Appendix 2) and scrutiny committees (Appendix 3) to ask.

This document is a good read and a useful resource for councils and councillors.

The UK Government document "*Expectations and Indicators of Good Practice Set for Category 1 and 2 Responders*" (Ref. 260) provides further guidance to category 1 and category 2 responders.

11. Nuclear off-site emergency planning

> *(1) Where premises require a planning zone under either or both of regulations 8 or 9, the local authority must make an adequate off-site emergency plan covering that zone or zones.*
>
> *(2) The plan required by paragraph (1) must be designed to mitigate, so far as is reasonably practicable, the consequences of a radiation emergency outside the operator's premises.*
>
> *REPPIR-19 Regulation 11(1) and 11(2)*

This section discusses the process of writing an off-site plan.

11.1. Introduction

A nuclear emergency is a situation in which radioactive material is imperilling, or may imperil, members of the public. In these situations, the members of the public may need the conventional support required by people caught up in an emergency such as warning, safety advice, transport, a safe haven in which to rest and shelter, reuniting with family and friends, sustenance, immediate and longer-term financial support, legal advice, health (including mental health) monitoring and support. They may also have special concerns about radiation and contamination and the fear of short term and long-term health effects.

11.2. Writing an off-site plan

REPPIR-19 regulation 11 puts a duty on the local authority to write an "*adequate off-site emergency plan*" where an operator's risk assessment (regulations 4 - 7) has led to the determination of a detailed emergency planning zone or where the regulation 9 requires outline planning. This plan must be designed to mitigate, so far as is reasonably practical, the consequences of a radiation emergency outside the operator's premises. Refer to Section 9.8 for an outline of the REPPIR regulations.

The ONR expectations for producing off-site plans are expressed in Ref. 228. This document contains the statement that "*the process of producing off-site plans should be embedded within the local authority's management systems. There should be arrangements for the review, consultation, amendment and formal approval of the off-site plan with clear governance arrangements and oversight. Arrangements to test the plan and implement lessons learned following exercises, events, and other learning opportunities (e.g., other local authority's exercises) should be evidenced including root cause analysis and prioritisation where appropriate*".

Part 2 of REPPIR schedule 6 contains the minimum information the plan must contain and schedule 7 lists some principles and purposes "*to which the plan must have regard*".

One process for preparing an off-site plan might be (Figure 30):

- **Management system** -Ensure that you have a management system that ensures that the process for checking that the off-site is consistent with the regulations, guidance and with the ONR expectation expressed above. This management system would have to identify the person or role responsible for the writing of the plan, the steps taken to produce the plan and the various parties that should be consulted or informed a progress and the person or role responsible for the formal approval of the plan.

  One aspect of the management system would be an action tracking system to capture and monitor progress of any actions agreed, particularly in the exercise debrief process, to ensure that any additional training or updates to documents, training materials or facilities are completed.

- **Risk assessment/Scope** (see section 11.4) - This would entail a review of the Operator's consequence report as an absolute minimum but should also review the range of accidents that are reasonably foreseeable and those that are less likely. It should look at aspects such as the presence or lack of warning time, the potential off-site dose rates and the potential duration of the release and how any of these might vary as the particular characteristics of the event reveal themselves. It also important to identify how you could know the magnitude of the fault as early as possible i.e. prepare to respond to the particular characteristics of any event. This process would probably entail detailed discussions with the Operator.

A useful addition to this section of the work would be a short list of scenarios that the plan may be required to cope with. This could include:

- A representative range of the faults suggested by the risk assessment including some that have no warning and some that have a warning period; a range in terms of the amount of radioactivity released including a release that just triggers the need for prompt protective actions, a release that requires prompt protective actions within the DEPZ range but no further, one that is a bit bigger and a significantly bigger one;
- A site based false alarm where members of the public misinterpret activities taking place on site;
- A transport accident somewhere in the county;
- A dirty bomb or threat of one in a built up area;

These are often presented as "planning assumptions". The potential to be facing multiple challenges at the same time should be considered i.e. the nuclear emergency might have been triggered by an earthquake or severe storm that has wider effects, changes priorities and diverts resources and attention.

- **Functional specification**
  Having considered the range of faults that require planning (either detailed or outline) the planners should consider the challenges facing a response; in particular what set of protective actions need to be applied as quickly as possible and are therefore candidates for automatic application on alert. They should also consider how they could learn that the automatic actions are insufficient and how quickly they would have to reach an agreement to extend the protective actions (see Section 11.3).

  This functional specification would be informed by the scenarios produced in the previous stage.

  See response times objectives in GS-G-2.1 (Ref. 190) for international guidance on this topic.

- **Consultation plan**
  Having made a proposal for automatic protective actions and considered how they might need to be extended based on circumstances the emergency planners need to discuss how these protective actions will be delivered and what resources will be required to monitor the situation, make recommendations, make decisions and implement those decisions.

  An important step is determining who to consult and how to consult them. Regulation 10(5) and the associated guidance lists some that must be consulted but as for most lists of this nature it ends with "*such other persons, bodies or authorities as the operator considers appropriate*".

  Consultant can take many forms from phone calls and emails to meetings of various sizes, workshops and exercise. It is unlikely that the plan can be prepared based on a single meeting. The important thing to ensure is that you consult those that are needed to deliver the plan and that the consultation ends with agreement about the plan and how it will be delivered.

  The approved code of practice requires (paragraph 293) that the local authority must ensure that the plan "*can be put into effect without delay when required ....*" . This is interpreted in guidance (paragraph 299) as "*the local authority should seek confirmation, to the extent possible, from other responding organisations that the underpinning capabilities required to implement the plan are in place and readily available for example by asking for written confirmation of this when consulting on and reviewing the plan*". This is an important step in the consultation and one of the key elements tightened by the 2019 revision of the REPPIR regulations.

- **Gap analysis**
  It may be worth reviewing the requirements of regulation and the associated ACOP and guidance, other relevant guidance and good practice/best practice guides and trying to identify areas in which you need to improve. Peer review is another potential tool.

  This is also the time to review the learning gained from any exercising or real deployment of any component of the plan to see if potential improvements have been identified.

- **Draft plan**
  The plan is likely to be a collection of documents that may include;
  - A scoping document summarising the risk assessment, the functional specification and the base scenarios;
  - A concept of operations document detailing how system is expected to respond to an alert. This would capture the roles and responsibilities of the various responders in the preparedness phase (maintaining facilities in a state of readiness and training sufficient people to meet the requirements of the plan) and in the response phase (responding in a defined timescale and in a defined manner). This could usefully describe a full range of scenarios with different initiators, different time structures, and different outcomes.
  - A collection of action sheets for each role, outlining their role and providing any data or information that they need to meet their duties;
  - A training plan detailing what experience and training each role needs and how it will be delivered and recorded;

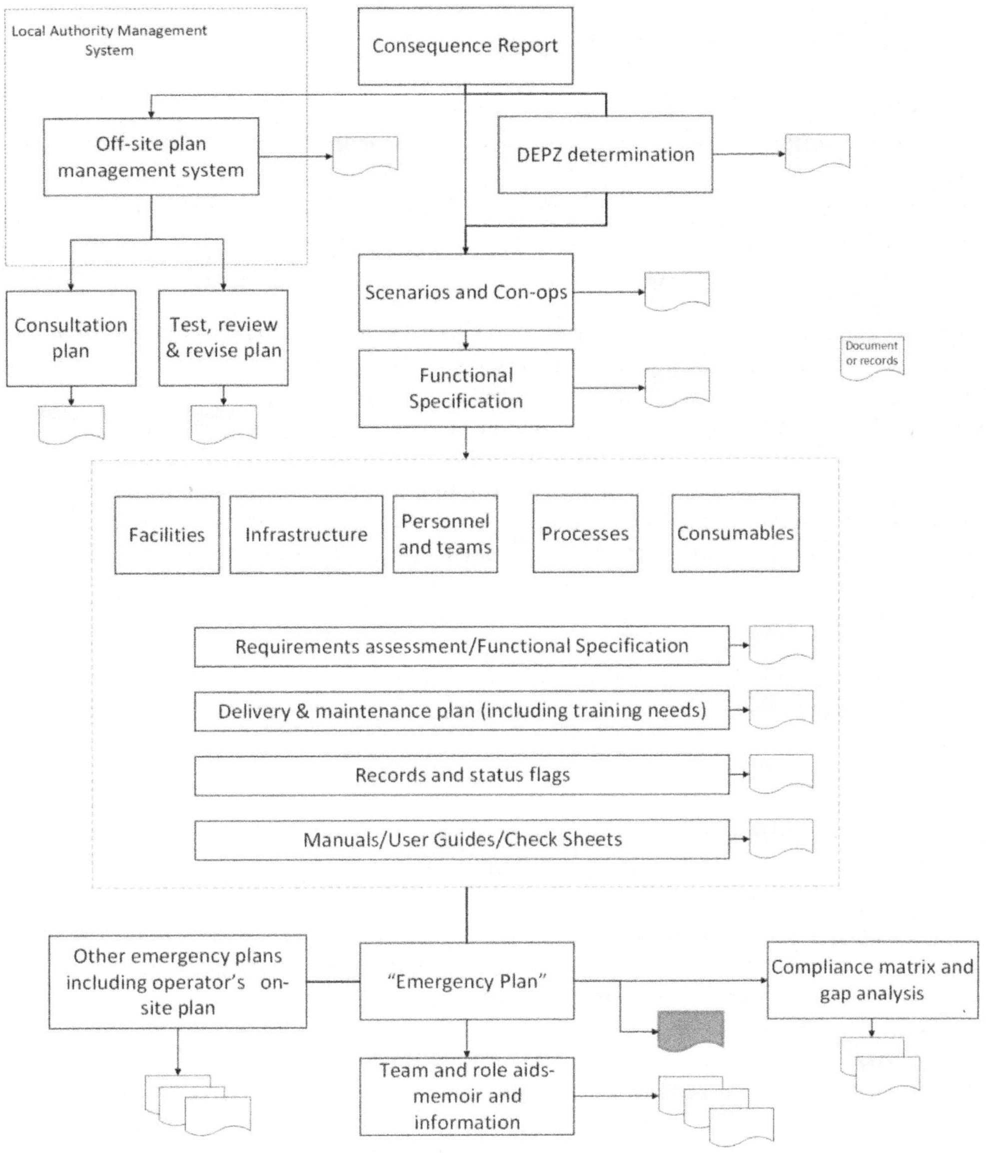

Figure 30 Plan writing process (Idealised)

- **Test Plan**
  If the plan is new or heavily revised it may be sensible to have a series of exercises or workshops to walk through the plan and ensure that the plan is complete and coherent and that everyone is comfortable that they understand and can deliver their organisation's role.

  At least once every three years a review and test of the plan is required under regulation 12.

- **Compliance matrix**
  Although not an absolute requirement it may be considered a good idea to have a document that shows how you intend to comply with regulations and guidance. This would point to items of planning and to processes or records against each identified requirement. This is useful when managing time used to maintain the plan and helps in accountability reviews with management and with audits from regulators.

Note: Most of the licensed sites in the UK are several decades old and have had off-site emergency plans since the aftermath of the Windscale fire. These sites may not have recently gone through the steps outlined above in the clear and logical order presented but they all have plans that have passed repeated testing and close scutiny.

11.3. Emergency planning challenges

The perceived challenges for off-site nuclear emergency plans have been:

- Can you get people in a defined area, or in the downwind sector from the site, into shelter within an hour or two for a no-notice event?
- Can you get people in a defined area to take stable iodine tablets within a few hours for a no-notice event?
- Can you use TV and radio to inform and reassure the public?
- Can you extend your response if the event is more severe than your automatic protective action strategy assumed?
    - Can you gather and explain the data required to support decisions on protective action extension?
    - Could you evacuate certain areas given more notice, say 24 hours?
    - Could you extend shelter and stable iodine distribution out further? and
    - Could you and the general public cope with an event which continued for several days?

The first three bullet points were considered appropriate for the worst set of "reasonably foreseeable faults" and the remainder for faults that could be imagined but were thought to be of too low a probability to maintain "detailed" plans to manage them should they occur (the term "reasonable worst case" is used in some circles).

Further challenges in the short term are:

If you have an area, say the DEPZ, subject to shelter and/or stable iodine advice could you manage to promulgate the message to those in the affected areas and respond to any questions or concerns raised by the public (including those outside the affected zones)?

These may include:

- Families split between inside and outside the affected zones;
- Concern for other members of family and friends (and pets or farm animals);
- Vulnerable people;
    - (a) Those requiring home visits for food, hygiene, medicines & other routine help;
    - (b) Those with physical or learning difficulties, less able to cope independently;
    - (c) Those with health issues (chronic and/or acute);
    - (d) Those who are worried;
    - (e) Those who cannot find or do not have sufficient iodine tablets;
- Normal emergencies such as fires, road traffic accidents etc.

In the longer term (in the days and weeks after the event) the challenge for the local authority becomes: "how are you going to support community recovery if your area has been subject to a nuclear emergency where people have had to shelter, take tablets and stop eating local foods and where the public are aware that the environment is now contaminated with radioactive material?"

11.4. Scoping the emergency plan

Before writing an off-site emergency plan it is sensible to pause for a while and consider what the plan has to be able to deliver in order to be successful or worthwhile. The principles and purposes of an emergency plan are discussed in REPPIR-19 (See Section 8). This gives the purposes as:

a. to reduce or stop the effects of the radiation emergency;

b. to reduce the exposure to individuals and to the environment resulting from the release of ionising radiation;

c. if necessary, to ensure that provision is made for the medical treatment of those affected by the radiation emergency; and

d. to prioritise the implementation of the plan in relation to a person exposed to a dose consequence in excess of the reference levels set out in regulation.

Given that (a) is the remit of the operator's on-site plan the local authority must produce a plan capable of satisfying (b) - (d) of the above list. Item (b) the reduction of exposure requires the consideration and implementation of protective actions, principally shelter, stable iodine and evacuation in a timely manner.

The Vulcan off-site plan (for example) gives the aim as "*to set out the initial arrangements*[10] *for dealing efficiently with an incident involving the nuclear reactor or the handling of nuclear fuel at Vulcan NRTE*" and the principal objectives as:

- *To clarify the roles of the participating agencies through defined responsibilities;*
- *To provide an initial response document for all the agencies;*
- *To ensure procedures are in place and remain relevant and appropriate and, therefore, meet both the relevant regulations and responding agencies requirements in order to achieve the aim of this plan.*

The AWE off-site plan gives the aims as to "*provide detailed information for all responding agencies to work to in order to facilitate the protection of the public and/or environment following an event involving an on-site accident at either of the Atomic Weapons Establishments". With objectives:*

- *Provide details about the sites and their hazards;*
- *Provide details of the roles and responsibilities of each responding agency;*
- *Provide details of activation, command & control and coordination;*
- *Provide details of countermeasures;*
- *Provide details of communications;*
- *Provide details of recovery;*
- *Provide details of where to find more information.*

The Torness plan (Ref. 261) has the purpose being to:

- *provide a framework for the initial response to an emergency at Torness Power Station, to mitigate the effects of a site incident or off- site nuclear emergency.*
- *outline the management structure at Strategic and Tactical levels established in response to any such emergency.*
- *highlight the Roles and Responsibilities of the participating Agencies when responding to an emergency affecting the Site.*
- *establish in broad terms the initial anticipated actions of the participating Agencies in responding to an emergency as specified above.*
- *provide a source of information from which relevant agencies can progress a 'procedures' manual for internal use by the staff of that Agency. It is strongly recommended that partner agencies do so.*
- *describe the various counter measures which can be taken to minimise the effects of an incident, and to identify how these counter measures can be applied.*
- *set out the arrangements for providing the public with specific information relating to the emergency.*
- *outline a post emergency recovery strategy for reinstating the off- site environs to a condition as close as possible to their pre-emergency status.*

It can be seen that there are a range of aims and objectives stated in UK off-site plans but they all cover similar ground.

Historically emergency planning in the UK has tended to be based on a *Reference Accident* which was towards the top end of the consequence scale without being ridiculously unlikely. (This might be considered to be similar to assessing the emergency planning needs for aviation by considering the required response to a passenger aircraft crashing in an inhabited area but not on to Wembley Stadium during the FA cup final). The concept of emergency planning was to have a flexible system, demonstrably able to cope with the Reference Accident and flexible enough to cope with different events whether they be of lower or greater severity.

REPPIR-19 requires the operator to provide the local authority with a *consequence report* which recommends a minimum distance for the prompt imposition of protective actions and a recommended minimum size of the DEPZ. The standard information provided by these reports is totally inadequate for the preparation of a local authority off-site emergency plan and the local authority needs to make good use of regulation 7(5) which requires that "*the operator must comply with any reasonable request for information made by a local authority, following receipt of the*

[10] I particularly like the recognition that the plan is the "initial arrangements". It would be expected that the people involved and the command and control structure may morph to respond to the particular characteristics of the event and the resources available should the response be protracted.

consequences report, required by the local authority to enable it to prepare the off-site emergency plan which it is required to prepare under regulation 11, within 28 days of the date on which that information was requested'.

When scoping an off-site emergency plan the planner would want to know:

- How likely it is that the site would suffer an event severe enough to require urgent action to be taken to protect the public?
- For each type of fault;
  - The likelihood of the fault (expressed as the relative likelihood of occurring in the next five years in REPPIR-19 but more usually as a probability of occurrence per year)[11];
  - How much radiation dose members of the public might receive as a result of the accident, broken down by age group, how much of that dose might be averted by prompt protective actions and how the avertable dose reduces as the delay in achieving the protective action increases;
  - The warning time that might be available between the initiating event and the start of the release;
  - The variability in the amount of radioactive material released and how the particular characteristics of the event will be determined;
  - How long the release is likely to last for and what factors might affect that (this has an important effect on the avertable dose);
  - What environmental pathways could lead to exposure of the public to an additional radiation dose or chemo-toxic hazard (this helps you choose the protective actions to include);
  - What factors, such as the weather conditions, the state of the plant at the time of the event, the actual details of the event, the possible actions of the operator or emergency services etc. might affect any of the previous points.

This information must come from the operator. It may not all be in the consequence report provided so the Authority may need to use the meeting with the operator provided by the regulations to gather the remaining information.

Some operators offer the local authorities a "composite fault" which has characteristics that bound the characteristics of faults in the REPPIR-19 DEPZ planning area of the risk framework. This would, for example, have the worst-in-class early releases, the worst-in-class total releases and a range of possible durations. Any plan that could credibly manage the composite fault should be able to manage any of the faults considered in its construction. This was essentially the philosophy of the "reference accident" concept used in years gone by; a fault, not always a "real" fault, at the upper end of consequences for the lower end of chosen likelihood range. This runs the risk that the plan will not be particularly well designed to cope with the particular characteristics of the release on the day.

Figure 31 shows the REPPIR-19 risk matrix with some made up faults shown in terms of their consequences and likelihoods. It is the sort of information that should be in the operator's consequence assessment and is really needed in the consequence report although not mandated by the regulations.

In this example fault A would not be tolerated by the operator's safety management process nor the regulator. Its likelihood is "very high", that is greater than I in 2 over the next five years and its consequences are "moderate" with an interruption to normal life (people in the area would be advised to shelter). Further engineering or procedural work would have to be done to reduce the likelihood or consequence of this type of event; preferably both.

The faults labelled 0 have no serious off-site impacts and are therefore not relevant to the off-site plan. The faults of immediate interest to the planner are those labelled B - G. They would want to know more about them as discussed above.

Fault B is the most likely of those that matter. This will potentially lead to shelter advice being appropriate as the projected dose is above the 7.5 mSv threshold at which 3 mSv, the lower ERL for shelter, is considered to be avertable. The next question is how quickly does shelter need to be imposed? Very different problems arise for a "no-notice explosive dispersion" event than for one that gives a generous warning period or is drawn out. The former is very difficult to prepare for, it would require a near instant response from the people downwind and their ability to get into

---

[11] In these schemes 1 in 20,000 likelihood of occurring in the next five years is the same as a probability of 1×10^{-5} per year.

cover quickly. At 4 $m.s^{-1}$ (about average wind speed in the UK (Ref. 279)) the plume could cover the first 100 m in less than half a minute and 2 km in just over 8 minutes. The only practical plan would seem to be "*if you hear an explosion and you see a plume heading your way then run (either downwind or cross wind), ideally into shelter*".

If fault B gives a generous warning time, or has a duration measured in hours rather than minutes then an emergency plan delivering shelter becomes practical. This could comprise of a warning system such as, for example, sirens, a man-in-a-van with a loud hailer or a phone system that calls all phones in the area with a warning message. The phone system could work well in an area used for businesses and homes which would be generously provided both with known phones and with nearby buildings in which people could take shelter. It may not work so well in open areas used by leisure walkers and the like who are harder to pre-inform so they may not know how to react, harder to contact and may struggle to clear the area or take shelter if the alert gets to them.

If we assume that fault B has a generous time structure and the area is well supplied with buildings suitable to shelter in then we could develop an off-site plan that credibly attempts to contact and warn people in the area and where they have a credible chance of modifying their behaviour, i.e. taking shelter, to reduce their exposure to radiation exposure.

But what if fault C, which is much less likely, but with similar projected dose is forward loaded in its consequences giving less time to react? Do we need a faster acting plan or do we propose that the difficulties of enacting a plan quickly enough put this fault into upper ERL rather than lower ERL considerations and initiate the fault B plan as the best practical, but not very effective option?

Fault D is at the bottom end of "very low likelihood" putting it at about 1 in 20,000 likelihood of occurring in the next 5 years. This is at about the threshold of needing an off-site plan within REPPIR-19. The consequences are shown as "significant" which denotes additional doses in the region 100 - 1000 mSv for those near to the site. 100 mSv is the default reference level. It is a level of dose that represents the boundary between being acceptable under very rare circumstances and being considered unacceptable for an emergency plan not to make every effort to reduce. Advising shelter may be sufficient to result in residual doses of below 100 mSv. If not then evacuation might be considered. In which case the local authority and police would have to work out how that might be achieved in the time allowed and how much effort would be appropriate to plan for evacuation and even test that it could work. That would be a matter for discussion.

Fault E is a similar fault but with a likelihood outside the REPPIR-19 emergency planning range. You would need an outline plan for that fault.

Always at the back of our minds we have the faults like G. Incredibly unlikely, but not impossible, and with devastating effect. The local authority planners know that their resources would be inadequate to respond to this scale of event and national and international aid would be required. Developing some idea about how to start a response (how many people live and work in the area? who to call for additional resource? what resources are available? what hurdles would the responders face? how could they be overcome? what practical help can be given quickly? and how can this be ramped up) would be prudent.

You don't need different plans for each type of fault but this sort of analysis reveals the range of protective actions that might be appropriate on different occasions and removes the risk of a one size fits all mentality.

This style of analysis allows the emergency planners to determine which protective actions should be automatic on initial promulgation of the alarm and over what range and which might need to be imposed or removed when a greater understanding of the situation is achieved. This would normally be based on recommending automatic protective actions out to the range where the averted dose from worst fault in a given frequency range under average conditions would equal the emergency reference levels (ERLs) set by the PHE. The impact of less likely but more severe faults (resulting in larger releases) or unusual weather conditions (resulting in less dilution of the radioactivity as it spreads downwind leading to higher doses for each distance downwind) would be used as a sensitivity study to inform the plan's ability to manage the "variable factors" that cause the actual consequences on the day to differ from the planning assumption.

Guidance warns that "*an excessively large area could divert important resource from affected areas which require the most attention*" (guidance para 197) and the PHE principles require that protective actions should be justified. Making the DEPZ bigger than appropriate should not necessarily be seen as a conservative decision to be applauded. It could be

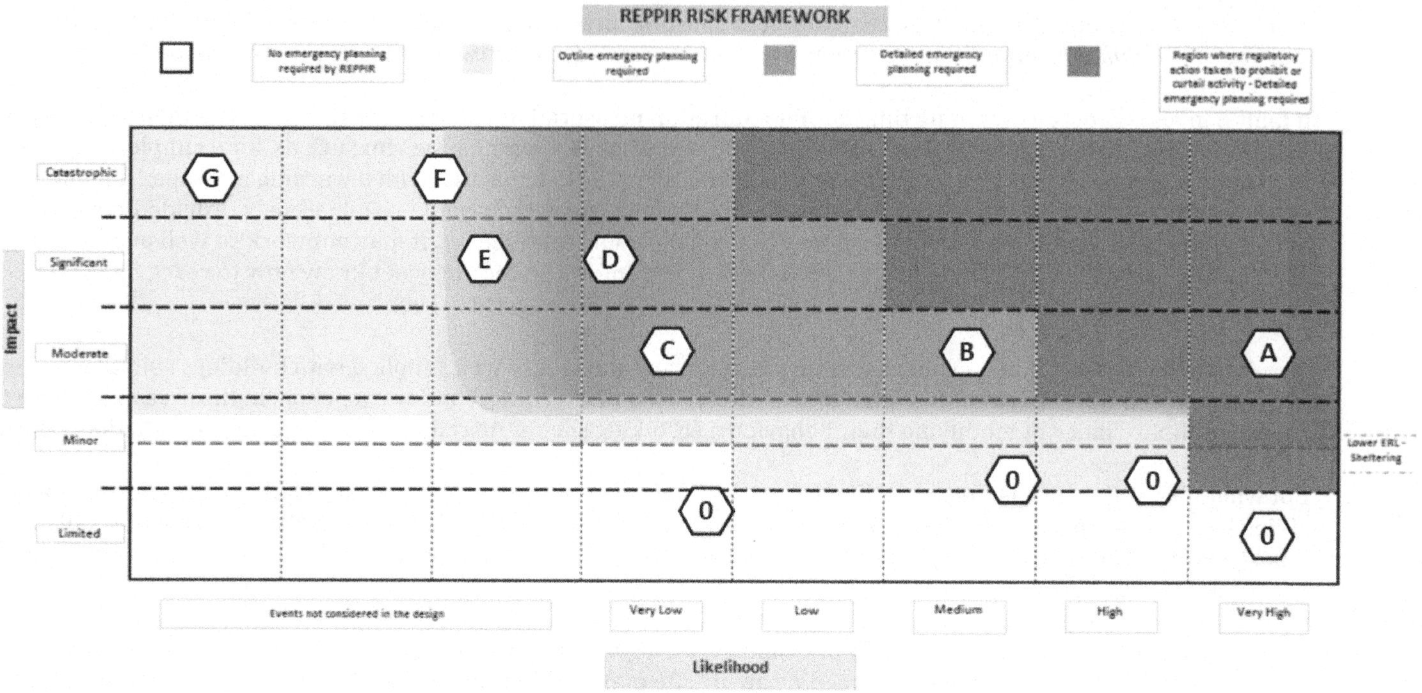

Likelihood descriptor	Relative likelihood of occurring in the next 5 years
Events not considered in design	less than 1 in 20,000
Very low	1 in 20,000 - 1 in 2,000
Low	1 in 2,000 - 2 in 200
Medium	1 in 200 - 1 in 20
High	1 in 20 - 1 in 2
Very High	Greater 1 in 2

Impact descriptor	Dose exposure (main quantifier, see table for additional potential factors
Limited	Background radiation levels
Minor	1 mSv - 10 mSv
Moderate	10 mSv - 100 mSv
Significant	100 mSv - 1000 mSv
Catastrophic	1000 mSv or more

Figure 31 REPPIR risk matrix (fictional)

a mistake.

The local authority is obliged by the regulation to set the DEPZ range at least to the distance recommended by the operator but can increase it and can also add "lumps and bumps" to base it on geographic features or to include nearby communities, or parts of communities, should that seem sensible. The local authority also seems obliged to take the operator's advice on the minimum distance for the prompt application of protective action advice.

11.5. REPPIR Detailed Emergency Planning Zone

Going back a few years the Detailed Emergency Planning Zone around a nuclear licensed site was a circle centred on the facilities with the potential for the largest releases on site and over which detailed emergency plans were written. The radii of the DEPZs were largely historical with Magnox sites being a few kilometres, AGR sites generally 1 km and the others depending on the perceived risks at some time in the past.

For the reactor sites the protective action strategy (in those days called the countermeasure strategy) was to declare an off-site nuclear emergency, call in the responders to the site and the Local Emergency Centre (LEC) (forerunner of the

Strategic Co-ordination Centre)[12] and send out off-site survey vehicles. If they detected airborne concentrations above a trigger value the operator recommended evacuation of that sector to a judged distance downwind. This was discussed in the Local Emergency Centre and, if agreed, an evacuation order with given and the police asked to evacuate an area with messages to the public being transmitted by radio. Members of the public were then to be herded to a reception centre where they would be given stable iodine tablets on arrival.

A significant improvement to these arrangements arrived with the introduction of automatic protective actions in which, on declaration of an off-site nuclear emergency, members of the public within the DEPZ were warned using an automatic telephone dialling system delivering a recorded message asking them to shelter and to take stable iodine tablets which, by this time were now pre-distributed to homes and businesses. Text alerts to mobiles may be the next step improvement.

The introduction of REPPIR-01 had little immediate effect on emergency plans, it was not until 2014 that ONR published revised principles for determining detailed emergency planning zones (DEPZs) (Ref. 263) which looked "*beyond the technical assessment of the hazards on site and also consider local practical and geographical issues in the operation of the off-site plan*". This was the introduction of "lumps and bumps" and the end of the circular DEPZ although some authorities had already started to define their DEPZ in terms of postcode region in the expectation that people would find it easier to judge if they were in the zone or out of it.

> REPPIR 10.5 (1) The local authority must determine the detailed emergency planning zone on the basis of the operator's recommendation made under (paragraph 2) of Schedule 4 and may extend that area in consideration of.
>
> (a) local geographic, demographic and practical implementation issues;
> (b) the need to avoid, where practicable, the bisection of local communities; and
> (c) the inclusion of vulnerable groups immediately adjacent to the area proposed by the operator.

Whereas in the past the DEPZ was the automatic protective action zone, this is no longer true. It is now acceptable for a site to have ranges of automatic protective actions that are smaller than the DEPZ, for example the pre-distribution of stable iodine tables around Sizewell was reduced to 1 km after the shutdown of the Magnox reactors at Sizewell A while their DEPZ stayed at to 2.4+ km (Ref. 264).

Operationally DEPZs tend to be divided into sectors using standard concentric circles and spokes. These make it easy for the different centres to communicate the areas where activity has been detected or where the avertable dose for a protective action may be above the appropriate ERL. These sectors are not obvious to anyone without the right map so protective action communication to the public tends to refer to geographic features such as roads, rivers, parks etc. or to post-codes for clarity.

In order to scope the required capability of an off-site emergency plan the operator must make some assumptions about the accidents it is seeking to provide protection for. This requires an understanding of several parameters about the site's risk profile and it is these that the consequence report and subsequent meeting are expected to communicate.

11.6. Concept of Operations

An outline of the UK's nuclear emergency planning from the view point of central government is given in the 2015 document Nuclear Emergency Planning and Response Guidance: Con-ops (Ref. 265) which outlines the structure shown in Figure 32 (more detail is given in the original diagram). At the local level co-ordination of the response is achieved by strategic, tactical and operational structures including the Strategic Co-ordination Centre. The tactical and operational structures allow the emergency services to co-ordinate their response to the site while the strategic level provides direction and resource to these responders. The strategic level, which includes a wider range of government bodies, regulators and experts also discuss the protection of the public including agreeing the protective action strategy

---

[12] The LEC was the forerunner of the SCC but was usually a facility financed by the operator, usually hosted by the local police service, often with a day job such as a training centre or office suite, and used as a command centre only for nuclear events.

and the media management strategy and media messages.

It is noticeable that this document makes no mention of the scenarios it has been planned against which maybe reveals a weakness in the UK planning process.

The structure is designed to "grow upwards" if the situation demands it. A smaller event (if you can have a "small" nuclear emergency) may only need the local components. There is, however, an apparent ambiguity in some of the relationships. The connectors are not carefully defined so could be taken to be "chats to". More clarity could be achieved by differentiating between information flow and instruction and how this may change with time and severity. Generally, information and requests for resource go upwards and instructions and advice go downwards from command and technical structures respectively.

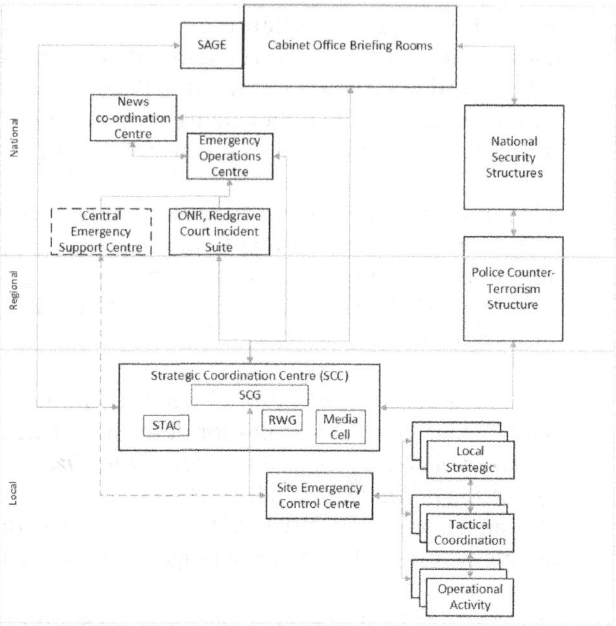

Figure 32 Nuclear emergency response Con-ops

The UK Nuclear Emergency Planning and Response Guidance Concept of Operations (Con-ops) document (Ref. 265) identifies the core characteristics of an effective response as:

- *Anticipation: Risk assessment and risk communication.*
- *Preparedness: Clarity of roles and responsibilities, supported by training, exercising and the availability of equipment and agreed procedures.*
- *Continuity: It is helpful if individuals and teams are asked to carry out similar roles, using the same tools as they employ in their normal work rather than expect them to do something very different or use unfamiliar systems with occasional training.*
- *Subsidiarity: Decisions should be taken at the lowest appropriate level, with co-ordination at the highest necessary level.*
- *Direction: Clarity of purpose should be delivered through a common awareness of the strategic aims and supporting objectives for the response agreed at the Strategic Co-ordinating Group (SCG).*
- *Integration: Effective co-ordination should be exercised between and within organisations both at the local and national levels of a response. For the emergency services this is second nature based on the JESIP principles. (Joint Emergency Services Interoperability Principles (Ref. 267).*
- *Communication: Good two-way communications are critical to an effective response. This is both about physical communications technologies and about situational awareness and shared understanding.*
- *Co-operation: Positive engagement based on mutual trust and understanding will facilitate information-sharing and deliver effective solutions to arising issues.*

It is interesting to consider the French plans as published in 2014 which discuss eight different scenarios including accidents at power stations with different release profiles, releases abroad or at sea and transport accidents (Ref. 272, kindly translated into English). Each scenario has a list of initial actions and objectives. The plans are broadly similar although the French government seems to have been more systematic in comparing their plans to a range of scenarios than is evident in the UK plans.

A review of Chinese emergency plans can be found in reference 273. Admittedly this is more of a public relations document rather than a technical document, but it allows us to understand that their system is more alike ours than different. They use the concept of defence in depth, have clear lines between the operator's responsibility on-site and the local authority and government responsibly off-site. They train at three different levels: state (nationwide), provincial (local and regional government) and operator. They exercise their plans regularly.

On protective actions it says "*Organizing personnel to implement emergency protection actions. If radioactive isotope of iodine has already been released or may be released arising from an accident, a professional team shall organize the public within a given area to take stable iodine pills to minimize radiation exposure to the thyroid gland. The people in the affected area shall be duly organized to take emergency protection measures deemed necessary, including hiding, evacuating, seeking temporary shelter or permanently moving out in order to prevent or minimize injuries from radiation exposure. Psychological assistance shall be provided in a timely manner to minimize social anxiety and panic*".

It does suggest a slightly different relationship with the general public. Under the heading maintaining social order it has *"Illegal or criminal acts, such as spreading rumours to create panic, shall be dealt with rigorously"*. Although it does undertake to provide accurate and timely information to the public and to the international community.

The document also suggests that they are further advanced on integrating emergency data on reactor state, drones, vehicle and vessel mounted detection devices and weather information to give a fuller picture of the off-site situation.

A high-level review of emergency preparedness in China can be found in reference 274. Canadian nuclear emergency plans are signposted at Ref. 275 and a concise review of Japanese nuclear emergency plans can be found in Ref. 276.

The output from the Con-ops document is an overview and justification of the proposed emergency plan with performance expectations for several linked facilities and consideration of the communications capabilities required between facilities and between facilities and those in the field. This then leads to functional specifications for the facilities and the identification of the individual and team roles that the responders will be expected to fill.

11.7. Staffing Emergency Operations Centres

Having agreed a concept of operations it is necessary to consider the people that will be needed to deliver the required response from each facility. Most such facilities will require the cooperation and coordination of people with a range of expertise and experience to quickly determine what is happening, reach conclusions and initiate actions.

There will be leadership of some description, determining what will be discussed when and how and also keeping an eye on how individuals and teams are coping; technical experts providing briefings and opinion (for nuclear emergencies these are likely to include health physics and facility personnel); technical teams researching and explaining the situation; operational people explaining what response teams are doing on the ground; support staff keeping information and communications technology running; the facility security personnel keeping every one safe, operational and administrative personnel recording meetings and tracking actions; support personnel keeping every one fed, watered and supplied with stationary and other needs, a veritable hive of activity.

Emergency plans generally come with a list of roles required to achieve the mandate of a facility. Exercises show if this list is complete or if new roles are required or assistants needed to make workloads more manageable.

Each such role should have a brief description of the role and the authority and experience likely to be necessary to be successful in it.

For almost all staff in local authority and company emergency operations centres attendance at the centre for training/exercises or for responding to a real event takes them away from their day job. Some are more willing than others to do this. This is an important consideration when selecting people, they are of little use to you if you cannot train them. Consideration also needs to be given to business continuity. If you may need to operate an emergency response of any nature for more than a few hours you need to understand what happens to the responders' day jobs; there is a limit to how long the demands of both roles can be satisfied.

The key roles within the process will probably have a duty rota to ensure that there will always be at least one person to fill the role at short notice if an alert is promulgated. Other roles may be on a "best endeavours" basis where several people are trained and express a willingness to respond, and a mechanism is in place to find and nominate one of them after the alert.

11.8. Warning and informing the public

There is little point in lots of experts coming together and devising a clever protective action strategy if you cannot get the message to the public that there is a threat to their wellbeing and give them advice on how to minimise the harm it can do to them.

The requirement to be able to warn and inform the public runs through many regulations and expectation sets, for example both REPPIR-19 and the Civil Contingencies Act. The UK government advice is that (Ref. 277):

Immediately when an emergency occurs, and during the first hour:

The public needs:

- basic details of the incident - what, where, when (and who, why and how, if possible);
- to know the implications for health and welfare;
- advice and guidance (eg stay indoors, symptoms, preparing for evacuation) and reassurance (if necessary).

The public wants to know:

- other practical implications such as the effect on traffic, power supplies, telephones, water supplies, etc;
- a helpline number;
- what is being done to resolve the situation.

Broadcasters will require:

- well-thought-out and joined-up arrangements between the emergency services, local authority and other organisations, capable of providing agreed information at speed;
- an immediate telephone contact;
- a media rendezvous point at the scene (adapted from the BBC's Connecting in a Crisis).

It also suggests that's some warning methods include:

- mobilising officers to go round on foot and knock on doors;
- from car or helicopter, by loudhailer or other amplified means;
- media announcements;
- electronic/variable message boards, eg at the roadside or on motorways;
- direct radio broadcasts to shipping (in maritime incidents);
- PA announcements in public buildings, shopping centres, sports venues, transport systems, etc;
- automated telephone/fax/e-mail/text messages to subscribers;
- site sirens.

To this list could be added drones (which are clearly moving from the development stage to mature deployment in many areas) with loudspeakers and possibly with two-way communications capability.

For fixed sites with off-site plans under REPPIR it is less likely that the police will be prepared to plan to mobilise officers to go round on foot knocking on doors. Site sirens, telephone-based warning systems, local radio messages and social media messages, are rather more appropriate.

Additional government advice on warning and informing can be found in "The Ten Step Cycle - an informal guidance note" (Ref. 278). The National Steering Committee on Warning & Informing the Public (Ref. 262) seems to have vanished without a trace.

11.9. Dose limits for responders

It is recognised that if there is an uncontrolled release of radioactive material from a site then members of the public and personnel from services supporting the public may be exposed to additional exposures to ionising radiation. These exposures must be controlled. The three radiological protection principles of justification, optimisation and limitation apply more to planned exposure situations but the last two: doses should be as low as reasonably foreseeable and significant effort is expected to reduce doses above or near stated thresholds apply during an emergency.

The situation for the public and for responders is not entirely equivalent. Whereas public would be expected to avoid radiation exposure we may ask responders to undertake tasks that entail additional exposure. An issue for the health physics profession is that in this context they are dealing with people with less understanding and experience of ionising radiation and radiological protection and working in a less controlled environment.

There is a requirement in REPPIR-19 regulation 18 (see Section 9.8.13) for employers to identify any of their personnel who may be exposed to radiation in the response to an emergency and to suitably train and equip them. They are also required to declare the upper bound of dose that they think appropriate for each role and to have a system to formally move into "emergency exposure" situations.

Each employer must have a management system that ensures that on the day of the accident emergency exposures are used wisely; that all activities that might lead to an IRR-17 dose limit being exceeded are undertaken in accordance with REPPIR regulation 18 by suitably trained, equipped and supervised volunteers who are appropriately briefed before the task commences. All doses must be as low as reasonably foreseeable.

The 2016 Torness plan (public version) (Ref. 261), for example lists the dose limits set for some of the responders[13]. These include:

Police Scotland have an operational dose limit of 5 mSv but in an emergency they have a range up to 100 mSv (Dose limit 1) for "*simple non-life saving rescues, maintaining important plant and reducing further doses to responders and the public*" and the range up to 500 mGy (Dose limit 2) for "*lifesaving intervention only and also includes intervention to save critical infrastructure which, if not attended, may yet threaten public and / or responder life*".

The Scottish Fire and Rescue Service work to:

- *Male: 20 mSv per year or single incident;*
- *Female: 13 mSv in any consecutive period of 3 months or single incident with a maximum of 20mSv per year; (Dose limits for the abdomen of women of reproductive capacity are more restrictive than those for men, and are intended to protect the fetus which, at some stages of development, is particularly sensitive to ionising radiation).*
- *Anyone: 100 mSv in exceptional circumstances and on a voluntary basis for life saving purposes.*

The Ambulance Services have:

- *Reference level one: 1 mSv total for all staff per event. This Dose Reference Level (DRL) applies to all general ambulance staff that have not been identified for specialist roles such as Special Operations Response Team (SORT) in Scotland and Hazards Areas Response Team (HART) in England and Wales;*
- *Reference level two: 5 mSv maximum dose per event. This DRL applies to those staff who have volunteered to be involved in CBRN decontamination roles;*
- *Reference level three: 100 mSv the maximum dose (for life saving operations) where the casualty cannot be immediately removed from the area of high dose rate or contamination. All ambulance staff can volunteer to be exposed to this level provided that they have been fully briefed and understand the implications.*
- *The SORT and HART teams work to an annual whole body dose limit of 20 mSv. These teams are generally the first choice for operations within the affected area.*

Interestingly ICRP 103 sets a medical exposure limit for comforters and carers of 5 mSv per episode suggesting that this level of exposure is acceptable for members of the general public who are providing valuable support to a vulnerable person (in this case a patient).

11.10. The co-ordination and decision-making process

An important part of emergency preparedness is determining how the efforts within and between the responding organisations are going to be co-ordinated and then providing the facilities and the training and testing that the system works. The emergency services are used to working together in a wide range of circumstances and, in the UK, have a formalised process for co-ordination, the Joint Emergency Services Interoperability Principles (JESIP) (Ref. 267). This recognises a three-tier chain of command; strategic, tactical and operational also identified as Gold, Silver and Bronze respectively. Figure 33 shows the UK Ambulance Service Chain of Command (Ref. 268).

The strategic (gold) commander is in overall control of their organisation's resources at the incident but will not be at

---

[13] Note these may be subject to change. Always check the current Plan.

the scene. The role is to determine the strategy and provide resources to enable the strategy to be realised. The gold commander sets the strategic direction, co-ordinates responders and prioritises resources. In the UK, the Civil Contingencies Act nominates the police to co-ordinate the strategic level of a multi-agency response.

The tactical (silver) commander co-ordinates the efforts at the scene determining the tactics required to meet the strategy. The silver commander can be in a distant control room, a command vehicle near the scene or at the scene, depending on the service and the situation. The silver commander interprets strategic direction, develops a tactical plan, and co-ordinates activities and assets.

The operational (bronze) commander(s) (there can be several for each organisation working on different aspects of the response) will be at the scene and is in direct control of their organisation's resources at the incident. The bronze commander executes the tactical plan, commands the single-service response, and co-ordinates actions.

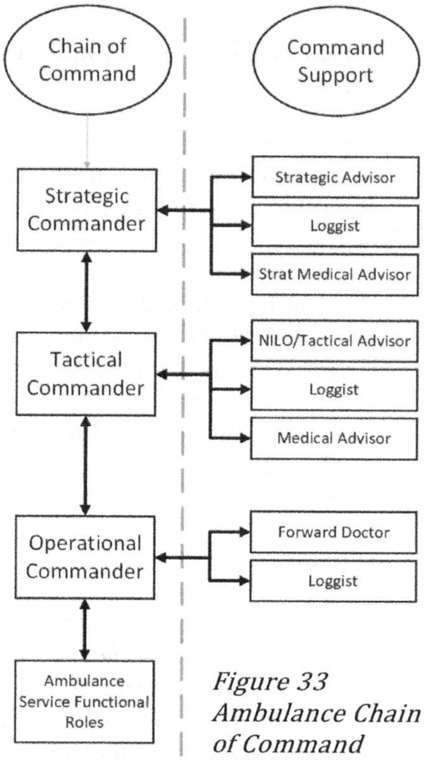

Figure 33 Ambulance Chain of Command

Cooperation is supported by communications systems such as Airwave (Ref. 269) (to be progressively replaced by ESN (Ref. 270)), a secure and resilient telecommunication system (voice and data) used by the emergency services and some operators and Resilience Direct (Ref. 271) an online private network which enables civil protection practitioners to work together across geographical and organisational boundaries during the preparation, response and recovery phases of an event or emergency. However, much of the co-operation is managed by face-to-face interaction in a number of control rooms.

A local authority view of Command and Control of an incident is shown in Figure 34 (Taken from the Devonport off-site emergency plan (DOSEP)(Ref. 266). In alignment with the con-ops this identifies the Strategic Co-ordination Centre (SCC) as the focal point for strategic co-ordination.

There are many organisations who are represented at the SCC. For example, the 2016 version of the Torness Off-site Plan identifies the following organisations expected to attend the Torness Strategic Co-ordination Centre (TSCC) in the event of an off-site nuclear emergency:

- EDF Energy
- Government Decontamination Service
- HM Coastguard
- Office for Nuclear Regulation
- Network Rail
- Scottish Government
- British Telecom
- City Edinburgh Council
- Police Scotland
- Midlothian Council
- NHS Lothian
- Scottish Borders Council
- Scottish Ambulance Service
- East Lothian Council
- Scottish Fire and Rescue Service
- Public Health England/ Chemical and Environmental Hazards
- Food Standards Scotland
- Animal Health Plant Agency
- Scottish Water
- Scottish Environment Protection Agency
- Scottish Power Energy Networks

The co-ordination and decision-making process in the event of an off-site emergency takes place in the Strategic Co-ordinating Group (SCG) within the SCC which would set strategic objectives at its first meeting and review them thereafter. The Torness plan, for example, suggests:

- *preserve life;*
- *prevent escalation;*
- *ensure safety of personnel;*
- *co-ordinate the response; and*
- *provide public reassurance and maintain community confidence.*

The emergency plan may have suggested agenda for the Strategic Co-ordination Group and suggestions about who should be on it. During a response these would be kept under continuous review and adapted according to need.

Agenda items for the Strategic Co-ordinating Group could include:

- the situation at the scene of the incident, projections about the future and resources required and committed;
- discussion of the radiation levels around the area and how they might change over time - this is intended to inform discussions about the potential future doses and the dose that might be avertable;
- health issues including early and late protective actions and casualties taking advice from the STAC (this could include a brief discussion of the vulnerable people in areas subjected to protective action advice and their need for assistance);
- access to affected areas - this is important should there be a requirement to enter the area to provide support to the public;
- thoughts about how the event will end and any actions that are required to support and hasten the return to normality;

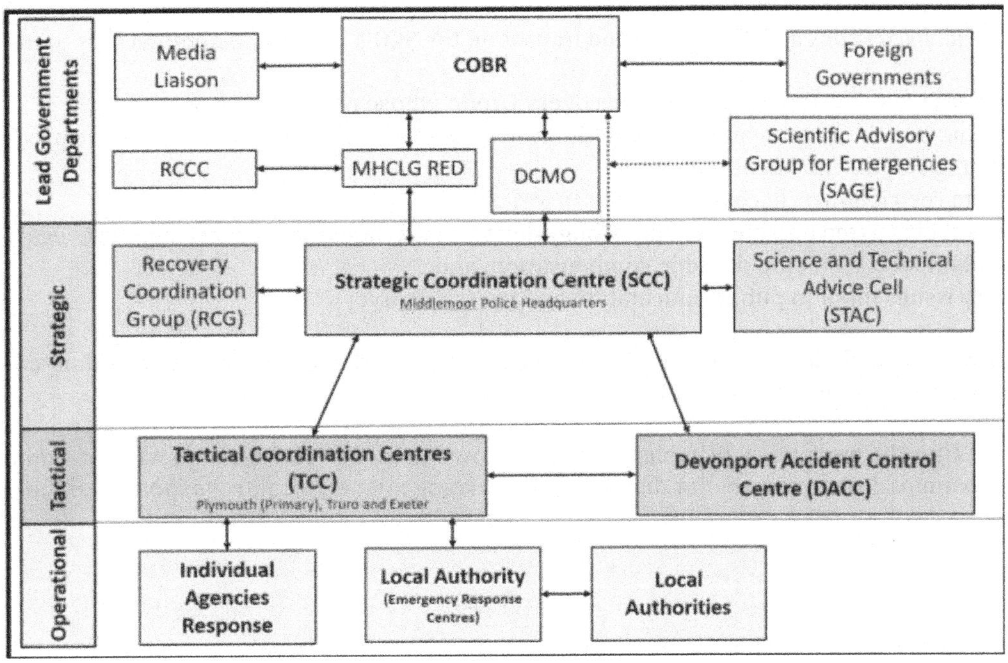

Figure 34 Groups and Facilities that communicate with SCC (Devonport)

Meanwhile the STAC will be meeting to discuss:

- a summary of the incident and actions taken. This should include a discussion of the likely and possible progression of the event (most likely outcome and worst reasonable outcome and how you will be able to tell which is nearer the truth);
- the protective action strategy; the options, the uncertainties, the effectiveness
- risk analysis including the safety of responding agencies' personnel.

There is an implicit assumption within the decision-making process, and in particular within the science advice, that there is an objectively correct answer to where protective actions should be recommended; where they will clearly be of a net benefit when weighting the disparate costs and benefits including taking into account the inevitable untidiness of application. It is expected that the STAC or SAGE will discuss the situation as reflected in their information feeds, reach a learned consensus and identify the one right answer with all other options being wrong. This is simply not true. The situation may be changing. The release prognosis will be subject to uncertainties about the effectiveness of steps taken by the operator to regain control in addition to the inherent uncertainties about how a damaged system will behave. Different people will have a different understanding and will react to these uncertainties differently. The actual dispersion, which is complex, and the modelled dispersion, which is subject to simplifications and parameterisation, may not match particularly well. Whatever protective actions are recommended it is likely that some members of the public will not get the message, some will misinterpret it, some will be unable to comply, and some will ignore it.

We also know from experience of Covid-19 that the media and social media will find "experts" of their own to second guess the SCC advice, to "explain" it in their own way and to express contrary opinions. This is likely to be unhelpful. A tension to be aware of in the media circle is that some organisations or individuals will want a rapid briefing of the public while others will want a more cautious approach where care is taken to check and coordinate facts and media lines.

It is in STAC that the discussions about the costs and benefits of personnel entering the protective action zones to provide services to the public might be discussed. These discussions would consider

- the situation the member(s) of the public find themselves in;
- how stressful, uncomfortable and damaging to mental and physical health this could be if support is delayed (this would be related to the discussion of the release prognosis and weather forecasts);
- the options for providing the required/desired support;
- the dose and risk to the personnel asked to provide support if that is the decision;
- how to communicate the risks to the support personnel and the affected members of the public; and
- how these discussions are affected by and impact on the SCG's discussion about access to the affected areas.

Longer term issues would be discussed in the Recovery Group whose remit includes:

- continuing health issues including restrictions on consumption of agricultural/ marine produce;
- extent of environmental contamination;
- long term restriction on access;
- support of evacuated persons including long/short term re-housing, financial support, legal advice etc,
- financial impacts including possible compensation; and
- recovery issues input to public information and press releases;

It should probably also include the need for a Humanitarian Support Centre to provide co-ordinated support to those affected (and those who think they were affected).

After the event (after the emergency plan has been stood down) a ministerial decision will determine the reference level (target maximum dose to member of the public) and arrangements to control exposure to the public in affected areas (Ref. 88).

11.11. IAEA

The IAEA is not visible at the local level to a response in the UK but we do have responsibilities to inform them of nuclear emergency alerts and provide situation reports. The IAEA considers that it has five roles during the response to a nuclear or radiological emergency, irrespective of its cause (Ref. 286).

- Provision of notification, exchanging and sharing official information with States and international organizations;
- Performing assessment of potential consequences and prognosis of likely emergency progression;
- Timely co-ordination and provision of assistance or advice, upon request, to States/international organizations;
- Provision of timely, clear, factually correct, objective, consistent and easily understandable public information;
- Co-ordination of the interagency response.

11.12. The emergency planning community

This section briefly reviews the players in the UK emergency planning community, looking at their role in the preparedness phase.

The Civil Contingencies Act (Ref. 216) divides emergency responders into category 1 and category 2 responders and others and places demands on category 1 and 2 responders as discussed in reference 287.

11.12.1. The local authority

The principle local authorities are Category 1 responders in accordance with the Civil Contingencies Act (Ref. 216). They have clearly defined responsibilities in relation to civil emergencies and will typically lead the recovery from any emergency in their area. Under the Act, an emergency is defined as "*an event or situation which threatens serious damage to human welfare in a place in the United Kingdom, the environment of a place in the UK, or war or terrorism which threatens serious damage to the security of the UK*".

The ONR maintain a list of the local authority associated with each licensed site (Ref. 288).

As a category one responder the council must:

- assess the risk of emergencies occurring and use this to inform contingency planning;
- put in place emergency plans;
- put in place business continuity management arrangements;
- put communications arrangements in place to make information available to the public about civil protection matters and maintain arrangements to warn, inform and advise the public in the event of an emergency;
- share information with other local responders to enhance co-ordination;
- cooperate with other local responders to enhance co-ordination and efficiency;
- provide advice and assistance to businesses and voluntary organisations about business continuity management.

Councils should maintain a set of fully developed, tested and up to date plans covering a range of different scenarios, based on locally identified risks, to enable them to play a full and effective part in the response to and recovery from an emergency. Within this remit, nuclear emergencies are often seen as a low priority as risks such as flood, animal disease, fire, pandemic etc. are seen as much more likely. In fact, the local authority emergency preparedness team is often responding to current events rather than planning future one.

The local authority has a number of duties under REPPIR-19. The first key duty is under regulation 8.1, translating the operator's consequence report into a decision about the size and shape of the DEPZ and the prompt protective actions that will be applied in the event of an off-site nuclear emergency being declared. This usually involves an officer writing a briefing paper for one of the council's committees after discussions with the Public Health England and the operator (see Section 9.8.6).

The local authority must prepare the off-site plan in accordance with regulation 11 and schedule 6 & 7 and associated ACOP and guidance. This is a major undertaking requiring consultation and detailed work. The plan must be reviewed and tested in accordance with regulation 12, and, in accordance with regulation 17(3), must take "*reasonable steps to put it, or such parts of it as are necessary, into effect without delay ...*"

As the author and holder of the off-site plan, the local authority will be a key player in the EPCC (Section 11.12.2) and the LRF (Section 11.12.3).

11.12.2. The Emergency Planning Consultative Committee

An Emergency Planning Consultative Committee (EPCC), or a close equivalent, can be found at many of the licensed sites. It provides a forum to consult on, discuss and develop emergency planning issues related to the site.

Its objectives may include:

- keeping the adequacy of emergency arrangements under review;
- Setting an exercise programme;
- Considering exercise reports;
- Providing advice and information to a Site Stakeholder Group (SSG);
- Considering issues raised at Site Stakeholder Group which could affect emergency planning;
- Keeping under review the information provided to the public within the Detailed Emergency Planning Zone.

Membership would include the range of agencies involved in planning and response including the Emergency Services, Local Authorities, Government Departments, and public utilities together with specialist staff from the licensed site(s). It would meet as often as necessary to discuss any changes to the site's hazard evaluation, any learning from exercises or real events, any changes to the component organisations, the testing, review and updating of the plan.

11.12.3. The local resilience forum

The Local Resilience Forum (LRF) is responsible for the overall direction and policies for emergency planning and preparation in their police area to ensure the effective delivery of duties under the Civil Contingencies Act 2004 (CCA)

(Ref. 216). A Local Resilience Forum (LRF) is not a legal entity, nor does a Forum have powers to direct its members. Nevertheless, the CCA and the Regulations provide that responders, through the Forum, have a collective responsibility to plan, prepare and communicate in a multiagency environment in areas defined by police areas.

> *The purpose of the LRF process is to ensure effective delivery of those duties under the Act that need to be developed in a multi-agency environment and individually as a Category 1 responder. In particular the LRF process should deliver:*
>
> - *the compilation of agreed risk profiles for the area, through a Community Risk Register;*
> - *a systematic, planned and co-ordinated approach to encourage Category 1 responders, according to their functions, to address all aspects of policy in relation to:*
>   - *risk;*
>   - *planning for emergencies;*
>   - *planning for business continuity management;*
>   - *publishing information about risk assessments and plans;*
>   - *arrangements to warn and inform the public; and*
>   - *other aspects of civil protection duty, including the promotion of business*
>   - *continuity management by local authorities; and*
>   - *support for the preparation by all or some of its members of multi-agency plans and other documents, including protocols and agreements and the co-ordination of multiagency exercises and other training events.*
>
> The role of Local Resilience Forums: A reference document (Ref. 289)

See *The role of Local Resilience Forums: A reference document* (Ref. 289) for more details.

Expected standards are published in *National Resilience Standards for Local Resilience Forums (LRFs)* (Ref. 290)

11.12.4. The emergency services

The police will normally co-ordinate the activities of those responding to a land-based sudden impact emergency, at and around the scene. For the police, as for other responders, the saving and protection of life is the priority. There are however exceptions, for example, a fire and rescue authority would co-ordinate the response at the scene for a major fire.

In a nuclear emergency the police will chair the strategic co-ordination group, will provide traffic control and cordons within affected areas, and will be collecting evidence of any criminal activity. They will also attend the site to gather information and provide advice.

Where necessary, the police process casualty information and have responsibility for identifying and arranging for the removal of fatalities. In this task, they act on behalf of HM Coroner, who has the legal responsibility for investigating the cause and circumstances of any deaths involved.

The primary role of a fire and rescue authority in an emergency is to extinguish any fire and rescue anyone trapped by fire, wreckage or debris. On-site they will work closely with the operator's emergency teams. They may also assist ambulance services with casualty-handling, and the police with the recovery of bodies.

As part of the NHS, Ambulance Trusts have the responsibility for responding to and co-ordinating the on-site NHS response to short notice or sudden impact emergencies. This includes identifying the receiving hospital(s) to which injured people should be taken, which depending on the types and numbers of injured, may include numerous hospitals remote from the immediate area where the incident has occurred.

11.12.5. Health Services

The UK's NEPLG Guidance (Ref. 291) lists the Department of Health's role as:

- Provide the focal point on public health issues at national level;
- Activate DH Incident Response Team and, if necessary an Operations Centre; and

- Alert Regional Directors of Public Health and Clinical Commissioning Groups (if required) and provide authoritative information to regional office and health authorities.

And the National Health's as:

- Coordinate the NHS capability needed to support the local health response to a radiation emergency;
- Identify and initiate NHS mutual aid to deliver an effective health response; and
- Deploy staff to the STAC as required.

Noting that "*In England and Wales, Clinical Commissioning Groups deliver this function. In Scotland it is delivered by NHS Boards and the local health board will take the lead in establishing the STAC and providing advice to the SCG and other responders*"

NHS Emergency Planning is signposted from Reference 292. The high-level National Plan is at Reference 293. The radiological planning within the CBRN section is a bit thin.

The NHS Scotland "Preparing for Emergencies" (Ref. 294) has this to say about preparing for radiological accidents "*Radioactive material is widely used across industry, healthcare and research and may be released at or whilst in transit to or from such sites or accidentally released from a nuclear reactor. Nuclear incidents generally result from accidental leaks at nuclear sites or CBRN/malicious acts with potentially widespread effects from blast, heat and radiation. The response to the effects of an ionising radiation release from a radiological or nuclear incident and the measures required to mitigate them are broadly similar although management of the consequences would differ significantly*".

Detailed expectations regarding the ability of NHS Scotland to respond to casualties from an accident at a nuclear site can be found in Reference 295. This document lists seven particular scenarios with a brief outline of which organisations have responsibilities to prepare for and respond:

- Overseas reactor accident;
- UK civil nuclear reactor site;
- Military nuclear reactor accident;
- Nuclear fuel or waste in transit (civil or military);
- Nuclear weapon transport accident;
- Satellite with nuclear power or material returning to earth;
- Incident arising from industrial/agricultural/scientific/educational/medical use of radioactive material.

By comparison, the Ontario Canada Radiation Health Response Plan dated 2014 lists 18 different scenarios (Ref. 297) covering accidents and deliberate acts and ranging from nuclear reactor accidents, to radiological spills and satellite re-entry and from dirty bombs to deliberate contamination of food supplies or individual poisoning.

11.12.6. The Civil Contingencies Secretariat

Civil Contingencies Secretariat (CCS) sits in the Cabinet Office as part of the National Security Secretariat. It supports the Prime Minister and Cabinet, and leads the wider government effort, on civil emergency planning and response.

According to the Cambridge University Centre for Science and Policy (Ref. xxx) the particular policy responsibilities of the CCS are:

- The National Risk Assessment and National Risk Register (identifying and assessing risks to national safety and security arising from terrorism, major industrial accidents and natural hazards, over 5 years);
- The National Security Risk Assessment (identifying global risks to UK security interests, in a 5 to 20 year timeframe);
- Leading a cross-government resilience capabilities programme to improve public sector response to such emergencies;
- Contingency planning and capability building for the risks of catastrophic emergencies;
- Policy for secure and resilient national infrastructure, and corporate resilience in the private sector.

Assessment of capability is overseen by the Resilience Capabilities Programme Board (RCPB) and, ultimately, by the Ministerial Sub-Committee on Resilience, NSC (THRC), which is chaired by the Chancellor of the Duchy of Lancaster.

NSC (THRC) is the National Security Council Ministerial Sub-Committee on Threats, Hazards, Resilience and Contingencies.

11.12.7. Department for Business, Energy & Industrial Strategy

The Department for Business, Energy & Industrial Strategy (BEIS) is responsible for co-ordinating civil nuclear emergency planning in England, Scotland and Wales at a national level.

BEIS would act as the lead government department in response to an emergency at a civil nuclear site in England, Scotland or Wales. The department would assemble up to date information on the emergency, including measures being taken to protect the public and make the site safe again. This is done by liaising closely with other government departments, the Welsh Government, agencies and organisations close to the affected site, such as police, fire service and local authority.

As the lead government department, BEIS's role in a response includes:

- ensuring that information is available to the public and media;
- alerting the International Atomic Energy Agency, the European Commission and various neighbouring countries of the accident;
- providing the Secretary of State for Business, Energy & Industrial Strategy, who is responsible to Parliament for civil nuclear safety matters, with briefing on the course of the accident and measures taken to protect the public.

11.12.8. NEAF

The Safety Directors' Forum's (SDF) purpose is to help the industry to work together to identify the key strategic Health, Safety, Security, Safeguards, Environment and Quality issues facing the nuclear industry and identify the appropriate industry response (Ref. 298). They have a number of subgroups including the Nuclear Emergency Arrangements Forum (NEAF) through which are organised a number of face-to-face and teleconference meetings each year.

NEAF helps to maintain national best-practice advice in the areas of on-site emergency preparedness and emergency response and was particularly busy post-Fukushima responding to the event and to the UK government review of nuclear emergency planning that followed the event (Ref. 299).

The regular NEAF meetings allow the operators and regulators to discuss learning from the exercise programme and from real events.

11.12.9. Regulators

The Office for Nuclear Regulation (ONR) independently regulates nuclear safety and security at 36 nuclear licensed sites in the UK, regulates transport and ensures that safeguards obligations for the UK are met. ONR was established as a statutory Public Corporation under the Energy Act 2013 (Ref. 198) which provides the framework of responsibilities and the powers of the organisation.

The ONR's and HSE's primary roles during emergencies are to provide advice to support the local and national EP&R arrangements, in addition to the regulatory roles of investigating and collecting evidence. Both competent authorities are Category 2 responders and are not front-line responders in implementing protective actions.

In the emergency preparedness stage ONR may play close attention to emergency arrangements including attending site Emergency Planning Consultative Committee meetings (See Section 11.12.2) and witnessing on-site shift exercises and training. It is ONR who formally assesses the adequacy of the operator and local authority tests of their emergency plans (every three years for the off-site plan under REPPIR-19 regulation 12 (See Section 9.8.10)).

The ONR discharge their responsibilities by:

- carrying out inspections of local authorities' compliance arrangements;

- by assessing emergency plans and arrangements, and/or;
- assessing emergency exercises that demonstrate the implementation of the off-site plan.

The ONR expectations for off-site planning are summarised in Ref. 228.

> KEY DOCUMENT ONR GUIDE Off-Site Emergency Arrangements
>
> This guide is written for ONR inspectors. Its purpose is to "*promote a consistent approach for compliance inspections of off-site emergency preparedness arrangements and tests of those arrangements (emergency exercises, Regulation 10(1)), under REPPIR*".
>
> It explains what the inspector should expect to see when inspecting a local authority off-site plan and provides, at annex A, an aide-memoir for the inspector.
>
> This document should be read by any local authority emergency planner responsible for an off-site plan.
>
> ONR Guide, Off-Site Emergency Arrangements (Ref. 228)

In an emergency response, the Redgrave Court Incident Suite (RCIS), supports the collation of information, provides the interpretation of the collected information, and is the central hub of information exchange for ONR. The RCIS coordinates the overall ONR response.

ONR fulfils its advisory role in many locations. This includes sending inspectors from ONR and other regulatory bodies to the Strategic Coordination Centre (SCC) to provide timely and authoritative advice to the central and devolved government. In addition, a team, led by the ONR Chief Nuclear Inspector and other regulatory bodies, are sent to the Scientific Advisory Group for Emergencies (SAGE), which advises the Prime Minister in COBR. Other ONR teams are deployed to the affected nuclear site and to the licensee or operator's off-site facility established to provide technical support for their on-site emergency response.

The Environment Agency/Scottish Environmental Protection Agency/ Natural Resources Body for Wales have a number of roles in a nuclear emergency including the supporting and interpreting of environmental monitoring, the minimisation of environmental harm, provision of advice to the public, STAC and SCG on the environmental impacts of the emergency and how to minimise them, providing advice on waste management and checking for breaches of discharge authorisations.

As a statuary consultee (REPPIR regulation 11(5)) they will be involved at all stages of planning and response, with a heightened role in the recovery phase.

11.12.10. Director of Public Health

The Director of Public Health is a chief statutory officer of their local authority and is an independent advocate for the health of the population. They should provide leadership for health improvement and protection and should provide the public with expert, objective advice on health matters (Ref. 229). For more information about this role see "What is a DPH?" (Ref. 230).

The Director of Public Health local to a licensed nuclear site is responsible for ensuring that there are appropriate arrangements for the prompt distribution of potassium iodate tablets and for authorising their administration. This is usually done in consultation with the nuclear operator, local authority and emergency services within the framework of existing planning procedures.

The Director of Public Health is responsible for the health of the population within their geographic area and therefore for ensuring the public receive appropriate advice on what to do in the event of an emergency.

The Director of Public Health is a potential chair of the STAC.

11.12.11. PHE-CRCE

PHE's Centre for Radiation, Chemical and Environmental Hazards provides radiological protection advice, services and training (Ref. 231). In a radiological or nuclear emergency, CRCE would provide public health advice and information relating to the radiological protection aspects of the emergency to colleagues at PHE Centres and other responding organisations including:

- the Scientific and Technical Advisory Cell (STAC) and the Recovery Advisory Group (RAG);
- the Military Co-ordinating Authority (MCA) at a Ministry of Defence (MoD) emergency;
- the relevant lead government department, such as the Department of Business, Energy and Industrial Strategy (BEIS) or Home Office (HO), as well as Department of Health and Social Care (DHSC);
- other health organisations such as NHS England.

CRCE can also send teams to carry out radiological monitoring of the environment or the affected population, analyse samples and undertake specialist biological dosimetry. Data from these monitoring activities is used in CRCE's assessments and subsequent public protection advice.

CRCE's emergency response is co-ordinated from the Chilton Emergency Operations Centre which gathers information from its teams in local and national centres, assesses monitoring information and the weather conditions, and provides advice based on its analysis.

CRCE also has trained radiation protection experts who can attend press briefings and give media interviews to inform the public. This resource would be used as part of a wider multi-agency media strategy co-ordinated by the police local to the incident.

In the event of a UK radiation or nuclear emergency, CRCE would be responsible for national co-ordination of radiological monitoring but not the tactical operation of any other organisation's monitoring assets.

PHE is to become part of the new National Institute for Health Protection (NIHP) in the near future (Ref. 232).

11.12.12. The Meteorological Office

Procedures And Communications in the event of a release of Radioactive Material (PACRAM)

PACRAM is a facility operated by the Met. Office which provides predictions of the trajectory of possible contamination plumes following an incident at a UK nuclear power plant. The operator sends them an alert giving the location of the release and any details they know, and the Met office returns an "area at risk" prediction. The value of this can be enhanced once more information is available.

The National Radiation Monitoring Network and Emergency Response System (RIMNET)

RIMNET is the UK Government's national radiation monitoring and nuclear emergency response system set up following the Chernobyl accident as part of its arrangements for dealing with the effects on the UK of overseas nuclear accidents. It was designed to aid the assessment of levels of radioactivity across the UK resulting from a nuclear accident overseas. It is not intended as an early warning or first alert system.

RIMNET has a network of 94 fixed gamma dose rate monitoring sites across the UK, automatically measuring, analysing and informing on background radiation levels 24/7. All measurement and reference data is stored in the UK National Nuclear Database. The original design intent of RIMNET was to detect clouds of radionuclides originating from nuclear accidents from overseas but provides usual background measurements for comparison with levels experienced during a nuclear emergency.

RIMNET has recently been upgrade around some of the UK's nuclear sites by the addition of one or two rings of detectors at distances from the site. These are more directly useful to responders as they indicate the location of the plume, its direction of travel and give a measure of the external dose rates and dose rate trends.

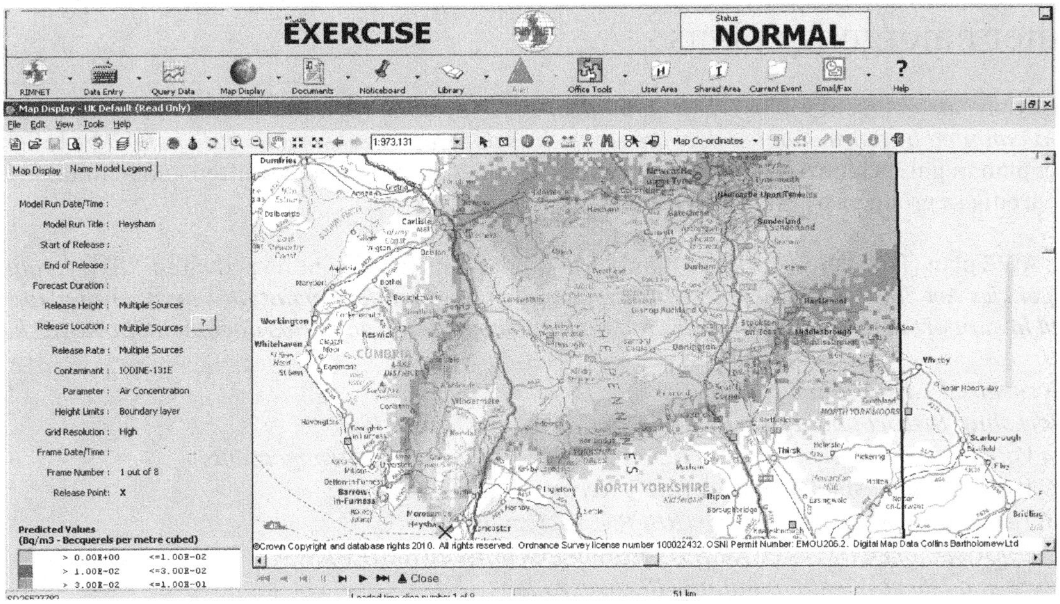

Figure 35 Area at risk plot from an exercise

RIMNET is managed by the Met Office, working in partnership with DECC and Defra on behalf of all government departments and agencies who would be involved in a radiological/nuclear incident, including those leading on non-overseas nuclear events.

Joint Agency Modelling

The Joint Agency Modelling (JAM) programme was one outcome of the review of nuclear safety and emergency response following the Fukushima event. It was described by ONR thus "*JAM (when complete) will provide a range of potential scenarios on how an incident might develop to be used at national and local level to inform medium/long term response. JAM will support strategic decision making at national level*" (Ref. 300). The design intent being that the nuclear operators would provide a series of source terms to the ONR and Met. Office which would be retained on file. In the event of an off-site nuclear emergency the operator would very quickly tell the ONR which accident they were having and which source term to use. The Met. Office would run this through their dispersion codes and PHE-CRCE would then interpret the results in terms of areas at risk and avertable dose and use it to help formulate, or at least confirm, protective action advice (Ref. 305).

A recent presentation suggests that JAM provides (Ref. 306: ((Figure 36).

- Assessments of the impacts and potential areas for protective actions;
- based on scenarios of what has been and/or what could be released (best estimate & reasonable worst case);
- to central UK Government to aid strategic decision-making during an emergency;
- by modelling AND reaching a consensus on the key points and uncertainties.

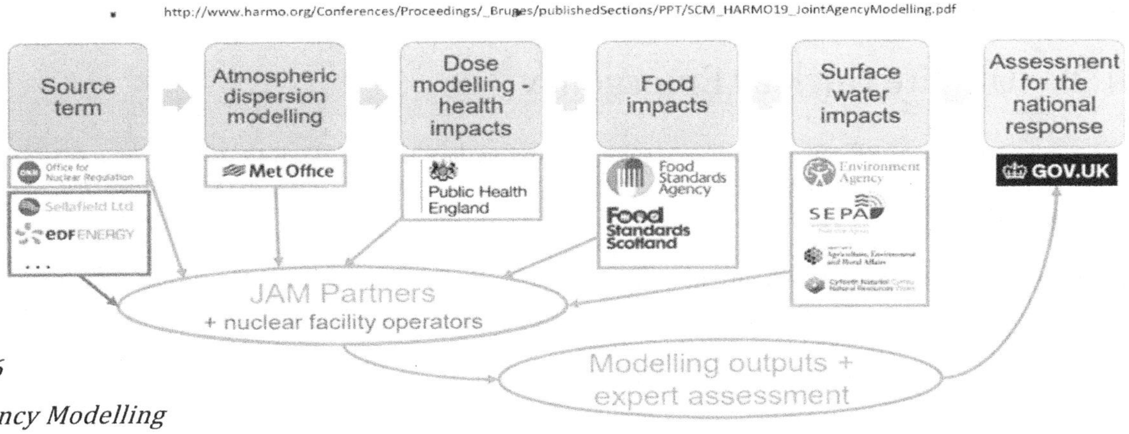

Figure 36

Joint Agency Modelling

11.13. Other emergency plans

REPPIR-19 states that "*an emergency plan is a document, or set of documents, that describes roles and responsibilities and may be supplemented by more detailed documents such as detailed event specific guidance*". The totality of a local authority off-site plan might include a range of capabilities described in documents written and maintained by the emergency preparedness group or by other responders.

For example the AWE plan (Ref. 307) refers to a number of other plans that might be activated. "*As a requirement of the Civil Contingencies Act 2004 (CCA) (Ref. 216), local Category 1 responders maintain a number of plans which may also be activated in support of this plan. There are too many plans to detail all in this document, however key plans include:*

- *West Berkshire District Council (WBDC) Major Incident Plan*
- *West Berkshire District Council (WBDC) Reception & Rest Centre Plan*
- *Thames Valley Local Resilience Forum (LRF) Emergency Response Arrangements*
- *Thames Valley Local Resilience Forum (LRF) Recovery Plan*
- *Royal Berkshire Fire & Rescue Service (RBFRS) Tactical Plan*
- *South Central Ambulance Service (SCAS) Incident Response Plan (IRP).*
- *Hampshire and Isle of Wight Community Recovery Plan*
- *Although not a Category 1 Responder, AWE also maintains On-Site Emergency Plans for the Aldermaston and Burghfield sites."*

11.14. Official declarations

In the UK there are two official "declaration states" that might be initiated at a nuclear power station that require a response from the local authority. Chemical plant and weapons establishments also have pre-agreed declaration states as discussed below.

11.14.1. Site Incident

At nuclear power stations site incident is defined as "*A hazardous condition which is confined in its effect to within the boundary of the site security fence*".

The operator may declare a site incident in response to events on site that threaten the safety of site personnel or may result in the failure of one or more of the layers of protection. The Operator will continually review the wisdom of remaining at Site Incident rather than escalating to an Off-Site Nuclear Emergency.

A Site Incident poses no real threat to the public. There may be a minor increase in the radioactivity released from the site but many site-incident scenarios release no radioactivity to the environment.

On declaration of a Site Incident the operator alerts the emergency services and the regulators as well as the UK and/or Scottish Government and the Local Council. There is no need for the SCC to set up on receipt of this declaration, but a heightened state of readiness might be considered.

11.14.2. Off-site Nuclear Emergency

At nuclear power stations an off-site nuclear emergency is defined as "*hazardous condition, which results, or is likely to result, in the need to consider urgent protective actions to protect the public outside the site security fence from a radiological hazard.*"

The Royal Navy add qualifiers to the off-site nuclear emergency declaration to provide more information. These are:

Radiation Hazard Confirmed	An off-site nuclear emergency in which a radiation hazard has been detected in the DEPZ
Release of radioactive material confirmed (RRMC)	An off-site nuclear emergency in which a release of radioactive material to the environment has been detected in the DEPZ
Recovery	An off-site nuclear emergency has transitioned into the recovery phase

The emergency notification chain for an off-site nuclear emergency reaches many more organisations than that for a site incident.

The operator will quickly recommend that members of the public in the DEPZ take shelter and, if it is an operating reactor event, take stable iodine tablets (pre-distributed at commercial sites or distributed on the day at submarine sites). This message is transmitted over a telephone warning and informing system that the public have been invited to register for and over local broadcast media. Thus, a number of members of the general public will be quickly alerted to the off-site nuclear emergency and will be advised to take protective actions. We can expect that very soon after the alert there will be a heavy demand for more information, advice and assistance that will last at least as long as the emergency.

Arrangements to set up the SCC and Media Briefing Centre (MBC) start as soon as declaration is made. The SCC takes over responsibility for advising on protective actions to protect the public and liaison with external agencies.

11.14.3. Cloudburst and Hexaflow

These code words are used in Cheshire.

Cloudburst is the generic codename for a chemical emergency while Hexaflow is specific to the Capenhurst Site and covers the operational response of the Police, Fire and Ambulance Services and the Local Authorities to an accident condition arising on the Capenhurst Site of sufficient severity to warrant the declaration of a Hexaflow Emergency.

The criteria for declaring a Hexaflow / Major Accident Primary Cloudburst Emergency will be the measurement of levels of HF at the site fence in excess of 3 parts per million (ppm) or visible HF fume being observed crossing the site perimeter fence.

Key declaration states in the USA for comparison:

Site Area Emergency (SAE) – Events are in progress or have occurred which involve actual or likely major failures of plant functions needed for protection of the public or hostile action that results in intentional damage or malicious acts; 1) toward site personnel or equipment that could lead to the likely failure of or; 2) that prevent effective access to, equipment needed for the protection of the public. Any releases are not expected to result in exposure levels which exceed EPA PAG exposure levels beyond the site boundary.

General Emergency – Events are in progress or have occurred which involve actual or imminent substantial core degradation or melting with potential for loss of containment integrity or hostile action that results in an actual loss of physical control of the facility. Releases can be reasonably expected to exceed EPA PAG exposure levels offsite for more than the immediate site area.

US NRC

11.15. Contents of an off-site plan

The information that should be in an off-site plan to satisfy the requirements of REPPIR are outlined in Schedule 6 Part 2 of the regulations with chapter 1 covering the DEPZ and chapter 2 the OPZ.

These cover the arrangements for initiating the plan (detection, declaration and alarm promulgation) and the arrangements to coordinate the response.

The first requirement is usually satisfied by explaining and defining the declaration categories, such as site incident and off-site nuclear emergency, and describing the call down chain and initial responses of those alerted.

The second requirement is usually met with a paragraph or two on the roles and responsibilities of the responding organisations and some diagrams showing how the different facilities and roles interact to reach decisions, set actions and monitor progress. This may include maps showing how to find facilities, details of access requirements and facility layout plans. The committee structure and proposed agenda for meetings may be included.

There is likely to be a section on protective actions. This will explain and justify the planned protective actions and show how they will be implemented, including the description of warning and informing systems and media briefing processes.

There is also likely to be a section on radiation dose control listing any emergency exposure limits and outlining the approval process.

There will also be data on the area such as various scale maps and pertinent demographic data that decision makers or tactical teams may want to refer to. This is likely to list schools, hospitals, care homes and any identified vulnerable groups.

The plan and associated documents serves a number of functions including:

- Providing evidence of the planning process and the agreements about the modus operandi between the responding parties;
- As a training tool to acquaint new responders and remind more experienced ones;
- As a working tool to remind responders of their role and interactions in real event.

12. On-site planning and response

12.1. Introduction

REPPIR-19 requires that an operator "*make as adequate emergency plan designed to secure, so far as is reasonably practicable, the restriction of exposure to ionising radiation and the health and safety of persons who may be affected by radiation emergencies identified by the evaluation*".

According to the ACOP the operator must ensure that the plan is a written document, or set of documents, and can be put into effect without delay when required. This latter requirement requires that the operator has personnel trained and ready to step into the plan quickly. For some roles this may require someone on site filling that role at all times with a formal handover at the end of a shift, for other roles the duty person can leave site but must be able to get back within some time frame (30 minutes, one hour and two hours are common targets depending on the role). Other roles, those in the "nice to have but not essential category" might have several trained personnel who do not go on a call rota but will respond if possible.

The site must also have any facilities and equipment that the responders are reasonably likely to need at hand. For some sites this means dedicated facilities and stores waiting for the alarm, for others it means meeting rooms or offices that can be converted to emergency use quickly and minimum stocks of those materials and equipment that might be required.

Maintenance of readiness is a role that should not be underestimated.

12.2. Emergency planning challenges

The challenges for emergency planning on a nuclear site are to keep the public safe (largely by preventing or reducing releases but also be providing timely warnings and advice), to keep the people on site safe including casualties, responders and those without a role, to protect the environment (again mainly by preventing or reducing releases), to understand the situation and to recover the site to a safe condition and then to facilitate recovery on-site and off-site. And to be able to do all of this at any time in the future, without warning.

12.3. Writing a plan

In principle the process of writing and maintaining an on-site plan might involve the following steps:

- Use the safety case to select a range of faults with different initiators and different developments;
- Work hard to reduce the probability of those faults and the consequences should they arise;
- Work out how you would recognise that those faults were in play, what information you would want to trace the progression and severity, what you would need to be done to reduce the consequences and bring the site back into a safe configuration and what people you would need with what training, equipment and co-ordination to make it happen. Summarise this in a "concept of operations" type document;
- Based on the analysis of faults, agree the "functional specification" of the on-site plan (What capabilities have to be bought to bear in what timescale and what conditions do they need to be able to work in?);
- Produce a concept of operations that shows how the site will respond to the full range of reasonably foreseeable events and how it will respond to some severe events. This should include the full progression from identification, alert, response, transition and recovery and consider corporate crisis management (strategic management of the company and its reputation during the crisis) and business continuity (to what extent, and how quickly, the site can recover and resume economic operation);
- Design, install, test and maintain an alarm system to alert you to the developing fault and a monitoring system to tell you how it is progressing;
- Identify, train and equipment the people you would need to put it right;
- Identify, develop, test and maintain the facilities that you would need;
- Ensure that your plan is well documented and that each role has all of the information it needs and any forms, tools, aids and action lists required to do the job;
- Ensure that everyone with a role in the plan, including external agencies, has the opportunity to discuss the plan, fully understands their role and maintains a state of readiness to respond in a time appropriate for their role;
- Maintain, test, challenge and improve.

The minimum contents of the operator's emergency plan are mandated in Schedule 6 of REPPIR-19.

The plan should explain how the initiation of an emergency will be identified (automatic alarms, monitoring instruments, call points), the arrangements made to ensure that these are always monitored, define any declaration states (such as *on-site incident* and *off-site nuclear emergency*) and state who is authorised to declare them and should describe how the alert will be promulgated. The call-out chain for a nuclear emergency is quite extensive and should be tested periodically to ensure that it works.

The plan should also detail the minimum number of responders and their roles and state the requirements on their availability. For a few roles there may be a 24 x 7 on-site requirement but for many of the emergency responders out of hours a target maximum time for return to duty station is stated. These tend to be in the 30 minutes to 2 hours range, depending on the urgency of the role.

For some operators, the nuclear emergency response organisation is at a constant state of readiness throughout the year, with in-hours and out-of-hours minimum manning levels. For others the state of readiness depends on the operational situation. For the nuclear submarine berths that are visited occasionally the organisation is stood down between visits and then re-established a few days before the submarine in due to dock.

The plan should flexible enough to cope with the range of potential radiation emergencies as identified in the safety case and any other unforeseen event of similar severity. Guidance (paragraph 690) states that:

"The plan should include:

(a) identification of the range and grouping of events with the potential to cause a radiation emergency as identified by regulations 4 and 5;

(b) the potential consequences of these events identified by the consequence assessment and the impact of any variable factors on the severity of the consequences (see regulation 10(2));

(c) the intended strategy for dealing with these events should they come about, including any relevant planning assumptions which have been made;

(d) details of any supporting procedures and documents in place to support the operator's emergency plan;

(e) *details of the employees who have roles to play in the emergency response, and their responsibilities;*
(f) *details of the facilities and communications equipment available to support the emergency response;*
(g) *details of the availability and function of special emergency equipment including fire-fighting materials, and damage control and repair items; and*
(h) *details of the availability and function of other resources."*

Arguably (a) to (c) are more about the planning process than the response plan.

It is not a great idea to have all of this information in one document and provide a copy of that to all potential responders. The scoping of the plan, elements (a) - (c) above with an outline of the remainder, could go into a "concept of operations document" or "emergency plan specification"; a more brief outline of the declaration conditions, the call out, the facilities, and the roles and responsibilities including command and control could constitute "the emergency plan" and then more details about what each role does, including check sheets, data sheets, forms and aide-memoirs could constitute the "emergency handbook" or "action sheets". An emergency directory could give all the required contact details.

Responders would be trained on the concept of operations and the emergency plan and their section of the handbook, looking at other handbook sections only so far as is needed to understand what the people around them in the response are trying to achieve.

In addition, the plan should contain "*arrangements for limiting the risk to persons on the premises*" including the warnings that are given and the expected responses. Usually alarms sound and personnel and visitors with no role in the response are herded into safe locations (for nuclear accidents, generally in-doors; for fire incidents in-doors or out-doors but evacuated from the building or zones affected) where they can be accounted for, informed and kept safe. At operating nuclear power stations stable iodine tablets are generally available at the muster points. Processes are generally in place to determine if anyone is unaccounted for by comparing registers of those on site against a roll call (section 16.2). Further systems seek to locate those unaccounted for.

The plan will contain an information management system that gathers pertinent information from the site and the surrounds, collates and interprets it and makes it available to the decision makers on-site and off-site.

The plan, or supporting documents, would provide each responder with an aide-memoir of their role. This could take the form of a check list of objectives and actions to be completed, a contacts directory of those they interact with, and any datasheets or forms that they require to undertake and record their contribution to the response.

The plan should show how the strategic and tactical structures change through the transition from the acute phase of the response to the recovery phase and how the recovery will be managed.

12.4. Protection of on-site personnel

There are a range of different types of sites at which events leading to exposures of on-site personnel could occur, including nuclear licensed sites, hospitals, and industries using radiographic sources. This means that there are a wider range of on-site plans than there are off-site plans. Events can include unshielded sources, spills of radioactive liquids, airborne releases into confined spaces and criticality events.

The requirement to keep radiation doses As Low as Reasonably Practicable (ALARP) remains a key consideration during any response to an accident. The three principles of justification (protective actions are expected to achieve more good than harm), optimisation (protective actions should be applied in a manner to optimise protection) and limiting doses to below dose limits and certainly below those liable to cause deterministic effects apply.

An NRPB report (Ref. 144) identifies "prior support measures", which are put in place in the planning stages to support an active response to the event, and "prior passive measures", which are put in place before the event to provide passive protection should the event occur. Prior support measures include the preparation of a response plan and the provision of equipment such as detectors, alarms, signed escape routes, PPE/RPE and equipped emergency control rooms and access control points. Passive prior measures include the provision of shielding.

For non-responders ALARP is achieved by moving them to relatively safe areas, giving them stable iodine if appropriate and keeping them informed of the situation until they can return to work or be dispersed from the site. To do this well requires evacuation routes to be determined, clearly marked and kept clear; instructions to be given in site inductions; and periodic practice evacuations. Evacuation Centres should be large enough to hold everybody in reasonable comfort for the maximum time they might be held there, and they should be supervised by trained personnel. Arrangements should be in place to account for all those on site at the time of the incident.

For sites where a criticality incident is reasonably foreseeable there should be processes to rapidly identify those who may have received a high radiation dose, monitor the onset of symptoms of acute radiation syndrome and prioritise them for treatment.

For responders, the defence against radiation sources, as opposed to contamination, are time, distance and shielding. Staying in areas with ionising radiation for as short a time as possible, staying as far away from the source as much as possible and putting something between the responders and the source, preferable something dense, thick and with high atomic number contents. Knowing the radiation levels is important so radiological surveys before and during entry are sensible. Dosimeters, ideally ones with sensibly set alarms on dose rate and total dose are essential.

For working in contaminated areas PPE and RPE are the first line of defence. RPE prevents you from inhaling airborne contamination thus cutting off what is often the principal dose route. PPE allows you to leave most of the contamination you get on yourself behind at the change area without sacrificing your clothes. It also provides useful protection against beta damage of the skin if beta contamination is high. Common PPE include gloves (because you generally touch more things with your hands than anything else), overshoes (to prevent you trampling contamination into clean areas when you leave), caps (it is hard to decontaminate hair and most people object to it being cut off by health physicists), eye protection (to protect the eyes) and coveralls (to prevent your skin and clothes getting contaminated). At higher levels of contamination, it becomes normal to seal the joints between coveralls, over shoes and gloves and between hood and respirator, add additional layers of PPE and use more robust materials. Different PPE is needed in wet environments.

See Ref. 176 for a comprehensive review of PPE in a nuclear emergency.

There will be a facility on site (there may be several options) that is used to control access to the affected area. All people working in the affected area are expected to enter and leave through this facility except in exceptional cases. Situation boards will be maintained to ensure that up to date information is being used. Forms will be used to guide briefs to the teams going in. There are likely to outline the situation, the team's mission, the conditions likely to be met, the PPE/RPE and dosimetry to be worn (including alarm levels on personal dosimeters), the expected dose rates and any dose limit for the entry and communications equipment and call signs. Great care will be taken to monitor teams' progress and time in the area. On exit the teams will be monitored, undressed and, if necessary, decontaminated. They will be carefully debriefed to learn what they achieved and what they learned about the situation.

Members of the emergency services who may be asked to enter the controlled area should be familiar with the layout of these access facilities and the procedures that are used within them. It is common practice for the emergency services to have representatives in these facilities as supervisors of their personnel and to aid co-ordination of the total effort.

13. Training

13.1. On-site responder training

REPPIR-19's ACOP (paragraph 248) requires that the operator's emergency plan can be "*put into effect without delay when required by providing the necessary information, instruction and training and making the necessary equipment available, in accordance with regulations 10(6)-10(8); and ensuring any other underpinning capabilities required to implement the plan are in place and readily available*". The underpinning capabilities include having enough trained employees.

The on-site response to a nuclear accident brings together two different populations. The first are the site personnel who are used to the complexity of the site and with working safely with the radiation and other hazards on site during normal operations (these include, for example, chemicals, confined spaces, working at height, and pressurised, hot and moving systems) but not used to working in emergency situations. The second population are the emergency services, chiefly the Fire and Rescue Service but also the ambulance and police services who may attend site. These are well versed in multi-agency response and working in emergency and hazardous conditions but less familiar with the complexities of the site and the hazards of radioactive sources and radioactive contamination.

Guidance 280 "*The information, instruction and training provided to persons with a role in the operator's emergency plan should allow them to perform their role effectively in the event of a radiation emergency. It will also enable their exposure to ionising radiation to be restricted, so far as is reasonably practicable*".

A Training Needs Analysis may be produced. This is a document that looks at each role in the emergency plan and identifies the "person type" required to fill the role. This covers the qualifications and experience required to be considered as suitable for the role. It also considers the additional training that the person may need to develop knowledge, skills and judgement[14] to ready them for the role. This will certainly include familiarisation with the emergency scheme, the role of their team within the response and their role within the team. It may also include particular skills such as the operation of specialist equipment or the analysis of data and they may be a requirement to show calm decision-making capability under emergency response pressures. Initial and periodic requalification training is mandated.

The TNA will also analyse the training opportunities that are available to the prospective and experienced responder. These include guided reading and practice, existing training courses, bespoke courses to fill gaps and team, facility and whole plan exercises. Some of the more general courses will be suitable for several roles and teams while some will be more specialist a suitable for fewer roles.

A good example of an emergency service's TNA is found in Ref. 301 which presents a summary of the UK Ambulance Service's training requirements and Safe System of Work (SSoW).

Finally, a training records system will be required to ensure that training records are retained and the need for refresher training flagged in good time.

The emergency services are all well trained in their own areas of operation; understand the risks and how to minimise them. Of particular importance is the need to train them to be safe and confident working in radiological environments. The EU BSS (Ref. 158) defines the training needs for "exposed workers" in Article 15.

The EU BSS (Article 15) "training of exposed workers and information provided to them" states that: *"Member States shall require the undertaking* [operator] *to inform exposed workers on:*

- *(a)* *the radiation health risks involved in their work;*
- *(b)* *the general radiation protection procedures and precautions to be taken;*
- *(c)* *the radiation protection procedures and precautions connected with the operational and working conditions of both the practice in general and each type of workstation or work to which they may be assigned;*
- *(d)* *the relevant parts of the emergency response plans and procedures;*
- *(e)* *the importance of complying with the technical, medical and administrative requirements*".

In the case of outside workers, their employer shall ensure that the information required in points (a), (b) and (e) is provided.

Emergency services personnel who may be called to respond to an affected site will be offered training that covers the key points identified above. This training may be delivered by the operator or by another organisation such as the Emergency Planning College or the PHE. They will also be offered site familiarisation events where they are shown around the site and the emergency facilities and will be invited to take part in shift and team training exercises as well as the exercises mandated by REPPIR-19.

---

[14] knowledge, skills, and attitudes in some versions

Guidance 279 also requires that "*the operator should ensure that all employees working on their premises, including those of other employers, are given suitable and sufficient information, instruction and/or training on what to do in the event of a radiation emergency*". This would include familiarisation with the alarms (both how to initiate them under the appropriate circumstances and responding to them if they sound), evacuation routes and evacuation centre procedures.

13.2. Off-site responder training

According to REPPIR guidance (Paragraph 293) the local authority must ensure that the off-site plan can be put into effect without delay. Among the steps required to do this is a requirement to seek confirmation, so far as reasonably practicable, from responding organisations that the necessary information, instruction and training has been provided and the necessary equipment for restricting exposure has been made available.

The training for the off-site responders such as those in the SCC and tactical centres and those, such as police officers and medical and social care personnel, who may be asked to perform tasks in areas subject to protective action advice depend on their role.

Those in the off-site centres will require training in the operation of those facilities, the role of their team and their personal role. If they are part of the decision-making process then, as well as the ability to participate in the decision making process, they will need a sufficient understanding of radiological protection, public dose exposure routes protective actions and emergency reference levels to take their part.

Personnel who may be asked to undertake tasks in the areas subjected to protective action advice will require training in the risks of radiation exposure, how to minimise their dose uptake with appropriate PPE/RPE and use of time, distance and shielding, how to use dosimeters and, possibly radiation monitors. They should also be familiar with the RPA advice process so that they can access advice and have confidence in it should the need arise.

Having identified the training team members need, and the maximum period before refresher training the Training Needs Analysis identifies the opportunities for training such as exercises, courses or self-training. There may be several different internal and external training courses, each with its defined objectives.

A record system will be required to show who has completed their training and when they need to complete refresher training. Local authorities can expect that the ONR will wish to audit the system and the training records periodically.

In the USA FEMA are introducing the OneResponder website (Ref. 302) to support their national qualification system for emergency responders (Ref. 303). This aims to provide minimum levels of skills for emergency responders to increase the confidence with which personnel trained in one area can support a response in another area should it be needed. Their philosophy is similar to that discussed above. There are prerequisites for those considered for a role. They then work through a role specific Task Book which can lead to certification and credentialing. This is a good model.

Ref. 304 outlines the wide range of IAEA training courses and signposts some useful training materials.

14. Support plans: Radsafe and NAIR

There are two schemes available within the UK to provide support to local emergency responders in the event of an incident involving radioactivity: Radsafe and NAIR.

Radsafe (Ref. 335) is a mutual aid association operated by and on behalf of the sections of the nuclear industry that send radioactive materials across the country by road or rail. If a qualifying shipment is involved in an accident or other event then Radsafe is alerted. It then sends immediate advice to the responding emergency services and arranges for an experienced and equipped health physicist to be dispatched from the nearest responding site to help control the scene until the dispatching company can send a recovery team.

Radsafe provides both awareness training and first responder training.

RADSAFE Awareness Training Course outline:

- Fundamentals of radiation protection;
- Instrumentation and monitoring;
- Overview of transport arrangements;
- RADSAFE arrangements;
- National Arrangements for Incidents involving Radioactivity (NAIR); and
- Desktop and practical exercises.

Objectives of the RADSAFE Responder Training Course are:

- to reinforce the understanding of RADSAFE;
- to expose the RADSAFE Responders to "incident" situations to improve their situational awareness;
- to provide an incident simulation to allow the RADSAFE Responders to practice their skills;
- to provide an understanding of an incident response structure; and
- to assess their knowledge.

A US equivalent to the Radsafe training can be found in Ref. 336 which provides a number of response flow charts. A Canadian Emergency response Guidebook gives advice on responding to a wide range of events including different levels of radiation (Ref. 337).

To aid them in incidents involving radioactivity that have no obvious "owner" the UK police have access to the NAIR plan (National Arrangements for Incidents Involving Radioactivity) (Ref. 280). The Centre for Radiation, Chemical and Environmental Hazards (CRCE) of Public Health England (PHE) co-ordinates the NAIR arrangements. This provides quick and widely available assistance to the police and other emergency services where no radiation expert is otherwise available. Assistance is drawn from hospitals, the nuclear industry and government departments. Reference 281 provides a two-page User Guide for Police Services. The Handbook (Ref. 194) provides a wealth of useful information.

Note: Companies that are not members of RADSAFE are not allowed to claim either RADSAFE or NAIR as their emergency or contingency plan for dealing with radioactivity under their control.

15. Transport Emergency Plans

The transport of radioactive material is highly regulated in most countries. The design specification of containers and the controls required before and during shipment are determined by the amount and type of radioactivity. Materials that have very low levels of radioactivity can be shipped with little or limited controls whereas highly radioactive material requires special containers, controlled routes, security, trained responders and plans to cover foreseeable events.

Approved packaging is tested to confirm that it can survive accidents, fore and water without losing its contents.

The UK package labelling rules are summarised in Table 18 with the signs shown in Figure 37. The signs may show the relevant UN number which can be found tabulated in Ref. 282 (which is a comprehensive "digital assistant" with a lot of useful information in it. It is well worth a quick look).

Transport Index	Maximum radiation levels at any point on external surfaces		Package category
	mSv/hr	μSv/hr	
0	< 0.005	< 5	I - White
0 -1	0.005 - 0.5	5 - 500	II - Yellow
1 - 10	0.5 - 2	500 - 2000	III - Yellow
> 10	2 - 10	2000 - 10,000	III-Yellow (conditional on being under exclusive use)

Table 18 Transport Index

Figure 37 International transport signage

Transport emergency plans must be written before shipment commences and must be available in the transporter. The information required in a plan is:

a) *the name or position of the person authorised to set emergency procedures in motion;*
b) *the name or position of the person in charge of and co-ordinating the mitigatory action;*
c) *for conditions or events which could be significant in bringing about a radiation emergency, a description of the action which should be taken to control the conditions or events and to limit their consequences, including a description of the safety equipment and resources available;*
d) *the arrangements for limiting the risks to persons likely to be affected by a radiation emergency including how warnings are to be given and the actions persons are expected to take on receipt of a warning;*
e) *the arrangements for providing early warning of the incident to the relevant local authority in which the radiation emergency has occurred, the type of information which should be contained in an initial warning and the arrangements for the provision of more detailed information as it becomes available;*
f) *the arrangements for providing assistance to the relevant local authority with mitigatory action;*
g) *the arrangements for providing information about the incident to the GB Competent Authority;*
h) *the arrangements for emergency exposures including the dose levels which have been determined as appropriate for the purposes of putting into effect the emergency plan;*
i) *the arrangements in relation to doses or reference levels as required by paragraph 9(1); and*
j) *the arrangements which the duty holder, or their agent, considers may assist in the transition from a radiation emergency to a situation where no further intervention is required.*

(From Carriage of Dangerous Goods (Amendment) Regulations 2019)

Each duty holder must ensure that suitable and sufficient information, instruction and training is given to all relevant personnel, and that any equipment necessary to restrict that employee's exposure to ionising radiation including, where appropriate, the issue of suitable dosimeters or other devices is available.

For example, the driver of a road vehicle, would be required to check the inventory of emergency equipment such as fire extinguishers, chocks, high-viz, touches, PPE before the journey commences. The vehicle should not be left unattended in public places and suitable placards should be in place. The plan may be to remain alert during the shipment, if there is a problem that could develop into a radiation emergency then:

- Apply the braking system, stop the engine;
- Avoid sources of ignition, in particular do not smoke or switch on any electrical equipment;
- The driver must notify Emergency Control Centre immediately if he suspects that:
    - any radioactive material has been lost, stolen, or has escaped from the vehicle;
    - any package containing radioactive material has been damaged in a road accident or otherwise; or
    - the vehicle and its load is in danger e.g. from fire;

  and give details of the radioactive material involved and any other relevant information;
- Put on the warning vests and place the self-standing warning signs as appropriate;

- Keep the transport documents readily available for responders on arrival;
- Do not walk into or touch spilled substances and avoid inhalation of fumes, smoke, dusts and vapours by staying up wind;
- Where appropriate and safe to do so, use available fire extinguishers to put out small/initial fires in tyres, brakes and engine compartment;
- Fires in load compartment shall not be tackled by members of vehicle crew;
- Where appropriate and safe to do so use any on-board equipment to prevent leakages and contain spillages;
- Move away from the vicinity of the accident or emergency, advise other persons to move away and follow the advice of the emergency services;
- Remove any contaminated clothing and used contaminated protective equipment and dispose of it safely.

(This list adapted from Association of University Radiation Protection Officers, Guidance note No. 6 (Ref. 283).)

For the first responders the initial response will follow the general methodology for approaching the scene of a potentially hazardous event. Including:

- assessing the situation from a safe distance (at least 30 m if possible) looking for
  - potential hazards, including radiological hazards;
  - people at risk; and
  - security concerns;
- looking for and identifying placards and labels or UN markings;
- identifying and talking to the vehicle driver or guard for briefing if possible;
- determining the inner cordon area and reposition personnel, vehicles and equipment accordingly;
- rescuing people in life threatening conditions;
- evacuating area;
- controlling access to inner zone;

Part 3 Emergency Response

16. Nuclear site response

This section reviews the on-site and off-site responses to an off-site nuclear emergency.

16.1. Introduction

The IAEA recognises a number of phases that the response to a nuclear emergency may go through (Figure 38) (Ref. 326). The *preparedness phase* ends and the *emergency response phase* begins when a formal declaration of an emergency is made. This phase ends when the situation is under control, the off-site radiological conditions have been characterized sufficiently well to identify whether and where food restrictions and temporary relocation are required, and all required food restrictions and temporary relocations have been put into effect. The *transition phase* may last from days to months, notwithstanding that for a small-scale emergency (e.g. a radiological emergency during transport or a radiological emergency involving a sealed dangerous source) the transition phase may last not more than a day. The termination of the nuclear or radiological emergency marks the end of the transition phase for a particular area or a site and the beginning of either an existing exposure situation or a planned exposure situation.

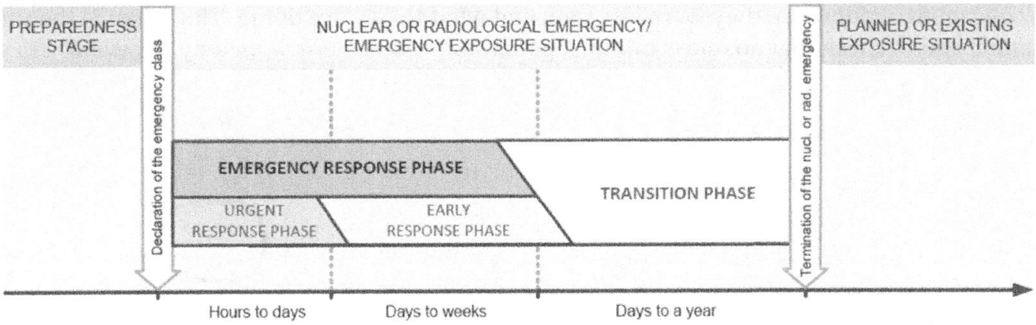

Figure 38 IAEA Phases of a nuclear emergency from GSG-11

That an emergency response will pass through several stages is a widely accepted piece of wisdom and obviously true. Labelling the "stages" and identifying when the response passes from one stage to the next is less clear because they merge into each other and different aspects of the response, or the response in different areas, can be in different stages.

The response will start with an alert and mobilisation phase where people respond to their duty stations and complete their early actions, which tend to be scripted. A battle rhythm will be established in the various control centres that will cycle between short meetings to share situational awareness and agree the way ahead and periods where participants complete actions and communicate with their teams outside the centre. At some stage the focus at the scene will move from preventing the situation escalating to making it safe and then to repairing damage and cleaning it up (recovery). Off-site the focus shifts from urgent actions to protect people from harm caused by the event in progress (protection from the airborne plume or external radiation from a source) to longer term actions protecting people and the environment from the lasting effects of the event (ground contamination and food contamination issues). Then effort turns to the future; attempting to restore home life and allow people to resume their working, home and leisure activities and to give support to those suffering longer term impacts.

For the operator and for some of the responders, there will be overlapping stages of emergency response (the prompt actions to make things better), crisis management (the strategic and tactical work required to enable work to resume in less affected parts of the business and to support the company reputation) and business continuity to rebuild and restore the company.

16.2. Alert and plan initiation

All nuclear sites have a comprehensive set of alarms. These include fire alarms and smoke detectors, radiation alarms inside the plant and on the site perimeter and process alarms monitoring temperatures, pressures, flow rates and levels. There are also several layers of security systems providing alarms in the case of intruders. In addition, there are alarm buttons around site and an emergency telephone number for personnel to report problems.

These alarms alert roles such as process operators, site duty managers or security personnel depending on the site and the circumstances. These are roles that are always present on site for the high hazard sites. The initial response depends on the combination of alarms sounding and eyewitness reports and can include making a formal declaration of an event (see section 11.14) and promulgating the alert.

Personnel on site will respond to the alarm sounding. Those with a role to play will report to their duty stations and commence the response. Those without a role will generally congregate in a safe "muster area" and a role call will be called to confirm that everyone is safe.

Different role call processes exist in different companies. Musters can be paper based with a register being called, based on cards or other tokens that denote when someone is in an area, or based on technology that requires a card to be shown to a reader or radiofrequency identification (RFID) chips at a longer range. Public address systems are often used to name those not accounted for in the hope that they will call in and report an error in the mustering system or to get reports from colleagues about when they were last seen and what they were doing. The intent is the same; to account for personnel and to check that no one is missing, potentially injured or trapped by the event.

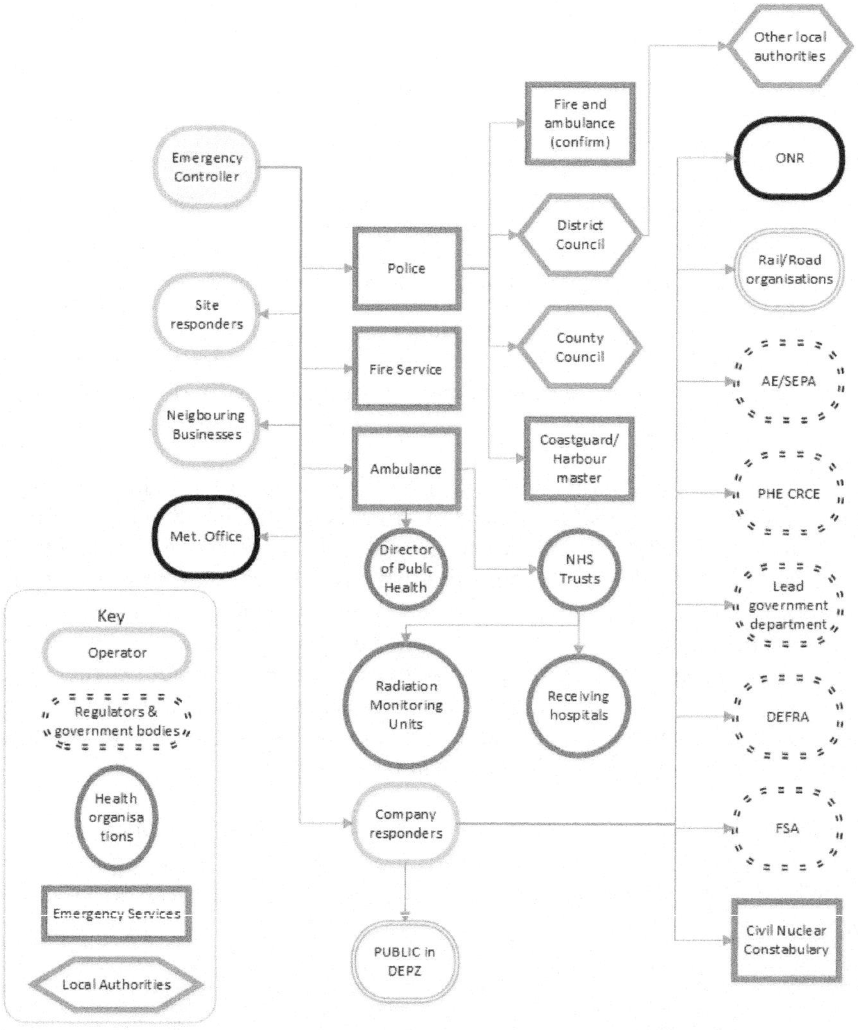

Figure 39 Callout chain (conceptual)

Once a declaration has been made an alert chain is initiated to contact the responding organisations and individuals. A cascade, where several of the people called then call further people themselves is common (See Figure 39, which is a composite of several real call out chains). Telephone based cascades are common, pagers are still common but other technology solutions are becoming more usual. Each site has a different call out chain. These are periodically reviewed and tested.

It should be noted that the callout contains little or no information beyond "the nuclear licensed site have declared a site incident/off site nuclear emergency please initiate the relevant plans and send people to their duty stations in the various response centres". This is for two reasons (1) there could be very little clear information to share at this time, the site are mobilising to determine and then report what has gone wrong and (2) the people running through the call out chain have other people to contact and then some other urgent actions so have no time to chat. The expectation is that once the response centres begin to fill with responders there will be structured information for them to work with.

16.3. Decision making on site

In normal times the nuclear industry tends to take a long time to make any important decision. There is a lot of technical work which is carefully defined and undertaken, gathering data, analysing it, discussing it, reaching conclusions and making recommendations or listing options. This is then verified equally carefully. It may then go to a Nuclear Safety Committee, consisting of up to a dozen highly experienced people given ample time to find any weaknesses in an argument and then interview the author and seeming to enjoy the sport. The regulators may witness this process and then may make their own decision to intervene in the process. It can take months if not years to reach a decision.

One thing you tend to lack in a crisis is time; you start with little or no information about the situation and how it might develop and you start with the clock running down on your ability to influence matters. This is not a comfortable situation. If you make a decision too soon it is likely to be based on incomplete information and possibly wrong, if you make it too late you may have lost an opportunity to prevent the situation worsening. There is a period of time, and it may be a short period, where there is sufficient information to make a decision and enough time for that decision to have a positive impact.

Some actions, such as advising urgent shelter in place for a few hours, are often agreed as defaults to be enacted without waiting for confirmation that it is the right decision as the potential benefit, if it is a right decision are significant and the costs, if it is the wrong decision, are manageable.

Some circumstances do not offer that comfort. In these cases, a wrong decision can cause more problems than it solves, while the right decision made too late does not deliver the potential benefits. If these potential dilemmas can be identified in the planning stage, then the plan can be written to give the responders the best chance of success by speeding up the support to decision making and improving the ability to carry out those actions that are reasonably foreseeable without undue delay enhancing the impact of decisions.

An example of this is that fixed reactor sites have gamma monitoring systems around their perimeter which constantly read out the gamma dose to various facilities on-site and off-site and have alarms set to alert operators to unexpectedly high levels which may indicate an atmospheric release of radioactive material. Thus, if there is a reactor fault, a release can be quickly confirmed - the right data is to hand and is readily interpreted.

Impact is improved by having thought through the actions that might be required and ensuring that the equipment required is to hand and in good working order and that trained teams are available to respond to the call and get the action completed quickly. An example of this is warning and informing systems, that contact registered members of the public resident or working in the DEPZ, and a tested plan to ensure that this triggered early in the response and that it has operated correctly.

The combination of radiation monitoring alarms including those on the perimeter, authorised and trained personnel able to trigger systems to phone the public or sound sirens, members of the public informed about the meaning of alarms and the expected response constitutes a credible response to an unexpected atmospheric release of radioactivity.

Because of the unsuitability of normal decision-making processes for crisis management it is formally suspended and "command and control" adopted for the duration of the crisis. In the nuclear industry this means that one person is given clear and unquestioned authority over the resources of the site. The system is designed to give this person the clearest view of the situation and the best advice possible within the time constraints and teams of trained people to act quickly and decisively on any instructions they give.

Emergency Control Centres across the industry come in a range of shapes and sizes but tend to have several similarities. There is generally a "top table" consisting of the Emergency Controller and their advisors (situation on site, technical options, radiological protection and administration support). In front of them are a number of display boards or screens possibly including:

- A "History Board" containing the key times of events that have happened;
- A "Current Situation Board" upon which one or two overworked people will try to summarise the situation, drawing particular attention to things that are important and things that are changing;
- A "Focus Board" upon which the emergency controller articulates the key areas where rapid progress is needed;
- An "Actions Board" where people have taken the Controller's foci and turned them into Actions with people identified as responsible for their completion and a challenging time for their completion.

Around the room may be other white boards or displays. This may include a "time line" showing events that occurred in the past and predictions about what will happen in the future. This might, for example, include reactor temperature. Comparing current and predicted temperatures against threshold temperatures at which reactor damage occurs helps estimate the time remaining to react. Comparing measured reactor temperature against the predictions can give an early warning that things are not going as well as hoped.

Beyond the top table will be support teams gathering data and interpreting data, determining options, feeding advice to the top table and working on the foci and actions coming back from the top table. This support network can extend beyond the control room and beyond the site boundary.

The effort at the scene of the event is generally controlled from another emergency centre sometimes called the Forward Command Point or the Access Control Point, this can be a fixed facility or a mobile facility based on vehicles, some sites have both. In this facility those going forward to complete a task are briefed, their PPE/RPE and dosimetry checked and they are dispatched. On return to the facility they are monitored for radiological contamination, decontaminated if necessary, helped to remove their over-clothes and debriefed. This facility can be a very busy area and training and discipline is required to keep it focussed, quiet and business-like.

The emergency services responding to site may have representatives in one or more of the on-site emergency centres and may have their own control vehicles to aid co-ordination and control at the scene.

16.4. Decision making in the Strategic Co-ordination Centre

The Strategic Co-ordination Centre would be opened as soon as possible after the declaration of an off-site nuclear emergency. It is usually in the headquarters of the local police force so they could be there to open it within minutes of the alert.

In many areas there are arrangements for the core of the Strategic Co-ordinating Group to hold virtual meetings, particularly in the very early stages of an event or where the site has declared a site incident which is not expected to develop into a full off-site nuclear emergency but might, or might be perceived by the public as such

The Strategic Co-ordinating Group would be getting information from:

- **The operator** who will be reporting what is happening on site, how they are responding, how it is likely to develop and a "worst reasonable case" of what could happen based on current understanding. They will also give protective action advice and the rational for that advice. This could be complex or quite straightforward depending on the circumstance;
- **The tactical centre (Silver Command)** who will be reporting on the attempts to control the event on the site and the work undertaken in the surrounding area to promulgate the advice to the public and support its implementation;

- **The ONR** who could have people on site and in COBR/SAGE (if activated) and will certainly have people in their emergency centre in Bootle (The Redgrave Court Incident Suite (RCIS)) trying to understand the situation and its possible consequences;
- **PHE** who will also be monitoring the situation from their own crisis centre and will have representation in COBR/SAGE (if activated) and may have people in the Operator's emergency centres, they will give advice on protective actions and, acting as a Radiological Protection Advisor (RPA), give advice to some of the responders about their dose uptake;
- **The emergency services** will be briefing about the situation on-site as they see it and off-site as reported by the police in the community;
- **The Local authority** will be reporting on the resources they have available and on the reaction of the population;
- **The media teams** will be reporting on the messages given out and on the responses of broadcast and social media.

Some information will be filtered through the STAC, some presented on information boards and information systems and some verbally at the Strategic Co-ordinating Group meetings. This section looks at a few of the roles in the Torness Off-site.

Generally, SCCs will have an 'Information Zone' which will be controlled by the Police and will use display boards and IT to collate and display information about the current situation and the working of the SCC.

The Company Technical Advisor (CTA) is a representative of the Operator and is usually a person trained and authorized to act as a Site Emergency Controller. This requires a firm grasp of reactor safety, accident modes and means of repair/recovery, radiation protection and public protective actions. They are in touch with the CESC[15] and so are well briefed on what is happening on the site and the expected changes in release rate, so far as they can be estimated. The CTA and their team is the person to go to for explanations about what is happening and what is likely to happen.

The Company Health Physicist (CHP) is a person with a good understanding of operational health physics and of the use of protective actions to protect the public. They are in touch with the CESC and in particular with the Emergency Dose Assessment Service (EDAS); a small team of people who are modelling the release and atmospheric dispersion to provide predictions of projected, avertable and residual doses under a range of circumstances and helping formulate the Company's input into protective action advice. The CHP is a good source of advice about radiological protection terms as well as the current situation with regard to off-site doses.

The Company Media Briefer is a person chosen for their wide understanding of the nuclear reactors the company operates, their safety cases and the health physics implications of the accident. They are trained to talk to the media. They are also in contact with the Company team within the SCC and the CESC and with the media briefing circuit. They are the go-to people for advice on media releases and explanations about events.

The ONR put a team into the SCC with a range of relevant skills that mirrors the operator's team. They are in contact with the CESC and also the ONR's own crisis centre in Bootle (the Redgrave Court Incident Suite (RCIS)). Between them they will be able to give technical advice on the state of the reactor, projections about what may happen next and outlines of the radiological situation and protective action advice.

Their information might be considered to be more independent than the operator's. A major problem with the Fukushima response was a marked lack of appetite in the site operator to accept and to report how bad the situation was. We do not believe that this would be such a problem in the UK, but the ONR is there as well and will be questioning and checking the validity of Company reports and projections.

---

[15] The Central Emergency Support Centre (CESC) is a facility based in the EDF Headquarters in Gloucestershire which co-ordinates the response of the wider company. It co-ordinates technical support to the site, manages the off-site radiological survey and provides information and advice to the on-site and off-site responders.

Public Health England (PHE) is an executive agency of the Department of Health and Social Care and is a Category 1 Responder as defined by the Civil Contingencies Act, 2004 (and subsequent amendments and revisions). PHE provides the UK's independent radiological protection expertise. They will be gathering data about the event from the CESC and from the off-site monitoring and forming their own opinion about the radiological situation and the wisdom of different protective action strategies.

They will be a good source of advice and interpretation, particularly of the radiological situation. They also provide an RPA service (Ref. 308) to some of the emergency services so provide advice on when it is safe to undertake work and what precautions to consider when planning work in contaminated areas.

Food Standards Scotland/Food Standard Agency (FSS/FSA) role is to help protect the public from risks to health which may arise through the consumption of food. They provide scientific advice relating to food to the STAC and to define the precautionary advice area in which relevant maximum permitted levels in food and feed might be exceeded.

They have powers to implement restriction orders under the Food & Environment Protection Act 1985 (FEPA) (Ref. 238) to restrict the supply, movement or sale of produce from the affected area.

The role of the Fire and Rescue Service is mainly focused on the on-site situation, but they do have the ability to call forward support from neighbouring areas if required and have statutory duties and capabilities that include the areas of hazardous materials and environmental protection incidents, urban search and rescue, dealing with CBRN incidents and responding to other emergencies. Their command officer and team in the SCC is likely to be a useful source of advice on such matters.

They may be called to an emergency within the affected areas.

The Police Service have a range of responsibilities which include:

- the saving of life in conjunction with other emergency services;
- co-ordination of the emergency services and other organisations during the emergency phase of the incident;
- the placement of appropriate cordons and the protection and preservation of the scene;
- traffic management;
- in conjunction with other agencies the restoration of normality at the earliest opportunity;
- co-ordination of the response to the media, and management of the media;
- liaison with friends and relatives of those affected by the incident; and
- application of protective actions to protect the public in conjunction with other services.

Their agreement and cooperation will be essential for any operations undertaken within their cordons and traffic management areas around the areas subject to protective action advice.

The NHS have a number of important roles with regard to the protection and reassurance of the public in the event of an off-site nuclear emergency. They will have personnel in the SCC and in their own Emergency Control Room. These people will have access to advice from skilled hospital radiological protection professionals and will be "*providing advice to local authorities, the public and the media on health aspects of a radiation incident, including those relating to foodstuffs and water in conjunction with FSA and other agencies*".

They will co-ordinate personal radiation monitoring at the Rest Centres if required and be able to give advice on the necessity for monitoring and the consequences of any results obtained. They will be providing skilled radiological protection personnel and services and will consider *the impact on health services to people not otherwise affected by the incident.*

In the event of an emergency involving the MOD the MOD Co-ordinating Authority (MCA) will:

- be the MOD's senior representative at local strategic level;
- provide authoritative and timely advice concerning the progress or development of the emergency and the potential implications, including mitigation of operations to make safe and recover the asset during a transport emergency;
- provide the MOD's Lead Government Department input to the Strategic Co-ordinating Group, liaising with the JMC as required;

- keep the Defence Nuclear Emergency Organisation Headquarters (HQ DNEO) informed on the status of the emergency and of the operational response to it;
- act on strategic direction from HQ DNEO and seek any additional military resources required.

The Scientific and Technical Advice Cell (STAC) has as its primary function the provision of health advice for public health protection. In the event of a nuclear emergency, it will co-ordinate advice in relation to the wider context of public safety, environmental protection and the results of sampling and monitoring of radiation levels and any radiological contamination.

The aim of the STAC is as far as is practicable to provide SCGs with authoritative information and agreed advice on the risk assessment of health and environmental hazards and technical failure, by (Ref. 285):
- bringing together or arranging contact with all the relevant specialist advisers through a single group;
- providing agreed recommendations on risk management action;
- providing agreed risk communication messages; and
- confining the main discussion on such issues to within the STAC itself (rather than at the co-ordinating group meetings).

The STAC may consist of:
- Cell Lead;
- Secretariat / Staff Officer support;
- Gold Liaison;
- Relevant emergency service technical advisors (e.g. Fire Service HAZMAT officer);
- Health Protection Agency;
- The Operator;
- Primary Care Trust and/or Strategic Health Authority Director of Public Health;
- Environment Agency;
- Food Standards Agency;
- Health and Safety Executive;
- Local Authorities (e.g. Environmental Health Officer);
- Met Office;
- Government Decontamination Service, if required;
- Operational partners of the Department for Environment, Food, and Rural Affairs (Defra);
- Other agencies invited to address sector specific issues, such as the Utilities or transport operators;

With so many "voices" and so much complex technical information that decision makers need to understand, it is important that a "battle rhythm" be established. This must allow the strategic co-ordination group time to meet to exchange information, discuss and make decisions without hampering the members' ability to spend time with their teams setting and chasing actions and receiving information.

The strategic group would meet as soon as possible after the alert, and thereafter as often as necessary, to exchange information, to make sense of the situation, to determine if any other organisations should be invited to join and to agree on and co-ordinate the response. Their meetings must be long enough to reach decisions on matters that need to be decided upon and to set actions but short enough to allow SCG members time outside the top table to talk to their teams to exchange information and commands. Meanwhile there are other groups such as STAC, the recovery group and the media group acting in parallel that must be co-ordinated.

If this all sounds calm and orderly then it has misrepresented the situation. In the SCC, as in all crisis centres, there is immense pressure to make important decisions quickly against a complex, changing and barely understood situation. Larken warns that "*Time pressure and incomplete information are uncomfortable but often unavoidable fellow travellers; they heat decision making and are stressors*" (Ref. 284).

16.5. Setting protective actions

The art of advising protective actions is to identify the point at which the avertable dose compares to the relevant ERL. This is not easy. Some of the complexity is shown by Figure 40 which shows the centres and teams involved for an AGR (EDF) event.

Dose avertable on the day depends upon:
- The radioactivity release from the event;
- The dispersion of that radioactivity in the environment;

- Where the exposed person is in relation to the release and the weather;
- The person's age and activity including indoor/outdoor;
- What the person would do if they didn't hear the protective action advice;
- What the person would do once they heard any advice;

To be sure of getting protective action advice right you need a very good crystal ball!

We don't have a crystal ball. Instead we have a system that aims to integrate the knowledge and experience of a number of different people with different skills and in different places with different information available to them and make the best collective educated guess possible. The people involved include:

- Damage assessment/repair teams at the scene;
- On site technical support and decision making personnel;
- Off-site survey teams and Health Physics teams;
- Atmospheric dispersion modellers (Operator, PHE and Met. Office);
- Off-site dose expertise (Operator, PHE);
- Expertise within STAC and SAGE;
- Decision makers with SCC and COBR.

Shelter prior to controlled evacuation is a sensible set of protective actions if the projected dose is high enough. For many sites shelter is a default action on declaration of an off-site nuclear emergency.

It should be noted that changing protective action advice, as the situation becomes more serious or as resource become available, can be misconstrued by the media and the public as indecision, dithering or confusion; presentation is key.

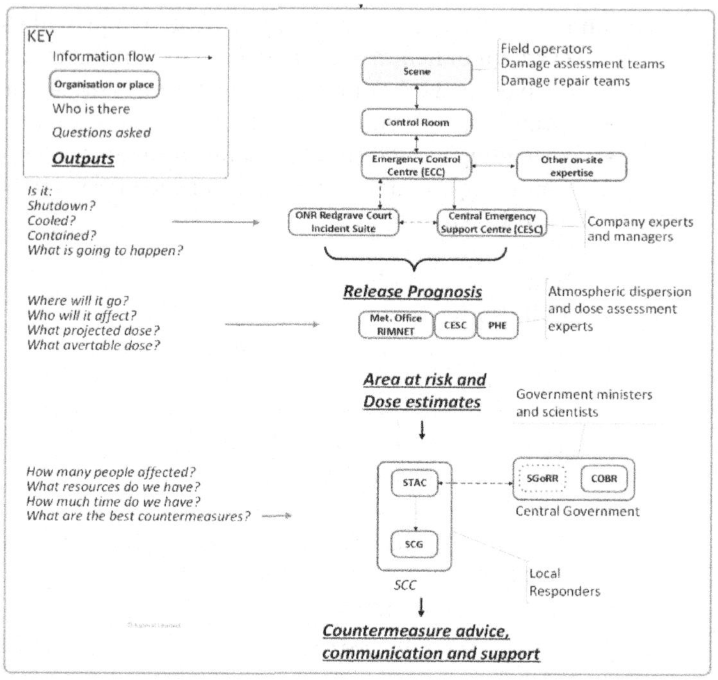

Figure 40 Centres involved and questions asked (EDF reactor accident)

16.5.1. Why isn't it easier to determine the source term?

This section is designed to explain to the reader why no one can provide a clear and unambiguous assessment of the amount and nature of the radioactivity already released from the reactor nor how much will be released in the future. It is included because of the unrealistic expectations in many responders that the operator will know exactly what will be released once the initiating event has occurred. ***This is important as it describes the variability in the source term and thus affects the "particular characteristics" of the release.***

It is quite common to hear impatience in units such as STAC and the SCG when they are told that the projected doses to members of the public are uncertain. "*Surely it is just a matter of putting the source term (the list of nuclides and amounts released) into a dispersion and dose computer code!*" (Which is what JAM is intended to do (See Section 11.12.12)).

However, the real world is more complex and uncertain. There are two ways to estimate the source term. The first is to look at what is happening on the site and refer to the nearest safety case. The second is to take some measurements off-site and work out how much had to have been released to give that measurement.

The limitations of safety case source terms

Safety cases produce estimates of source terms that rely on a chain of calculations with a number of estimates and a number of assumptions:

- They estimate the inventory of radionuclides (the amounts of radioactive nuclei that are there) by running computer codes which model the chain reaction and the effects of neutrons on the fuel and surrounding material. There are a lot of parameters in these calculations that depend on complex models of neutron behaviour and the constituents of the fuel and reactors. Assumptions are made about how long the reactor has been running and at what power ratings. For gas cooled reactors this is made more uncertain by the variation in fission product inventory in fuel pins around the reactor caused by the refuelling cycle and the difference in neutron distribution across the core.
- They estimate the changes in temperature caused by the event. This depends on the assumed configuration and effectiveness of the cooling circuits, the power density of the fuel elements, how things change after shutdown and some complex physics. In an AGR different parts of the core experience different power densities and different cooling rates. Safety cases tend to be pessimistic about all variables; cooling is as bad as it can reasonably get and power density as high as it is likely to be. This is unlikely to be the case in a real event.
- They estimate the damage done to the fuel and the reactor internals by the temperature transient, by the pressure transient and by changes in the coolant composition (in a gas cooled reactor it should be CO_2 but under some circumstances can be contaminated with steam or with air).
- They estimate the release of radionuclides from the fuel and reactor internals. This depends on the chemical and physical forms of the radionuclides and the temperatures and levels of damage - more volatile chemicals such as iodine and caesium come out more rapidly and at lower temperatures than others such as plutonium and uranium. Complex chemistry is involved.
- They estimate the behaviour of the released radionuclides in the cooling circuit which offers a high surface area with surfaces at different temperatures (the inner surface of a boiler tube with water feed is a lot cooler than the surface of the fuel, for example). Complex chemistry and surface chemistry are involved.
- They estimate the release of radionuclides suspended in the coolant flow to the atmosphere. This depends on assumptions about the coolant flow and the location and nature of any leaks.

It is very unlikely that the safety case for a site contains a set of calculations that exactly match the real event so the safety case source term can be taken as a guide but attempts are needed to make better estimates on the day.

The limitations of measurement

On the day it is possible to make measurements of the radiation levels and radionuclides present. Why can't this be used to give an accurate source term?

The radioactive material released from a reactor in an accident is likely to consist of several different radionuclides, maybe several hundred. Each radionuclide has a characteristic decay process which will give off radiations that may be alpha, beta, gamma or neutrons. The alpha, gamma and neutron radiations are emitted in a small number of well-defined energies that are characteristic of their parent nuclide. Beta radiation is emitted in a distribution of energies with a characteristic maximum energy.

Thus, a sample of the radioactive material released from a reactor in accident conditions will contain a mixture of radionuclides and will be emitting alpha, beta, gamma and neutron radiations with different energies. The mixture released from the reactor depends on the nature of the event and will vary with time. It may change suddenly as a critical temperature is passed at some location within the reactor or as a barrier to release is breached. The mixture will also change after release as the effects of radioactive decay depletes some radionuclides much faster than others and as the constituents of the plume deposit on surfaces at a rate dependent on their chemical and physical form.

Dose rate measurements can be taken by fixed monitors such as the Emergency Perimeter Gamma Monitoring System (EPGMS) and the RIMNET system, by the company's off-site survey vehicles and by many other organisations and individuals. All such detectors measure the energy deposited in their sensitive volumes in one way or another and this can be related to the external dose that a human tissue would receive if it was in the same place. This is a valuable measurement, but it does not tell us a great deal about the isotopic composition, we just see the total energy deposition summed across all radiations present. It also tells us nothing about the chemical and physical form of the release.

Unfortunately dose rate measurements such as these do not directly tell us the current total radiation dose rate to an individual at that point since they do not measure inhalation dose or ingestion dose. Nor do they give us any information about how the dose rates at that point were different in the past or will vary in the future.

Off-site survey vehicles can suck air through a specialist filter (a "Maypack" filter) and capture the particulate activity and some of the vapour activity in the air. Measuring this on a doserate meter gives an indication of the particulate activity in the air and is used by EDF to check their default action levels. Knowing the particulate activity in air gives a better guide to the inhalation dose. Off-site survey vehicles do well to do much more than two such measurements an hour each. The recent introduction of gamma spectroscopy capability into survey vans greatly improves performance here (Ref. 339).

High resolution gamma spectroscopy of samples collected in this way can be used to identify some of the characteristic radiations and therefore identify individual radioisotopes and estimate their concentrations in the air. Not all radioisotopes can be identified in this manner, but the gamma spectroscopy of Maypack filters is a major step forward in dose assessment. From this stage the dose assessors are using measured, but incomplete, radionuclide mixtures rather than those estimated before the details of the fault were known.

Radiochemical analysis of samples can improve the ability to detect other radionuclides and measure their quantities but these processes take time.

Thus, the dose assessors may have some continuous reading of dose-rate at fixed points, some measurements of airborne activity taken at different times and different locations, some gamma-spectroscopy derived isotopic analyses (usually with an hour or so delay), and some radiochemical analyses (usually with about a day delay). They are using this to try to understand a radiological situation resulting from a site releasing a varying mixture of radionuclides at varying rates into an atmosphere where the weather is changeable and where wind speeds, wind direction, temperatures and feature-induced turbulence (the impact of hills, trees, buildings etc. on the wind) all have an impact on dispersion.

To make it harder the customers for their work are less interested in what has happened and much more interested in what will happen in the near and further future. It is not easy and it is not exact.

EDF have a unit within the CESC called "the Emergency Dose Assessment Service" (EDAS) which is able to hear updates from the Technical Support Team about what has happened on the site and what is expected to happen next and can see all of the Company's fixed radiation monitors and off-site survey reports. They use computer codes and spreadsheets to try to assimilate all of this information and weather reports into a consistent picture which is used to predict doses to the public in order to support protective action advice. Transition from emergency exposure to existing exposure situation

The emergency exposure reference level will apply when making decisions about the advice to give to members of the public about urgent protective actions to reduce their exposure to a temporary plume of radioactive material.

Existing exposure situations exist when considering the longer-term reduction of doses resulting from deposited radioactivity after the release is confirmed as terminated.

An awkward transition phase exists when the release is newly terminated and the radioactivity in the environment has not been adequately characterised but if the discussions have moved on from airborne concentrations and ERLs and onto deposition, maximum permitted levels in food stuffs and doses over periods of more than about a week then existing exposure rules are applicable.

PHE-CRCE are expected to alert the SCG to the requirement to discuss changeover at the appropriate time. A formal decision by the SCG will be made and recorded.

Any change in the extent reference level due to change of status or reassessment will be noted and recorded in the Operator's report of the event (Regulation 17(6)) as required by regulation 20(8).

16.6. Responding in an area affected by a Nuclear Emergency

It will be important to continue to identify, monitor and consult with affected individuals as the emergency unfolds. This will help those co-ordinating and delivering the activities to adjust plans and support activities to ensure that individual needs are being met in the most appropriate and effective way.

HM government (Ref. 333) use the term "Humanitarian Aid" to describe "*those activities [that are] aimed at addressing the needs of people affected by emergencies; the provision of psychological and social aftercare and support in the short, medium and long term*".

16.6.1. Warning, informing and confirming

Having agreed upon a set of protective actions, which might be "*evacuate area A, advise those in area B to shelter and those in area C to take stable iodine tablets*" (these areas may overlap), the requirement is to ensure the maximum benefit of that advice is achieved. The concept of operations and the off-site plan should explain how this is to happen and detail actions for a number of organisations.

The planning phase for warning and informing is discussed in 11.8. On the day, the operator may initiate their component of the warning system which could be, for example, a telephone system or a siren. The police may set up control points on the edges of these areas to advise people trying to drive or walk into the area to turn around, and the local authority may cause signage to be put up on minor roads and pathways alerting people to the issue.

The mainstream media and social media will also be used to promulgate the alarm and give further information. We cannot be sure that the media will transmit exactly the information we provide to them without their own layers of interpretation and comment. This is potentially important as the media has a role in the production, circulation and reproduction of social interpretation and, as such, they have a strong impact on public behaviour.

Predictable issues include:

- Those advised to take stable iodine tablets but without stocks or unsure of dosage of method of talking;
- Those advised to shelter but lacking a suitable building;
- Those not sure which, if any, protective actions are relevant to them;
- Those advised to shelter who would rather evacuate;
- Those in the area unaware of the issue (either familiar with the prior advice or not);
- Those outside the area concerned with family and friends inside the area an vice-versa (less a problem now that mobile phones are much more common - but, even now, not everyone has them)
- Those subject to shelter advice but who normally benefit from in-home care.
- Those who work peripatetically in the area including carers concerned about their charges;

Ideally mechanisms will be in place to manage all these issues.

The SCC will probably seek feedback on how well the alert has been spread, how those affected have responded and how many requests for help have been received. This aspect of the response is notoriously difficult to simulate in an exercise while being absolutely fundamental to the success of the response.

16.6.2. Radiation protection on the day

It the event of a major atmospheric release of radioactivity there will be a concerted effort to understand the off-site situation. This will include:

- Information feeds from the affected site with their release prognosis (is it leaking? Do we expect the leak to reduce, increase or stay the same over the next few hours? What is the worst reasonable scenario for development from here? When will we know more?)
- Information from on-site and off-site radiation monitors. There are a range of radiation monitors that will be deployed. These include fixed monitors within the site, on the site perimeter and beyond the site and mobile monitors in vans which will take readings within the plume.

- Information from mathematical modellers using computer codes and weather data to predict where the activity will be blown by the winds to generate "area at risk" plots.

These will inform the protective actions that are recommended to the general public.

They will also be used to predict the radiation doses that might be accrued by individuals and teams working in or near the plume. These could include, for example, police officers on road checks or other activities associated with establishing protective actions, Fire and Rescue and Ambulance personnel responding to conventional emergencies within the affected areas, or social services providing direct in-house support to vulnerable people.

If there were a plea for help from somebody within such a zone there would be a number of factors to take into account when deciding how to respond:

- Who was asking for help, particularly the nature of their vulnerability;
- The nature of the help required;
- The urgency of the help required;
- The potential duration of the exposure time for the responders;
- The dose-rate predictions for the response area and how they may vary with time;

Each "mission" into the DEPZ will have to be judged by the employers of the people going in to determine if the benefits of the mission merit the additional radiation dose. In this they will be supported by the Strategic Coordination Centre and by their own Radiation Protection Advisor.

It may be that the release is expected to stop within the next few hours and that the service required can be delayed that long without causing too much distress. It may be possible to find another way to provide the service such as asking a neighbour to pop next door to provide support, or at least better judge the issue.

If it is decided that a mission should go ahead the employer should ensure that their RPA has given advice to the entry team on PPE/RPE usage and on stay time limits and that the team are comfortable with the brief and prepared to go in. If there are traffic controls to pass on the way in then a clearance process will have to be in place to allow the responders access.

Individuals being asked to work in an area subject to protective action should ensure that they are adequately briefed. Advice should be sought from their RPA about the value of the task in hand compared to the dose implications of entering the area and about how to measure and minimise their dose.

- Plan the quickest way in and out and to spend the minimum time exposed as possible;
- Consider other ways of providing the help:
  - Ask a family member or friend to make the visit, if they have the right skills;
  - Use neighbours if they are available and able;
  - Delay the visit until dose rates have dropped (depends on release prognosis and situation);
  - Provide reassurance and/or instructions over the phone; or
  - Arrange for the person to be bought out of the area (becomes more sensible if release could be prolonged or get more serious and the person is reasonably mobile).

Setting	Dose or dose rate	Alarm sound
Level 1	100 μSv/hr	Intermittent double fast chirp
Level 2	5 mSv	Slow two-tone alarm
Level 3	100 mSv	Continuous tone
[LOW BATTERY]	-	Intermittent slow tone

Figure 41 Ambulance service EPD alarm levels

Responders should remember that any radiation meter or Electronic Personal Dosimeter (EPD) (sometimes referred to as a Radiation Detection Pager) provided with only shows external radiation dose and does not include any inhalation dose. Different Services may have different alarm levels set. It is important that responders understand the meaning of alarms on their dosimeters. Figure 41 from NARU *Guidance for Dealing with Radiological Incidents*, 2020 (Ref. 301).

Purpose: Level 1: Detecting above background radiation levels, Level 2: Default operation dose limit reached, Level 3: Emergency Exposure dose limits reached.

16.6.3. Working in contaminated areas

It is unlikely that personnel untrained in contaminated area work would be sent into a significantly contaminated area without a health physics escort. Decision making is discussed in more detail in Section 16.8. Occupational Health Services/Radiation Protection Advisors within each responding agency are responsible for the provision of health and safety and welfare advice to their respective personnel.

Avoiding contamination involves wearing appropriate PPE so that, if someone does get contaminated, it will be removed when they (carefully) remove their outer layers. Contaminated areas should only be entered with good reason and responders should try to minimise touching of potentially contaminated surfaces. However, these steps only tend to become important in areas of very high contamination.

Control of contamination involves trying not to spread contamination into previously clear areas. Mechanisms such as over-shoes (essentially over large socks that go over the person's shoes and are sealed at the top by elastic) which can be removed at a boundary (ideally a barrier) between contaminated and clean areas to prevent contamination being trodden into the clean area. Similarly, coveralls, gloves and head covers, removed at a barrier, will prevent contamination being spread on clothes and hair.

It is important not to eat or smoke while in contaminated areas. Drinking should be discouraged and only done with care. Open wounds should be covered before entering the area. Wounds sustained in the area should be bought to the attention of a Health Physics Monitor if possible (otherwise wash in running water being careful not to get the run-off water into eyes, ears, nose or mouth).

Finally, if people, rather than their clothes, are contaminated then the activity can usually be removed by mild washing processes, continued with patience if not immediately successful.

Simple advice on undressing techniques that avoid contamination on clothes contaminating the skin can be found in an infographic (Ref. 178). Much more detail on contamination and contamination control can be found in the Fire and Rescue Service National Operational Guidance (Ref. 310).

16.6.4. Process to set a new reference level

The local authority would take advice from their RPA, PHE-CRCE and/or the STAC about the advantages and disadvantages of introducing a new reference level, either for a particular task or for the response in general. This would be discussed by the SCG and, if the co-ordinator agreed, enacted.

The Secretary for State may also set a reference level whether applicable locally or nationally in addition to any reference level set under paragraph (5) (Regulation 20(7). If the Secretary of State were to set a reference level the instruction would be promulgated through the SCG and the SCC'c communications systems and the STAC would discuss the implications to the response and recovery processes underway.

16.7. Communicating

In this section the communication about the situation, about protective action advice and about the potential long-term implication of radiation doses is discussed. It is divided into three main sub-sections. The first discusses where information can be found and who can help interpret the more complex knowledge for the use of personnel and members of the public. The second section looks at communications with the responders and the third with communications with the public.

16.7.1. Communicating with responders

There is a requirement that people who may receive high emergency doses as a result of carrying out their duties are informed of the potential for exposure and are "informed volunteers". While it is not clear that these regulations and guidance documents apply to local authority employees asked to work in areas subject to protective action advice the moral case to shadow them is strong.

Since it is unlikely that the situation will arise it is sensible to consider the effort that is appropriate to provide and refresh the information. In some cases, "just-in-time" training could be arranged for delivery on the day of the event. An example of just-in-time training for hospital workers prepared by the US Centres for Disease Control and Prevention is given at (Ref. 309), however, it could be as simple as a few pages of text reminding them of the key points gleaned in training.

Planned just-in-time training can be text based, computer delivered, or internet delivered. Ad-hoc training would probably be offered in the event of a serious nuclear or radiological accident as many people would require new knowledge and skills to play their part in the recovery.

It is important to consider what pathways the responders would potentially be exposed to. These include:

- inhalation dose while travelling;
- cloud shine while travelling;
- ground shine while travelling;
- inhalation dose while attending;
- cloud shine while attending;
- ground shine while attending; and
- skin contamination dose, potentially continuing after leaving the area.

On the day, responders should be able to expect an understandable briefing from their Radiation Protection Advisor (RPA) detailing:

- the estimated doses they might receive in any planned foray into or near the protective action zones;
- the dose pathways that may be significant;
- any steps they can take to minimise their dose;
- any ways they can measure their dose and make dynamic decisions about whether or not to proceed;
- the risks posed by the potential doses, putting them into context with background radiation and daily risks; and
- the benefit to individuals or groups of the planned action.

They should also be able to expect to be debriefed after the event should they wish for it.

16.7.2. Communicating with the public

It will be important to ensure that all communication with the public about the emergency, about the protective actions advised, and those not advised, and about the long term protective actions, remediation and risks is careful and effective. Communication with the public in a crisis is not easy. Short fallings were seen in the Fukushima response (Appendix A.7) and Grenfell Tower response (Appendix A.10) and both show how easily a trust deficit can build up during the response phase and immediate aftermath that makes recovery more difficult.

In this context the IAEA (Ref. 313) define the public as:

- *The population affected by or potentially affected by the emergency;*
- *The public and the news media (i.e. public information);*
- *Other interested parties.*

They also state that the "*key objectives of public communication for a nuclear or radiological emergency should be:*

- *To protect the public;*
- *To inform the public, both at the preparedness stage and during the response, of protective actions and other response actions, and of the nature of any hazards, and to facilitate emergency response actions;*
- *To gain and maintain public trust in the emergency response by means of transparent, timely, clear and accurate public communication;*
- *To address public concerns with regard to potential adverse consequences for human life, health, property and the environment;*

- *To prevent undue concern, to mitigate anxiety and long term psychological effects, and to help to ensure that actions taken do more good than harm;*
- *To respond to misinformation and rumours;*
- *To enable interested parties to make informed decisions.*"

There are some thoughts on the dangers of misinformation and rumour and a recommendation that processes are in place to combat these issues; probably easier said than done.

After the event risk communication will be required to support residents' decisions about where to live, where to work and play and what to eat; all of which will be affected by the perception of the radiation hazard. Such communications have been shown to have a complex effect on subjective wellbeing post-accident (Ref. 311).

An ANS report on Fukushima (Ref. 312) concludes that "*Ideally, in a crisis, a government would communicate effectively to its people and the global community. Risks associated with the crisis and ongoing efforts to manage the crisis would be clearly articulated. Efforts would be made to provide factual reassurances to the international community. All of this would be done with timely information provided by recognized authorities in a co-ordinated fashion.*"

A presentation (Ref. 316) made at an IAEA conference suggested that not clearly answering the question "is my baby safe?" to pregnant women has resulted in:

- *Voluntary abortions;*
- *Unsafe evacuation of patients (deaths);*
- *Refusal to treat patients;*
- *Stigma;*
- *Economic impact;*
- *Psychological distress.*

So it is clearly important to communicate well!

Communications channels

There are many channels of communication:

- official news and advice coming from the Media Briefing Centre;
- briefs from the organisations and companies involved direct to the media, including their websites and social media;
- messages on the Warning and Informing system;
- responses to the public and media on your usual phone lines and any special phone lines set up;
- social media messages "leaking out" from the many responders in the variety of centres;
- the appearance in public of personnel, particularly those working in the areas subject to protective action advice or controlling the boundaries; and
- traditional and social media including the full spectrum from informed comment and news to sensationalism and fake news.

Communications Plan

All local authorities should have a communications plan. The Torness Off-site plan, for example, states that "*The co-ordination of information to be released to the public will be the responsibility of the Public Information Co-ordinator appointed by Police Scotland. This person will attend strategic meetings where information pertaining to the emergency, which can be released to the public, will be clearly identified. All news releases produced by participating organisations should, where possible, be co-ordinated through the Public Information Co-ordinator prior to issue in order to ensure consistency.*"

Communication content

The IAEA have a publication dedicated to public communication (EPR-Public Communications *Communication with the Public in a Nuclear or Radiological Emergency"* Ref. 250) This discusses the basics of public communication, namely an incident command system including a role such as a "Lead Public Information Officer" with an adequately sized and experienced team and links to all the key players.

The document provides specimen task lists and hints on how to select and prepare public facing spokespersons.

Its list of "steps to be taken into consideration in the event of an accidental release" is quite daunting:

- *Describe the radionuclide and the type of radiation involved in the emergency;*
- *Describe the possible pathways by which people could be exposed to radiation, and how they can protect themselves;*
- *If possible, give estimates of radiation doses to people and how they might compare with doses from other sources of radiation, such as natural background radiation or medical practices;*
- *Explain the possible health implications of the doses received, and symptoms to be aware of;*
- *Where appropriate, describe how people might be able to reduce radiation doses, sheltering being a prime example;*
- *Make clear the areas where populations might be affected and those where people are not (or are unlikely to be) affected. (Address as appropriate the possibility of weather conditions changing and dispersing contamination in a different direction.);*
- *Clearly explain any restrictions on food, milk or water consumption;*
- *Clearly explain any travel or transport restrictions or advice;*
- *Provide consistent, concise and clear advice. During a prolonged emergency, issuing information at a regular time will help people cope with the effects;*
- *Choose spokespersons carefully to deliver information and messages;*
- *Clearly communicate the rationale for any public health recommendations, including showering, sheltering, evacuation and stable iodine tablets (if appropriate);*
- *Be clear about any uncertainties related to the precise nature of the release so that the need for precautionary measures can be more easily understood by the public;*
- *Clearly explain messages containing precautions for children specifically. During an emergency, people naturally have concerns for family members who are involved or they think are affected;*
- *Clearly explain the risks from radiation exposure, including acute and long term risks, in straightforward language; and*
- *Clearly explain the need, when appropriate, for people to attend special monitoring centres set up in response to an emergency so that a comprehensive monitoring programme can be carried out effectively.*

It provides some tools for communicating the basics of radiation and putting dose-rates into context. It explains the main types of radiation and exposure pathways of concern. Rather frustratingly the material is all subject to copy-right so it seems likely that local authority cannot copy tables and diagrams into their publications without permission and probably at expense.

On the day of a real event the best advice on content for public information and support would probably come from the Centre for Radiation, Chemical and Environmental Hazards (CRCE), Public Health England.

16.7.3. International Nuclear Event Scale (INES)

The International Nuclear and Radiological Event Scale (INES) (Ref. 314) is a tool for communicating the safety significance of nuclear and radiological events to the public. Member States use INES on a voluntary basis to rate and communicate events that occur within their territory.

Events are rated at seven levels (See Table 19 and Table 20). The scale is logarithmic – that is, the severity of an event is about ten times greater for each increase in level of the scale. The scale is quite complex to use since it considers impact on people and the environment; impact on radiological barriers and control and impact on defence in depth.

In this scale a level 5 event is one "with wider consequences" i.e. off-site consequences that might involve a limited

	People and environment	Radiological barriers and controls at facilities	Defence in depth
Major accident Level 7	Major release of radioactive material with widespread health and environmental effects requiring implementation of planned and extended countermeasures.		
Serious accident Level 6	Significant release of radioactive material likely to require implementation of planned countermeasures.		
Accident with wider consequences Level 5	• Limited release of radioactive material likely to require implementation of some planned countermeasures • Several deaths from radiation	• Severe damage to reactor core. • Release of large quantities of radioactive materials within an installation with a high probability of significant public exposure. This could arise from a major criticality accident or fire.	
Accident with local consequences Level 4	• Minor releases of radioactive material unlikely to result in implementation of planned countermeasures other than local food controls. • At least one death from radiation.	• Fuel melt or damage to fuel resulting in more than 0.1% release of core inventory. • Release of significant quantities of radioactive material within an installation with a high probability of public exposure.	
Serious incident Level 3	• Exposure in excess of ten times the statutory annual limit for workers. • Non-lethal deterministic health effects (e.g. burns) from radiation.	• Exposure rates of more than 1 Sv/hr in an operating area. • Severe contamination in an area not expected by design, with a low probability of significant public exposure.	• Near accident at a nuclear power plant with no safety provisions remaining. • Lost or stolen highly radioactive sealed source. • Mis-delivered highly radioactive sealed source without adequate radiation procedures in place to handle it.
Incident Level 2	• Exposure of member of public in excess of 10 mSv. • Exposure of a worker in excess of the statutory annual limits.	• Radiation levels in an operating area of more than 50 mSv/hr. • Significant contamination within the facility into an area not expected by design.	• Significant failure in safety provision but with no actual consequences. • Found highly radioactive sealed orphan source, device or transport package with safety provision intact. • Inadequate packaging of a highly radioactive sealed source.
Anomaly Level 1			• Overexposure of a member of the public in excess of statutory limits. • Minor problems with safety components with significant defence in depth remaining. • Low activity lost or stolen radioactive source, device or transport package.

Table 19 INES Scale definition

Description and INES Level	People and Environment	Radiological Barriers and Controls at facilities	Defence-in-Depth
Major Accident Level 7	Chernobyl, 1986 — Widespread effects. External release of a significant fraction of reactor core inventory.		
Serious Accident Level 6	Kyshtym, Russia, 1957 — Significant release of radioactive material to the environment from explosion of a high activity waste tank.		
Accident with Wider Consequences Level 5	Windscale Pile, UK, 1957 — Release of radioactive material to the environment following a fire in a reactor core.	Three Mile Island, USA, 1979 — Severe damage to the reactor core.	
Accident with Local Consequences Level 4	Tokaimura, Japan, 1999 — Fatal overexposures of workers following a criticality event at a nuclear facility.	Saint Laurent des Eaux, France, 1980 melting of one channel of fuel in the reactor with no release outside the site.	
Serious Incident Level 3		Sellafield, UK, 2005 — Release of large quantity of radioactive material, contained within the installation.	Vandellos, Spain, 1989 — Near accident caused by fire resulting in loss of safety systems.
Incident Level 2	Atucha, Argentina, 2005 — Overexposure of a worker at a power reactor exceeding the annual limit.	Cadarache, France, 1993 — Spread of contamination to an area not expected by design.	Forsmark, Sweden, 2006 — Degraded safety functions, common cause failure in the emergency power supply system.

Table 20 INES classification in use

release of radioactivity and the need to advice some protective actions. This level could include several deaths from radiation. On-site the situation would probably involve core damage (for a reactor facility) and significant failure of one or more radioactivity containment barrier. The Windscale Fire and Three Mile Island both fit into this category.

A level 6 event involves a "significant release of radioactivity" leading to protective actions. This would have applied to the Kyshtym disaster at Mayak (Ref. 315) in the USSR in 1957 which led to the eventual evacuation of several villages.

Level 7 is the highest level on the scale. It involves a major release of radioactive material with widespread health and environmental effects requiring implementation of planned and extended protective actions. Both Chernobyl and Fukushima are in this category.

It is likely that if an event in the UK was categorised at level 2 or above the police and local authority would be expected to react in some way and to give reassurance to the public via the media.

16.8. Working within a protective action zone

It will always be difficult to decide if work should be undertaken within protective action zones.

The NRPB (forerunner of PHE CRCE) cite a survey that most people claimed that they could shelter for at least one day without experiencing significant disruption (Ref. 317). However, that is most people. Some will find it difficult to be cut off from helpers, be they family or home care, for any period of time. Other people may suffer issues such as going into labour or medical issues such as sudden illness or collapse.

If there were a plea for help from somebody within an area subject to protective action advice there would be a number of factors to take into account when deciding how to respond:

- who was asking for help, particularly the nature of their vulnerability;
- the nature of the help required;
- the urgency of the help required;
- the potential duration of the exposure time for the responders; and
- the dose-rate predictions in the response area.

Some of these have considerable uncertainty attached to them. Some are obviously connected.

If the operators and ONR are confident that the release rate will drop over the next few hours and the people seeking assistance can manage with no real hardship for those few hours then it would be better to provide immediate support and reassurance by phone (if possible) rather than in person. Clearly nursing and healthcare would take precedence over housekeeping while personal care and cooking or help with meals would need a case-by-case consideration and would become more pressing as the period in shelter increased.

If suitable support can be obtained more locally, for example by a neighbour, then this option could be considered.

Another option might be for a family member, friend or neighbour to bring the person out of the protective action zone for care in a safe facility if appropriate.

If there seems no reasonable alternative to a visit then a dose assessment should be made for the responders and this compared to the value of the task.

The operator or STAC should be able to give a dose estimate for any area subject to a number of assumptions about forthcoming changes in the release rate and the weather conditions. They will simplify the dose assessment methodology but, in doing so, will try to ensure that they err towards overestimating the dose rather than underestimating it.

RIMNET (Ref. 318) can provide an "area at risk" plot (see Figure 35 for an example - the system allows you to zoom in for more local information) which shows the footprint of the plume as predicted by their atmospheric dispersion models (Ref. 319). These can be accessed via the PACRAM process (Ref. 320) or via Resilience Direct (Ref. 321). These can be used to determine if the person requiring assistance is in the area truly affected by the event (protective action zones are usually deliberately oversized to allow for uncertainties in the projected dose estimates) and help to plan a route into and out of the area.

Resilience Direct is an online private 'network' which enables civil protection practitioners in the UK to work together – across geographical and organisational boundaries – during the preparation, response and recovery phases of an event or emergency.

Having worked out a dose estimate for a job a decision has to be made about whether or not the job goes ahead. The estimated dose could be compared to the appropriate dose limit for the service in question (See Section 11.9). For example, Police Scotland have an operation limit of 5 mSv, an emergency range (Dose Limit 1) up to 100 mSv and 500 mGy for lifesaving and critical infrastructure work.

Again, remember the dose limitation i.e. keeping within the appropriate dose limits is not enough. It is also required by law and morality that doses be kept as low as reasonably practical. This means that invoking the 100 mSv limit does not give you the right to plan and deliver a mission in a disorganised manner resulting in a dose to an employee of (say) 60 mSv when smarter planning would have delivered it in (say) 40 mSv.

Ideally the responders' Radiation Protection Advisor (RPA) would be consulted before work involving a potential dose of radiation is undertaken.

It is important to remember that if a team is sent into an area within the airborne plume, then they will be inhaling radioactivity and any doses or doserates shown on their personal dosimeters will be underestimating their total dose or doserate.

Several systems have been proposed to overcome this issue. They all rely on understanding the ratio between a measured dose or doserate and the actual or total dose or dose rate resulting from all pathways.

EDF have provided a "traffic light" tool to assist the emergency services make this kind of decision. This is based on an estimated relationship between the measurable external radiation dose and the total dose to a person taking into account inhalation dose assuming an isotopic mixture from the site's safety case and based on the reference accident. The process allows an estimate of the projected dose over 48 hours of continuous occupation of the DEPZ to be deduced from a reading from the Emergency Perimeter Gamma Monitoring System (which is always on and remote reading) so that a very quick risk assessment of police work in the area can be given.

If green (projected dose below 1 mSv) then proceed with work.

If amber (projected dose above 1 mSv but below 5 mSv) "*consider the methods used and actions undertaken when carrying out tasks to protect the public e.g. optimizing the time spent within the DEPZ*"

If red (projected dose greater than 5 mSv) then "*Only critical public protection action should be taken directly in the DEPZ until independent advice has been received by the police from their radiological safety advisors*".

This is a very pessimistic system as it is based on 48 hours in the DEPZ with unchanging conditions and is used to provide a very quick "go - no go" decision before more considered radiological advice and more knowledge of the actual event is available.

The IAEA (Ref. 322) suggests a similar system based on 4 levels (Table 21) and a number of perspective charts. This system is for after the release has stopped. In the same way as the EDF traffic lights, it presents charts which relate something that can be readily measured, such as dose rate 1 m above the ground, with the health hazard. This methodology supports decision making about the long-term protective actions to protect the public but also provides a framework for controlling the dose uptake of local authority employed recovery workers.

The above framework is described and extended in the IAEA document EPR-NPP Public Protective Actions 2013 (Ref. 322) which would be an invaluable book to have and to understand in the early recovery phase. It talks about Operational Intervention Levels (OILs) which relate measurable quantities (ground gamma dose, activity in food, dose external to a person's thyroid etc.) to levels of risk or harm and actions to be taken. A document with just the charts can be found at Ref. 323.

It has always surprised me that the PHE has never recommended the adoption of OILs by the UK.

Health hazard perspective	Explanation
Possibly dangerous to health (red)	There is a possibility of radiation induced health effects that are life threatening or can result in a permanent injury that reduces the quality of life (severe deterministic effects). At this level there is also the small possibility of an observable increase in the incidence of cancer due to radiation induced cases, if the number of exposed people is more than a few hundred.
Health concerns (orange)	The danger to health is very low. However, there is a possibility of doses exceeding the international criteria that call for taking protective actions and other response actions to include medical screening in order to further assess: (a) the small possible risk to pregnant women (fetus), and (b) the small possible increase in the risk of radiation induced cancers.
Provisionally safe (yellow)	All members of the public are safe including the most sensitive (i.e. children and pregnant women) and there are no hazards to health due to radiation exposure if the specified limitations are followed, such as remaining in the area is limited to a specific amount of time and/or specified protective actions are taken (e.g. reduce ingestion of radioactive material).

Health hazard perspective	Explanation
Safe (green)	All members of the public are safe including the most sensitive (i.e. children and pregnant women) as international GC are met. The doses for the specified conditions and public behaviour (exposure scenario) are less than the GC at which protective actions and other response actions are justified to minimize severe deterministic effects or to reduce the risk of stochastic effects. Below this level there will not be any severe deterministic effects or an observable increase in the incidence of cancer, even in a very large, exposed group. Furthermore, the risk of radiation induced cancers is too low to justify taking any action, such as a medical screening.

Table 21 IAEA classification of areas from health perspective

16.8.1. Vehicle monitoring

It is very unlikely that a person or team working within the DEPZ will become significantly contaminated with radioactive material. It may be deemed appropriate to undertake a quick radiological survey of any vehicle leaving the area, concentrating on the tyres. Such surveys will require radiation monitoring instruments that can measure ambient dose rates in the range of local background (typically between 0.05-0.2 μSv/h). These should be wrapped in a plastic bag to prevent contamination. Monitoring the item for gamma contamination should be performed holding the monitor about 10 cm from the surface. Anything below 1 μSv/h is "clean" and can pass onwards. Anything above this level should be decontaminated using fire hoses, scrub brushes and detergents, ideally recovering any runoff. (Based on IAEA advice at Ref. 322, see also Ref. 324).

16.9. Recovery

This section considers the activities the local authority will become involved in after the incident itself is over. This may be a long term situation; a new normality.

If there were to be a significant release of radioactive gases and dusts from a Nuclear Power Station then there might follow a protracted period of recovery. In some ways recovery is easier in that the situation is not rapidly changing but there are more stakeholders, with widely different needs, to consult. It will be a complicated business to measure where radioactive material has deposited itself, determine where it can be expected to migrate to in the future, estimate the dose implications to the public and agree what target should be set. Expect a lot of debate, not all of it particularly well informed (but this will improve with time), but some of it highly charged with emotion.

The IAEA (Ref. 325) suggest that "*Recovery means the achievement of conditions within which society can again fully function, such as:*

- *Stabilisation of the accident scene leading towards eventual dismantling;*
- *remediation of affected areas to reduce radiation doses to people to an acceptable level;*
- *effective and safe management of contaminated material and radioactive waste leading to its ultimate disposal; and*
- *re-establishment of infrastructure and the revitalization of communities*".

It is important to try to prevent any promises being given to the public and businesses in the area that the clean-up will restore the area to "*no radiation*" (quite impossible since the environment is naturally radioactive anyway) or "*as low as it was before the accident*" (also totally impractical). The clean-up costs and the volume of waste generated depends heavily on the agreed level of clean-up whereas the dose savings can be small, particularly at the lower levels of contamination.

It will be important to discuss with the public the results of radiation monitoring, the pathways whereby they (and their children) could receive radiation dose, the estimates and measurements of individual dose, possible remediation action (cleaning the environment) and possible changes in their habits that would result in useful dose reductions.

The IAEA (Ref. 326) states that "*As the emergency evolves, the focus of the emergency response will shift from bringing the situation under control and taking public protective actions, to allowing the timely resumption of social and economic activity. At this time, radiological considerations will be only one of the many factors to be evaluated in the decision-making processes. Decision making at this time will require the involvement of additional organizations, with relevant responsibilities at different levels, which might not necessarily have been directly engaged during the urgent response phase. These organizations should gradually be involved, when appropriate, in the emergency response in order to discharge their allocated roles and responsibilities. This involvement should be arranged in a way that enables ongoing response efforts to continue without interruption on a routine basis in the longer term, after the emergency response organization has been relieved of its duties*".

The financial impact of a radiological emergency includes those associated with the loss of use of resources such as properties (shops, homes, industrial units and offices), agricultural land, infrastructure and recreational facilities. There will be immediate financial impacts during the emergency phase associated with emergency response actions and the long-term financial impacts from the cost of recovery.

If there is to be any level of decontamination of the environment, then waste management will be both complex and expensive. Desired endpoints may be out of reach due to complexity or cost and some difficult decisions will have to be made.

Sometimes pragmatic decisions must be made in the short term to allow progress, but these can cause their own problems years later. For example, The IAEA tell us (Ref. 327) that "*The Radioactive Waste Burial Facility (RWBF) Buryakivka was designed shortly after the Chernobyl accident and constructed in 1987. Forced by the lack of more suitable storage and/or disposal options, disposal of waste with dose rates of up to 5 R/h (50 mSv/h) was permitted for a restricted period, even though the facility was not designed for this type of waste. Information on the characteristics of the disposed waste was not retained and presents a major challenge for assessing performance of the facility.*

In addition to Buryakivka, hundreds of interim storage facilities were created after the Chernobyl accident in 1986 - 1987. However, the inventories and characteristics of waste disposed in these facilities is not well known. As a result, significant efforts were required to reassess the safety of these facilities, including waste retrieval or reconstruction of facilities".

16.9.1. Management

At the end of the acute phase of the emergency the police are likely to stand down from chairing the Strategic Co-ordination Group in favour of the local authority. It would be expected that this handover would be the object of a process in which all of the agencies involved in the co-ordination of the response were consulted and a set of pre-conditions agreed and then met. It is a strategic decision.

Restoring and maintaining public confidence in the affected areas will be one of the more important factors considered. Other stakeholders, notably residents, have a stronger influence at this stage.

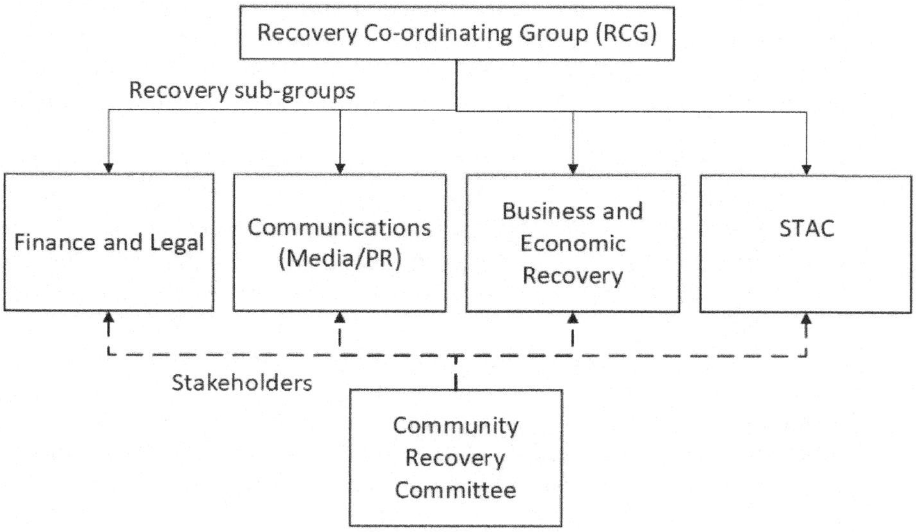

Figure 42 Recovery management

National Nuclear Emergency Planning and Response Guidance (Ref. 195) suggests the structure shown in Figure 42 but there may be local differences which would be described in the local off-site plan.

In the first stages of recovery the major issues would include the withdrawal of the shelter and evacuation protective actions and the determination of long-term advice for those with homes and business in the areas. This is likely to be a highly emotive process and, based on the experiences at Fukushima and following the Grenfell Tower fire, difficult to manage to everybody's satisfaction.

The target will be to achieve an acceptable "new normality". What is meant by 'normality' is not easily defined. The IAEA (Ref. 325) suggest that "*indicators of a revitalized infrastructure and community are:*

- *a place to call home and a sense of safety;*
- *community structures including facilities for health and aged care, education and leisure;*
- *stability and the certainty of governing structures;*
- *availability of employment, including opportunities for farming and local food production;*
- *the involvement of stakeholders in decision-making*".

16.9.1. Scene of crime work

It is likely that following any incident there will be a number of investigations about the causes of the event, how it was managed and the outcome. These may be undertaken by the Operator's or Carrier's Health and Safety organisation, by the civil, nuclear or MOD police forces, the HM Coroner or civil or criminal courts. There is a balance to be found between making safe the incident scene and preserving evidence for the various investigations.

> Radiological crime scene is the site containing records of past activities alleged to be a crime involving nuclear or other radioactive material.
>
> IAEA Radiological Crime Scene Management e-learning

While the police forces are used to working at scenes of crime and collecting evidence and health physicists are used to directing work to be as safe as possible in radiological controlled areas the two do not often have to work together and it is an area of nuclear response that is little exercised.

Whilst working in such an environment the investigators need to be aware of the need to minimise the time spent in the hazard control areas; to maximise the distance between the evidence contaminated with radionuclides and the individual collecting the evidence; to optimise the radiation shielding between the evidence and the individual collecting the evidence; and to minimise the spread of radionuclide contamination.

The IAEA suggest that the following personnel may be needed:

- *On-scene operational commander;*
- *Hazardous materials operations specialist;*
- *Safety specialist;*
- *Radiological advisor;*
- *Scene modeller;*
- *Evidence recovery personnel;*
- *Photographer;*
- *Evidence recorder/custodian.*

Which shows that this is not a trivial matter.

The IAEA advice on this matter is signposted via a web-site (Ref. 328) which gives access to a report (Ref. 329) and a half-hour e-learning course (Ref. 330) which may be useful as "just-in-time" training.

16.9.2. Monitoring

This section looks at the longer term recovery and the potential need for personnel to work in areas that have been contaminated with radioactive fallout from an off-site nuclear emergency.

A major release of radioactivity over the countryside and villages would require in a significant environmental surveying programme. As well as the areas known to be contaminated there would be a demand from worried members of the public to survey areas that the dispersion modellers were confident were clean.

A criticism of the authorities post-Fukushima was that "*during the initial weeks of the disaster in particular, citizens were given little actionable information concerning actual radiation levels, and what was presented both by official spokespersons and in the media was often incomplete and/or contradictory....... Also, there was no standardization or consistency in the available data in terms of detector type or processing methodology*". (Ref. 331). This is an area where technology has made significant advances over the last decade. Detectors linked to GPS and internet ready databases and maps are no longer on the limits of technology and, as well as being carried on helicopters and vans they can now be attached to drones. This means that the public can, and should, expect to see a comprehensive picture of local contamination and dose rates developing in days rather than in months. Whether or not it helps decision making remains to be seen.

Wide area monitoring such as airborne gamma spectroscopy would provide the "big picture" view of deposition, but more detailed ground-based monitoring would be needed to give a more accurate understanding. These days we might expect to see drones being used to perform wide area radiation surveys. A variety of drone and detector types can be used to give either the bigger picture over a wide area or more detail over a smaller area.

Dose rate measurements taken at 1m above the ground are used to determine the average ground gamma dose in the area so many of these will be taken. Post-Fukushima a lot of devices were developed that integrated positional information (GPS) with dose rate measurements to produce ground gamma maps. These included drone-mounted, vehicle-mounted and portable systems for different purposes.

High-resolution gamma spectroscopy measurements are also required to determine the "isotopic fingerprint" of the deposited activity. That is, the mixture of radioisotopes present. This helps the modellers determine both the rate of decay of the activity and the inhalation and ingestion doses that may result from resuspension of the deposits.

Experiments and measurements would be required to determine the mobility of the radioactivity in the environment to predict how it would behave in the days, months and years following the accident and how it might respond to different clean-up operations. This is necessary to provide data for the source > pathway > receptor model which is used to identify how people could become exposed to radiation, what the doses for each route might be and how best to balance dose reduction with quality of life.

Exposed individuals may also wish for personal monitoring during and shortly after the event. This may take place in Radiation Monitoring Units (RMU). The off-site plan should identify which organisation would be responsible for providing and managing any RMUs required.

Whole body monitoring and thyroid monitoring may also be used to measure internal burdens of radionuclides in the exposed population.

In the immediate recovery phase members of the public may wish to wear radiation dosimeters to measure and record their radiation doses. This would also be useful for the dose modelling effort. It should be noted that these devices are sensitive to external radiation only and do not assess inhalation or ingestion dose.

Post Fukushima the authorities erected a number of stations with gamma monitors coupled to large display screens so that passers-by could see the current dose rates. These were popular with the public as it gave them more information that had not been filtered by the government or the operator.

Local Authority personnel working on clean-up programmes should be adequately briefed and understand the, probably minimal, radiological risks associated with clean-up activities.

No matter how much monitoring is achieved in this situation it is unlikely to impress everyone. There will also be long arguments about the analysis of the data obtained and the meaning of it in terms of terms of the habitability of the area. School play areas can be particularly contentious.

16.9.3. Clean-up

There are a number of Management Options available for the decontamination of areas, buildings and facilities. The UK Recovery Handbooks for Radiation Incidents 2015, Inhabited Areas Handbook (Ref. 332) gives a table of options upon which the list below is based.

- Restrict access;
- Control workforce access (area classification and stay-time limits);
- Impose restrictions on transport;
- Temporary or permanent move of office accommodation;
- Restrict public access;
- Permanent or temporary relocation of residents;
- Remediation:
- Modify operation/cleaning of ventilation systems;
- Natural attenuation - with monitoring;
- Pressure and fire hosing;
- Vacuum cleaning;
- Water based cleaning;
- Treatment of walls with ammonium nitrate;
- Roof cleaning including gutters & downpipes;
- Use of fix and strip coatings;
- Use of reactive liquids;
- Surface removal (buildings & indoors);
- Surface removal and replacement (Roads);
- Demolish/dismantle and dispose of contaminated material;
- Treatment of waste water;
- Tie-down;
- Storage/cleaning of valuable equipment;
- Collection of leaves;
- Cover grass/soil with clean soil/asphalt;
- Top soil and turf removal;
- Grass cutting and disposal;
- Tree and Scrub pruning and removal;
- Mechanical and manual digging;
- Ploughing - various methods; and
- Snow/ice removal

Before an option was adopted it would be subject to analysis looking at the potential costs (materials, time, personal dose, waste generated, environmental harm, risks) and benefits (reduction in environmental radioactivity, reduced doses to the public, enhanced public confidence).

Optioneering will not be an easy process. Few people possess the knowledge to participate in the discussions and so rapid learning will be required. The recovery handbooks for radiation incidents (Ref. 332, but note that these are currently being updated (2020) and this reference may have been superseded) are a particularly good place to start. It is likely that the media will be paying a close interest. Inclusion of the public in the decision-making process is desirable and that will likely modify the terms of the discussion.

There may be training needs for the personnel undertaking the work. This training may include training in the use of the technique itself, conventional health and safety issues and radiological dose minimisation and measuring.

The managers and decision makers involved in the project will also require adequate training in these areas to allow them to make informed, transparent and defendable decisions. Most, if not all, of the training required could be given "just-in-time".

16.9.4. Human aspects

> *The human aspects of an emergency have been defined as "Those activities [that are] aimed at addressing the needs of people affected by emergencies; the provision of psychological and social aftercare and support in the short, medium and long term."*

Quote taken from: Human Aspects in Emergency Management Guidance on supporting individuals affected by emergencies. (Ref. 333).

It is important to remember that off-site nuclear emergencies are not only physical events requiring procedural approaches to planning and response; they are also psychological and social events. The needs of those affected by the emergency should be considered. Those affected include the responders, those members of the operator's personnel and members of the public who have had to take shelter, possibly take tablets, possibly move from their homes, either for a short while or for an extended period of time and who have been hearing a lot of terms that they don't fully understand and subjected to a hazard they cannot see, smell or feel and which they do not understand. Also affected are family and friends of those directly affected and those outside the areas delineated by the emergency responders but non-the-less worried about the invisible threat.

A "psychological" impact is one that affects a specific individual's mental state. "Psychosocial" impacts are those affecting a wider section of the population. Ideally both will be managed if there is a need.

Actions related to human aspects, or social care in emergencies, can include:

- emotional support;
- first aid;
- shelter, food, clothing;
- information updates about the incident and individuals directly involved;
- advice and support on financial, legal and insurance issues;
- support to restore social networks;
- opportunities for remembrance/memorialisation;
- input into any evaluation process following the emergency (for example a public inquiry).

Planning is expected to take account of vulnerable groups and diverse communities (considering factors such as culture, language and faith).

This endeavour is not unique to nuclear emergencies. Local emergency planners are expected to consider the impacts of the events postulated in their local risk registers and to put in place reasonable plans to manage the situation should it arise. (See regulation 15 of the Civil Contingencies Act 2004 (Contingency Planning) Regulations 2005).

It is to be expected that a casualty bureau, as usually constituted, will not be required as the expectation is that few, if any members of the public will become casualties in an off-site nuclear emergency. What will happen though is that families and friends will be separated with some within areas subject to prompt protective actions and some outside the area. This could be distressing. Schools would be expected to care for children until they can be reunited with their parents and most of the others would be expected to be in contact via phones and/or social media. A few may appeal to the police to make contact with sundered friends and family.

In the first 24 hours of an incident the local authority may be required to:

- To direct those people affected to places where they can shelter and recover;
- To ensure names and addresses of all those affected are recorded and fed into one central point (a function that might be taken by the police where relevant);
- For both survivors and worried family and friends – to offer a single point of information about what's happening and a single point of access to local responders, as the emergency unfolds.

A basic Rest and Reception Centre may be an adequate start but it is worth remembering there may be a reluctance on the part of the decision makers to allow movement within areas subjected to shelter advice for a number of days even if the release only lasted a few hours. The reason stated for this is the need to complete radiological surveys before declaring the area safe to resume normal activity.

If you cannot return to normality the needs of those affected become more complex with time and safeguarding children and vulnerable adults becomes more pressing.

In the following days and weeks there may be the need to meet with residents, provide them with information as it becomes available and then listen to them and respond to their support needs.

HM government guidance and advice is focussed on the Human Aspects of Emergency Management web page (Ref. 334).

17. Non-nuclear site response

Radiological emergencies can occur anywhere including:

- Uncontrolled (abandoned, lost, stolen or found) dangerous sources that could lead to deterministic effects and may be handled by people unaware that they are radioactive;
- Misuse of, or accidents with, dangerous industrial and medical sources (e.g. those used in radiography);
- Public exposures and contamination from unknown origins;
- Serious overexposures (i.e. exposures leading to deterministic effects);
- Malicious threats/acts; and
- Transport emergencies.

Clearly the local emergency services and local hospitals are likely to be the first responders with national expertise arriving in due course.

IAEA advice for first responders is given in Reference 187 which should be on the emergency planning bookshelf of emergency services, local authorities and hospitals, and ideally part of routine training. Associated E-learning tools are available in Reference 188.

The big difference between an event at a nuclear licensed site and an event elsewhere is likely to be the lack of radiological protection expertise, the ideal instruments for surveying (emergency vehicles will carry "all purpose" instruments which can generally do a useful job) and permanent command and control facilities.

The objectives of the response are familiar. The IAEA express them as:

- To promptly perform all reasonable actions to protect the public in order to
- minimize the radiological and non-radiological (e.g. psychological) health effects;
- To protect emergency personnel during response operations;
- To gather and protect information that can be useful in managing health effects, for law enforcement purposes and in preventing similar emergencies in the future;
- To establish and maintain public trust in the response;
- To provide a basis for an extended response.

Figure 43 shows the IAEA view of scene management.

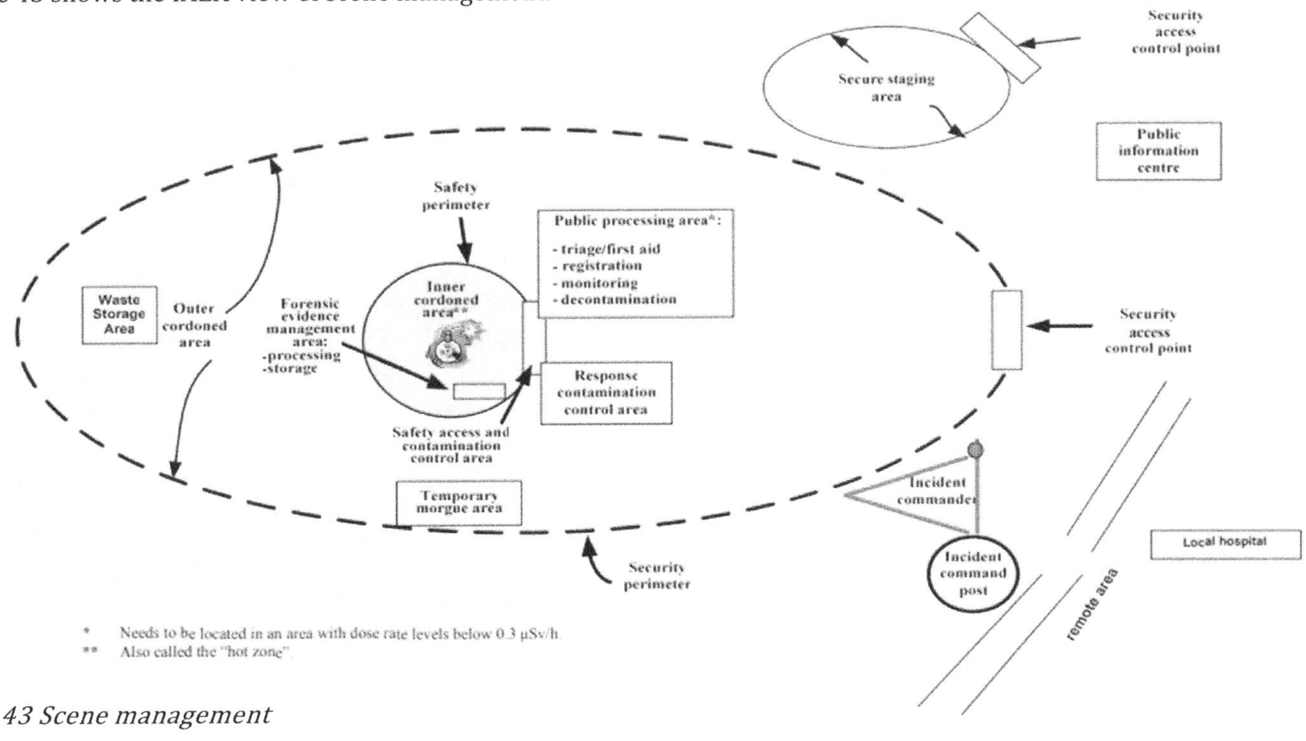

Figure 43 Scene management

Another aspect of events outside licensed sites is that members of the public who may be contaminated, especially following large incidents, may leave the scene and subsequently seek assistance at a nearby healthcare facility. All UK healthcare facilities are required to have arrangements in place to manage self-presenting patients. These arrangements include incident response plans, business continuity plans and lockdown plans which are designed to protect the wider community of service users as well as those primarily affected (Ref. 185).

18. References

1. HSE, The Tolerability of Risk from Nuclear Power Stations http://www.onr.org.uk/documents/tolerability.pdf.
2. The Office for Nuclear Regulation, http://www.onr.org.uk/.
3. Radiation (Emergency Planning and Public Information) Regulations 2019, http://www.hse.gov.uk/radiation/ionising/reppir.htm.
4. IAEA Training Course on Safety Assessment of NPPs to Assist Decision Making, Safety Assessment of General Design Safety Assessment of General Design Aspects of Aspects of NPPs (Part 1)(IAEA Training Course on Safety Assessment of NPPs to Assist Decision Making.
5. ERI Consulting, Perspectives on Reactor Safety, NUREG/CR-6042, Rev. 2 SAND 93-0971. https://www.nrc.gov/reading-rm/doc-collections/nuregs/contract/cr6042/.
6. Identifying people who are vulnerable in a crisis: guidance for emergency planners and responders, https://www.gov.uk/government/publications/identifying-people-who-are-vulnerable-in-a-crisis-guidance-for-emergency-planners-and-responders.
7. Remarks prepared by Lewis L. Strass, Sept 16, 1954. http://www.nrc.gov/docs/ML1613/ML16131A120.pdf.
8. International Atomic Energy Agency, World Uranium Geology, Exploration, Resources and Production, IAEA, Vienna (2020).
9. Chemeurope.com, United States Radium Corporation, https://www.chemeurope.com/en/encyclopedia/United_States_Radium_Corporation.html.
10. Frank Goldner, US DOE, Overview of the accident-tolerant fuel development, NEA/NSC/DOC(2013)9, https://inis.iaea.org/collection/NCLCollectionStore/_Public/47/013/47013906.pdf.
11. Gov.uk, managing nuclear materials and spent fuels, https://www.gov.uk/government/collections/managing-nuclear-materials-and-spent-fuels.
12. Fast reactors and related fuel cycles: next generation nuclear systems for sustainable development FR17, https://www-pub.iaea.org/MTCD/Publications/PDF/STIPUB1836web.pdf.
13. Westinghouse, Hex business at Springfields. https://www.westinghousenuclear.com/Portals/2/Documents/Brochure%20-%20Hex%20Businessl.pdf.
14. Westinghouse, Fuel Manufacture at Springfields. https://www.westinghousenuclear.com/Portals/2/Documents/Brochure%20Nuclear_Leaflet_04.pdf.
15. ONR, Determination of the Off-Site Emergency Planning and Prior Information Areas for the Springfields Fuels Ltd. Nuclear Licensed Site, http://www.onr.org.uk/pars/2017/springfields-17-001.pdf.
16. ONR, Letter of 20th September 2017, to Mr M R Gornall, Managing Director, Springfields Fuels Limited. TRIM Ref: 2017/0266899, Unique Number: ONR-EPR-17-3, http://www.onr.org.uk/pars/2017/springfields-17-001-letter-springfields-fuel-ltd.pdf.
17. Westinghouse UK Fuels Springfields – Radiation (Emergency Preparedness and Public Information) Regulations 2019 – Reg 7 Consequences Report, https://www.lancashire.gov.uk/media/918349/sfl-consequence-report.pdf.
18. Lancaster County Council, Detailed Emergency Planning Zone Determination Report: Local Authority DEPZ Determination Report, https://www.lancashire.gov.uk/media/918380/springfield-depz-determination-report.pdf.
19. HSE, Public information on establishments subject to COMAH 2015, https://www.hse.gov.uk/comah/comah-establishments.htm.

20. UUK Capenhurst Site, ENSREG Stress Tests Report (2012), https://www.urenco.com/cdn/uploads/supporting-files/uukstresstestsreportforuukwebsite150512_1.pdf.
21. Urenco ChemPlants, Tails Management facility, https://www.urenco.com/global-operations/urenco-chemplants.
22. Joint Cheshire Emergency Planning Team, Detailed Emergency Planning Zone Determination Report. https://www.cheshirewestandchester.gov.uk/your-council/emergency-planning/urenco-uk-ltd-capenhurst/documents/uuk-depz-determination-report.pdf.
23. Capenhurst Radiation (Emergency Preparedness and Public Information) Regulations 2019, Regulation 7 Submission Consequence Report, https://www.cheshirewestandchester.gov.uk/your-council/emergency-planning/urenco-uk-ltd-capenhurst/documents/capenhurst-reppir19-consequence-report.pdf.
24. ONR GDA Current reactors being assessed, http://www.onr.org.uk/new-reactors/assessment.htm.
25. WNA, Physics of Uranium and Nuclear Energy, https://www.world-nuclear.org/information-library/nuclear-fuel-cycle/introduction/physics-of-nuclear-energy.aspx.
26. W. R. Lloyd, M. K. Sheaffer, W.G. Sutcliffe, Dose Rate Estimates from Irradiated Light-Water-Reactor Fuel Assemblies in Air, (1994), https://www.osti.gov/servlets/purl/10137382.
27. Department for Business, Energy and Industrial Strategy, Advanced Gas-cooled Reactor (AGR) decommissioning (2019), https://www.gov.uk/government/publications/advanced-gas-cooled-reactor-agr-decommissioning
28. IAEA, General Design and Principles of the Advanced Gas-Cooled Reactor (AGR), https://nucleus.iaea.org/sites/graphiteknowledgebase/wiki/Guide_to_Graphite/General%20Design%20and%20Principles%20of%20the%20Advanced%20Gas-Cooled%20Reactor%20(AGR).aspx
29. AGR, Erik Nonbel, Riso National Laboratory Roskilde, Denmark, Description of the Advanced Gas Cooled Type of Reactor (AGR)
30. EDF, EU Stress Test –Torness, https://www.EDFenergy.com/sites/default/files/jer-srt-stt-pub-fin-008_tor_stress_test_v1.1.pdfhttp://www.iaea.org/inis/collection/NCLCollectionStore/_Public/28/028/28028509.pdf .
31. Torness Station Procedure, Hazard Identification and Risk Evaluation (HIRE) https://www.EDFenergy.com/file/1490/download.
32. EDF, 2017 REPPIR Report of Assessment Hunterston B Power Station, https://www.edfenergy.com/sites/default/files/2017_hnb_roa_npm_-_final.pdf
33. IAEA Safety Reports Series No. 30 Accident Analysis for Nuclear Power Plants with Pressurized Water Reactors, https://www-pub.iaea.org/MTCD/Publications/PDF/Pub1162_web.pdf
34. EDF, 2017, REPPIR Report of Assessment for Sizewell B, https://www.edfenergy.com/sites/default/files/szb_roa_final_2017.pdf
35. EDF, The Radiation (Emergency Preparedness and Public Information) Regulations 2019, Sizewell B Power Station Consequences Report https://www.suffolkresilience.com/uploads/EDF_SZB_CR_-_Jan_20.pdf
36. USNRC Technical Training Center, Pressurized Water Reactor (PWR) Systems, https://www.nrc.gov/reading-rm/basic-ref/students/for-educators/04.pdf
37. IAEA, Westenghouse AP600 Advanced Nuclear Plant Design, http://www.iaea.org/inis/collection/NCLCollectionStore/_Public/31/007/31007051.pdf
38. UK EPR - Generic Design Assessment, http://www.epr-reactor.co.uk/scripts/ssmod/publigen/content/templates/show.asp?P=57&L=EN
39. UK-EPR PCSR – Sub-Chapter 14.6 – Radiological consequences of design basis accidents, http://www.epr-reactor.co.uk/ssmod/liblocal/docs/PCSR/Chapter%2014%20-%20Design%20Basis%20Analysis/Sub-Chapter%2014.6%20-%20Radiological%20Consequences%20of%20Design%20Basis%20Accidents.pdf
40. HMG, New nuclear power stations: assessing reactor designs, https://www.gov.uk/guidance/new-nuclear-power-stations-assessing-reactor-designs
41. General Nuclear System, GDA Library, http://www.ukhpr1000.co.uk/documents-library/.

42. HPR-GDA-PCSR-0032-Pre-Construction-Safety-Report-Chapter-32-Emergency-Preparedness-Rev-001, http://www.ukhpr1000.co.uk/wp-content/uploads/2020/02/HPR-GDA-PCSR-0032-Pre-Construction-Safety-Report-Chapter-32-Emergency-Preparedness-Rev-001.pdf.

43. IAEA, Advances in Small Modular Reactor Technology Development: A Supplement to: IAEA Advanced Reactors Information System (ARIS) 2020 Edition, https://aris.iaea.org/Publications/SMR_Book_2020.pdf.

44. Rolls-Royce, UK small modular reactor: pioneering intelligent power, (2020) https://www.rolls-royce.com/products-and-services/nuclear/small-modular-reactors.aspx#/.

45. US Office of Nuclear Energy, The Big Potential of Nuclear Microreactors (2019) https://www.energy.gov/ne/articles/big-potential-nuclear-microreactors.

46. Market and technical assessment of Micro Nuclear Reactors, https://www.gov.uk/government/publications/market-and-technical-assessment-of-micro-nuclear-reactors.

47. Nuclear Emergency Arrangements Group (NEAG), NEAG Paper No 2, Radiation (Emergency Preparedness And Public Information) Regulations 2001, Naval Nuclear Propulsion Programme, Update Of Information Provided To Local Authorities (2009). https://www.nuclearinfo.org/sites/default/files/Naval%20Nuclear%20Propulsion%20Programme%20-%20Update%20Of%20Information%20Provided%20To%20Local%20Authorities.pdf

48. Clyde Off-Site Emergency Plan – May 2020, Published Version, https://www.argyll-bute.gov.uk/sites/default/files/clyde_off-site_emergency_plan_-_may_2020_-_published_version.pdf

49. NDA, End in sight for reprocessing nuclear fuel at Sellafield, https://nda.blog.gov.uk/2017/01/24/end-in-sight-for-reprocessing-nuclear-fuel-at-sellafield/

50. P. Hallington, Magnox Reprocessing - 50 not out, https://www-pub.iaea.org/iaeameetings/cn226p/Session5/INV06Hallington.pdf

51. WNA, Processing of Used Nuclear Fuel, https://world-nuclear.org/information-library/nuclear-fuel-cycle/fuel-recycling/processing-of-used-nuclear-fuel.aspx

52. Sellafield Limited, REPPIR Report of Assessment (RoA) to meet the requirements of the Radiation (Emergency Preparedness & Public Information) Regulations [REPPIR), https://wcssg.co.uk/wp-content/uploads/2015/01/SL-RoA-for-publication-vers-1.pdf.

53. Sellafield Limited, REPPIR Consequence Report, S&R-MSC(2019)020. https://www.cumbria.gov.uk/elibrary/Content/Internet/533/561/44148123654.pdf.

54. Cyclife UK, Workington facility, https://www.cyclife-edf.com/en/cyclife/governance/cyclife-uk#metal-treatment.

55. Lillyhall Landfill radioactive substances permit variation: Summary of consultation responses Version 1 February 2020 https://assets.publishing.service.gov.uk/government/uploads/system/uploads/attachment_data/file/873133/Lillyhall_summary_of_consultation_repsonses.pdf.

56. https://www.inutec.co.uk/.

57. Dounreay Emergency Plan, 2018, DSSWP(18)P26 Issue 11, https://assets.publishing.service.gov.uk/government/uploads/system/uploads/attachment_data/file/767321/Dounreay_Emergency_plan.pdf

58. MOD, The UK's nuclear deterrent: what you need to know, https://www.gov.uk/government/publications/uk-nuclear-deterrence-factsheet/uk-nuclear-deterrence-what-you-need-to-know.

59. Local Authority and Emergency Services Information (LAESI) Edition 11 (2017), https://assets.publishing.service.gov.uk/government/uploads/system/uploads/attachment_data/file/643525/20170831-LAESI-Ed_11.pdf

60. Joint Emergency Planning Unit, AWE Off-Site Emergency Response Plan, Version 1.8 January 2019, https://info.westberks.gov.uk/CHttpHandler.ashx?id=42490&p=0

61. Atomic Weapons Establishment Burghfield consequences report, 2020. https://info.westberks.gov.uk/ChttpHandler.ashx?id=48825&p=0

62. Atomic Weapons Establishment Burghfield consequences report, 2020. https://info.westberks.gov.uk/ChttpHandler.ashx?id=48824&p=0

63. HM Naval Base Devenport, Report of Assessment of HAZARD IDENTIFICATION and RISK EVALUATION (2017), https://www.plymouth.gov.uk/sites/default/files/HMNavalBaseDevonportReportofAssessment.pdf

64. ONR-OFD-PAR-18-028, Revision 0, Rosyth Royal Dockyard, Approval of Revisions to the Approved Emergency Arrangements under Licence Condition 11(3) http://www.onr.org.uk/pars/2019/rosyth-18-028.pdf

65. Mr. Archie Hamilton, UK Parliament, (1989), https://publications.parliament.uk/pa/cm198889/cmhansrd/1989-05-05/Writtens-3.html

66. Portland Consequence report (2020), https://www.dorsetcouncil.gov.uk/emergencies-severe-weather/emergencies/emergencies-pdfs/portland-port-consequence-report-20190920.pdf.

67. The Highland Council, Dounreay and Vulcan NRTE Emergency Plan, Version 7.0 – Public Version (Nov. 2020). https://www.highland.gov.uk/download/downloads/id/23061/dounreay_and_vulcan_nrte_offsite_emergency_plan.pdf

68. A L Jones and M P Harvey , Radiological Consequences Resulting from Accidents and Incidents Involving the Transport of Radioactive Materials in the UK – 2012 Review, PHE-CRCE-014, https://assets.publishing.service.gov.uk/government/uploads/system/uploads/attachment_data/file/345814/PHE-CRCE-014.pdf.

69. ONR, Transporting radioactive material Guidance on emergency planning and notification for the transport of class 7 goods, (2019), http://www.onr.org.uk/transport/emergency-planning-notification-class-7.pdf

70. The Carriage of Dangerous Goods (Amendment) Regulations 2019, https://www.legislation.gov.uk/uksi/2019/598/contents/made

71. HSE RP Community article - Transport Radiation Risk Assessments and the Carriage of Dangerous Goods - November 2020, https://webcommunities.hse.gov.uk/connect.ti/radiationcom/view?objectId=701797.

72. US REMM, Video: Emergency Response to Transportation Accident Involving Radioactive Material, https://www.remm.nlm.gov/tepp_er_youtube.htm

73. Video handout, Emergency Response to a Transportation Accident Involving Radioactive Material Accident Involving Radioactive Material, https://www.remm.nlm.gov/3.d.1TransportationAccidentRAMUserGuide.pdf
User Guide at https://www.remm.nlm.gov/3.d.1TransportationAccidentRAMUserGuide.pdf.

74. Carlos Rojas-Palma et al, Guidelines for first responders based on results from deploying a mockup radiological dispersal device, Carlos Rojas-Palma et al 2020 J. Radiol. Prot. 40 1205, https://iopscience.iop.org/article/10.1088/1361-6498/abb833/meta

75. HMG, Novichok nerve agent use in Salisbury: UK government response, March to April 2018, https://www.gov.uk/government/news/novichok-nerve-agent-use-in-salisbury-uk-government-response

76. HMG, Clean-up work underway in Salisbury in next phase of recovery, https://www.gov.uk/government/news/clean-up-work-underway-in-salisbury-in-next-phase-of-recovery

77. US NRC, Response to Dirty Bombs , https://www.nrc.gov/about-nrc/emerg-preparedness/respond-to-emerg/resp-dirty-bomb.html.

78. Radiological Dispersal Device (RDD), Response Guidance, Planning for the First 100 Minutes, https://www.dhs.gov/sites/default/files/publications/NUSTL_RDD-ResponsePlanningGuidance-Public_171215-508.pdf

79. Radiological Dispersal Device (RDD) Response Guidance: Planning for the First 100 Minutes, Youtube video. https://www.youtube.com/watch?v=qDOQApRmk5Q&list=PLPaVPPbYHq_i2MtwprSSdsjI9LcJsB6hZ&index=2&t=0s

80. FEMA Dirty Bomb Fact Sheet, https://www.fema.gov/media-library-data/20130726-1621-20490-3999/dirtybombfactsheet_final.pdf

81. IAEA, Guidance for Medical Physicists Responding to a Nuclear or Radiological Emergency, EPR Medical Physicist, 2020.

82. International Atomic Energy Agency, Lessons Learned from the Response to Radiation Emergencies (1945–2010), IAEA, Vienna (2012), http://www-pub.iaea.org\books/IAEABooks/8920/Lessons-Learned-from-the-Response-to-Radiation-Emergencies-1945-2010
83. IAEA, Goinana, http://www-pub.iaea.org/mtcd/publications/pdf/pub815_web.pdf
84. IAEA, EPR-Medical/T Medical Preparedness and Response for a Nuclear or Radiological Emergency, Lecture 33, https://www-pub.iaea.org\books/iaeabooks/8831/Medical-Preparedness-and-Response-for-a-Nuclear-or-Radiological-Emergency
85. IAEA, The radiological accident in Cochabamba, https://www-pub.iaea.org/MTCD/publications/PDF/Pub1199_web.pdf
86. C. Rojas-Palma et al, On the use of retrospective dosimetry to assist in the radiological triage of mass casualties exposed to ionising radiation, Carlos Rojas-Palma et al 2020 J. Radiol. Prot. 40 1286.
87. IAEA-TECDOC-1162, Generic procedures for assessment and response during a radiological emergency, https://www.iaea.org/publications/5926/generic-procedures-for-assessment-and-response-during-a-radiological-emergency
88. Ionising Radiation (Basic Safety Standards) (Miscellaneous Provisions) Regulations 2018, https://www.legislation.gov.uk/uksi/2018/482/data.xht?view=snippet&wrap=true
89. NRNL, Gamma Constants, https://www.orau.org/documents/ivhp/health-physics/ornl-rsic-45.pdf
90. Council Directive 2003/122/EURATOM of 22 December 2003 on the control of high-activity sealed radioactive sources and orphan sources, https://www.legislation.gov.uk/eudr/2013/59/chapter/ix/section/2
91. WNA, Nuclear Reactors and Radioisotopes for Space, http://www.world-nuclear.org/information-library/non-power-nuclear-applications/transport/nuclear-reactors-for-space.aspx
92. IAEA, Safety Series 119, https://gnssn.iaea.org/Superseded Safety Standards/Safety_Series_119_1996.pdf
93. CNSC, Operation Morning Light – "The year of the AECB's grand adventure", http://nuclearsafety.gc.ca/eng/resources/canadas-nuclear-history/past-presidents/alan-prince.cfm?pedisable=true
94. NASA, Multi-Mission Radioisotope Thermoelectric Generator (MMRTG), https://mars.nasa.gov/internal_resources/788/
95. NASA RPS, Presidential Memorandum on Launch of Spacecraft Containing Space Nuclear Systems, https://rps.nasa.gov/news/36/presidential-memorandum-on-launch-of-spacecraft-containing-space-nuclear-systems/
96. Frisch-Peierls Memorandum, March 1940, https://www.atomicarchive.com/resources/documents/beginnings/frisch-peierls-2.html
97. UK National Risk Register Of Civil Emergencies 2017, https://www.gov.uk/government/uploads/system/uploads/attachment_data/file/644968/UK_National_Risk_Register_2017.pdf
98. HM Government, National Risk Register, 2020 edition. https://assets.publishing.service.gov.uk/government/uploads/system/uploads/attachment_data/file/952959/6.6920_CO_CCS_s_National_Risk_Register_2020_11-1-21-FINAL.pdf
99. CPNI, Chemical, Biological, Radiological and Nuclear (CBRN) Threats, https://www.cpni.gov.uk/chemical-biological-radiological-and-nuclear-cbrn-threats-0
100. Responding to a CBRN(E) Event: Joint Operating Principles for the Emergency Services First Edition September 2016, https://www.jesip.org.uk/uploads/media/pdf/CBRN%20JOPs/JESIP_CBRN_E_JOPS_Document_On.pdf
101. Wikipedia, Nuclear weapon yields, https://en.wikipedia.org/wiki/Nuclear_weapon_yield
102. Top Sectret. Test of a clean hydrogen bomb with a yield of 50 megatons, https://youtu.be/nbC7BxXtOlo
103. FEMA Skin burn range, https://emilms.fema.gov/IS3/FEMA_IS/is03/REM0502100.htm
104. Homeland Security, Nuclear Blast, https://www.ready.gov/nuclear-blast
105. FEMA 7:10 rule, https://emilms.fema.gov/IS3/FEMA_IS/is03/REM0504050.htm

106. Quick Reference Guide: Radiation Risk Information for Responders Following a Nuclear Detonation, https://www.dhs.gov/sites/default/files/publications/Quick%20Reference%20Guide%20Final.pdf
107. Health and Safety Planning Guide for Protecting Responders Following a Nuclear Detonation, https://www.dhs.gov/sites/default/files/publications/IND%20Health%20Safety%20Planners%20Guide%20Final.pdf
108. Nuclear Attack, https://www.dhs.gov/sites/default/files/publications/prep_nuclear_fact_sheet.pdf
109. War Game (1960), https://vimeo.com/90896890
110. US Homeland Security Council, Planning Guidance for Response to a Nuclear Detonation, https://nepis.epa.gov/Exe/ZyPDF.cgi/P1004DE8.PDF?Dockey=P1004DE8.PDF.
111. UK atmospheric nuclear weapons tests Factsheet 1, https://www.gov.uk/government/publications/uk-atmospheric-nuclear-weapons-tests-factsheets.
112. Los Alamos National Laboratory, A Review of Criticality Accidents 2000 Revision, LA-13638 https://www.orau.org/ptp/Library/accidents/la-13638.pdf
113. Ian Jackson and Shehnaz Jackson, DTI, Siting New Nuclear Power Stations: Availability and Options for Government (2006), *https://aleo.org.uk/downloads/root/regions/uk/consultations/file39030.pdf*
114. Malcolm Grimston, William J Nuttall and Geoff Vaughan , The siting of UK nuclear reactors, JRP, https://iopscience.iop.org/article/10.1088/0952-4746/34/2/R1
115. Dr John Highton and Mr David Senior, HSE NII, Nuclear Safety Advisory Committee the Siting of Nuclear Installations in The United Kingdom, (2008). https://webarchive.nationalarchives.gov.uk/20190701154122/https://www.hse.gov.uk/aboutus/meetings/iacs/nusac/030708/p12-sittingpaper.pdf
116. National Policy Statement for Nuclear Power Generation (EN-6), https://assets.publishing.service.gov.uk/government/uploads/system/uploads/attachment_data/file/47859/2009-nps-for-nuclear-volumeI.pdf
117. Nuclear Installations Act, https://www.legislation.gov.uk/ukpga/1965/57.
118. ONR, Safety Assessment Principles (SAPs) (2020), http://www.onr.org.uk/saps/index.htm.
119. HSE, Reducing Risk, Protecting People, http://www.hse.gov.uk/risk/theory/r2p2.pdf
120. Lord James Douglas-Hamilton, Hansard, https://api.parliament.uk/historic-hansard/commons/1987/mar/24/nuclear-power.
121. U.S. Nuclear Regulatory Commission. Special Inquiry Group., Frampton, G. T., Rogovin, M. (1980). Three Mile Island: a report to the Commissioners and to the public. Washington, D.C.: Available from GPO Sales Program, Division of Technical Information and Document Control, https://catalog.hathitrust.org/Record/003177367.
122. IAEA, Safety Series No. 55, Planning for Off-Site Response to Radiation Accidents in Nuclear Facilities, https://gnssn.iaea.org/Superseded%20Safety%20Standards/Safety_Series_055_1981.pdf.
123. HMNII, Outline Emergency Planning for Licensed Nuclear Power Stations, 1990.
124. HSE, Arrangements for Responding to Nuclear Emergencies, (1994), http://www.onr.org.uk/arrangement-for-nuclear-emergencies.pdf
125. Council Directive 96/29/Euratom of 13 May 1996 laying down basic safety standards for the protection of the health of workers and the general public against the dangers arising from ionising radiation Official Journal of the European Communities (1996) 39, No L159, 1-114 ISBN 0 11 915263 0.
126. Radiation (Emergency Planning and Public Information) Regulations 2001, http://www.hse.gov.uk/radiation/ionising/reppir.htm.
127. Hansard (2002), https://api.parliament.uk/historic-hansard/written-answers/2002/jun/21/civil-contingency-secretariat#S6CV0387P2_20020621_CWA_264.
128. Cabinet Office, Dealing with disaster, Revised Third Edition, https://webarchive.nationalarchives.gov.uk/20050112000000/http://www.ukresilience.info/contingencies/dwd/dwdrevised.pdf.

129. CCS, Emergency Response and Recovery Non statutory guidance accompanying the Civil Contingencies Act 2004, (2013), https://assets.publishing.service.gov.uk/government/uploads/system/uploads/attachment_data/file/253488/Emergency_Response_and_Recovery_5th_edition_October_2013.pdf.

130. ASN, Technical guidelines for the design and construction of the EPR, http://www.french-nuclear-safety.fr/Inspections/Supervision-of-the-EPR-reactor/Resources/Technical-guidelines-for-the-design-and-construction-of-the-EPR.

131. UK-EPR, PCSR – Sub-Chapter 14.6 – Radiological consequences of design basis accidents, http://epr-reactor.co.uk/ssmod/liblocal/docs/PCSR/Chapter%2014%20-%20Design%20Basis%20Analysis/Sub-Chapter%2014.6%20-%20Radiological%20Consequences%20of%20Design%20Basis%20Accidents.pdf.

132. NRPB, Living with Radiation, 5th Edition, ISBN 0-85951-419-6.

133. Martin et al, An Introduction to Radiation Protection, 7th Edition (2019), CRC Press, ISBN 978-1-138-333307-9.

134. PHE, Ionising radiation: dose comparisons, https://www.gov.uk/government/publications/ionising-radiation-dose-comparisons/ionising-radiation-dose-comparisons

135. PHE, Utube video on background radiation, https://www.youtube.com/watch?v=VCsJcfmL7TE&feature=player_embedded

136. PHE, Ionising radiation exposure of the UK population: 2010 review, https://www.gov.uk/government/publications/ionising-radiation-exposure-of-the-uk-population-2010-review

137. World Nuclear Association, What is Background Radiation? http://www.world-nuclear.org/uploadEDFiles/org/Features/Radiation/4_Background_Radiation%281%29.pdf

138. PHE, Patient dose information: guidance, https://www.gov.uk/government/publications/medical-radiation-patient-doses/patient-dose-information-guidance

139. Radioactivity in Food and the Environment RIFE, https://www.food.gov.uk/research/radioactivity-in-food-and-the-environment

140. Radioactivity in Food and the Environment (RIFE) Report 2015, https://www.food.gov.uk/research/radioactivity-in-food-and-the-environment/radioactivity-in-food-and-the-environment-rife-report-2015

141. Nature 53, 274-277 (1896), https://www.nature.com/physics/looking-back/roentgen/index.html

142. John Macintyre and the world's first x-ray department, http://www.auntminnieeurope.com/index.aspx?sec=ser&sub=def&pag=dis&ItemID=608942

143. The First Fifty Years of Radiation Protection, https://sites.google.com/isu.edu/health-physics-radinf/history-of-radiation-and-radiation-protection/first-50-years

144. Protection of On-site Personnel in the Event of a Radiation Accident, Documents of the NRPB, Vol. 16 No. 1 2005. https://assets.publishing.service.gov.uk/government/uploads/system/uploads/attachment_data/file/425146/Documents_of_the_NRPB_Volume_16_Number_1.pdf

145. The Nagasaki Atomic Bomb Survivors' Tumour Tissue Bank, http://www.thelancet.com/pdfs/journals/lancet/PIIS0140-6736(15)00697-2.pdf

146. Miquel Macià i Garau, Anna Lucas Calduch, and Enric Casanovas Lópezc, Radiobiology of the acute radiation syndrome, Rep Pract Oncol Radiother. 2011 Jul; 16(4): 123–130., https://www.ncbi.nlm.nih.gov/pmc/articles/PMC3863296/.

147. Medical management of the acute radiation syndrome, Reports of Practical Oncology & Radiotherapy, Volume 16, Issue 4, July–August 2011, Pages 138-146 https://www.sciencedirect.com/science/article/pii/S1507136711000642.

148. PHE, Chemical, biological, radiological and nuclear incidents: clinical management and health protection, https://assets.publishing.service.gov.uk/government/uploads/system/uploads/attachment_data/file/712888/Chemical_biological_radiological_and_nuclear_incidents_clinical_management_and_health_protection.pdf.

149. IAEA, Medical Management of Radiation Injuries, Safety Reports Series 101. https://www.iaea.org/publications/12370/medical-management-of-radiation-injuries.

150. References for METREPOL: MEdical TREatment ProtocOLs for Radiation Accident Victims, https://www.remm.nlm.gov/ars_references.htm.

151. Pathway to a paradigm, http://radiationeffects.org/wp-content/uploads/2014/11/Kathren1996.pdf.

152. WHO, Cancer, https://www.who.int/news-room/fact-sheets/detail/cancer.

153. Cancer Research UK, https://www.cancerresearchuk.org/health-professional/cancer-statistics/mortality.

154. IAEA, Communication with the Public in a Nuclear or Radiological Emergency, https://www-pub.iaea.org/MTCD/Publications/PDF/EPR-Communcation_web.pdf.

155. Health Risks from Exposure to Low Levels of Ionizing Radiation: BEIR VII Phase 2 (2006) Chapter: 12 Estimating Cancer Risk, https://www.nap.edu/read/11340/chapter/14#278.

156. Fukushima Nuclear Power Plant Accident, ICRP ref: 4847-5603-4313, http://www.icrp.org/docs/fukushima%20nuclear%20power%20plant%20accident.pdf.

157. ICRP Publication 103, The 2007 Recommendations of the International Commission on Radiological Protection, http://www.icrp.org/publication.asp?id=ICRP%20Publication%20103.

158. (Council Directive 2013/59/Euratom of 5 December 2013 laying down basic safety standards for the protection against the dangers arising from exposure to ionising radiation) (Ref. Council Directive 2013/59/EURATOM, http://eur-lex.europa.eu/legal-content/EN/TXT/PDF/?uri=CELEX:02013L0059-20140117&from=EN).

159. Ionising Radiations Regulations 2017. http://www.hse.gov.uk/radiation/ionising/.

160. IAEA Safety Standards Series No. GSG-2, Criteria for Use in Preparedness and Response for a Nuclear or Radiological Emergency (2011), https://www.iaea.org/publications/8506/criteria-for-use-in-preparedness-and-response-for-a-nuclear-or-radiological-emergency.

161. PHE, Public Health Protection in Radiation Emergencies, PHE-CRCE-049, 2019.

162. SSK, Basic Radiological Principles for Decisions on Measures for the Protection of the Population against Incidents involving Releases of Radionuclides, https://www.ssk.de/SharedDocs/Beratungsergebnisse_PDF/2014/RadiologischeGrundlagen_e.pdf.

163. H.M. Government and Scottish Government, Nuclear Emergency Planning and Response Guidance, Part 2, Response, October 2015, https://www.gov.uk/government/publications/national-nuclear-emergency-planning-and-response-guidance.

164. CDC, Protective Actions for Radiation Emergencies - Get Inside, Stay Inside, Stay Tuned, https://www.youtube.com/watch?v=VXxIQLhL-XA

165. Wanyu R. Chan, William W Nazaroff, Phillip N. Price, Ashok J. Gadgil, Effectiveness of Urban Shelter-in-Place II: Residential Districts, https://www.osti.gov/scitech/servlets/purl/928232

166. J.J.Jetter, C.Whitfield, Effectiveness of expedient sheltering in place in a residence, https://cfpub.epa.gov/si/si_public_file_download.cfm?p_download_id=478508

167. Ready, https://www.ready.gov/shelter

168. Australian Clinical Guidelines For Radiological Emergencies - September 2012, Pulmonary Lavage, http://www.health.gov.au/internet/publications/publishing.nsf/Content/ohp-radiological-toc~ohp-radiological-16-app-b~ohp-radiological-16.1-pulmonary.

169. Board Statement on Emergency Reference Levels, Documents of the NRPB, Vol 1, No. 4,1990, https://www.gov.uk/government/publications/radiation-emergency-reference-levels.

170. WHO, Guidelines for Iodine Prophylaxis following Nuclear Accidents Update 1999, http://www.who.int/ionizing_radiation/pub_meet/Iodine_Prophylaxis_guide.pdf

171. Stable Iodine Prophylaxis Recommendations of the 2nd UK Working Group on Stable Iodine Prophylaxis, https://www.gov.uk/government/publications/stable-iodine-prophylaxis-recommendations-for-radioiodine-release.

172. IAEA, Overview of Nuclear Emergency Preparedness and Response, Iodine Prophylaxis, http://www-pub.iaea.org/MTCD/Publications/PDF/eprmedt/Day_5/Day_5-27.pps

173. McColl & Prosser, Emergency Data Handbook, NRPB-W19 (2001).

174. IAEA, General Safety Requirements No. GSR Part 7, Preparedness and Response for a Nuclear or Radiological Emergency, https://www-pub.iaea.org\books/iaeabooks/10905/preparedness-and-response-for-a-nuclear-or-radiological-emergency.

175. Council Regulation (Euratom) 2016/52 of 15 January 2016 laying down maximum permitted levels of radioactive contamination of food and feed following a nuclear accident or any other case of radiological emergency, and repealing Regulation (Euratom) No 3954/87 and Commission Regulations (Euratom) No 944/89 and (Euratom) No 770/90, https://eur-lex.europa.eu/legal-content/EN/TXT/?uri=CELEX:32016R0052.

176. REMM, Personal Protective Equipment (PPE) in a Radiation Emergency, https://www.remm.nlm.gov/radiation_ppe.htm.

177. PHE, Chemical, biological, radiological and nuclear incidents: clinical management and health protection https://www.gov.uk/government/uploads/system/uploads/attachment_data/file/712888/Chemical_biological_radiological_and_nuclear_incidents_clinical_management_and_health_protection.pdf.

178. CDC Decontamination graphic, https://emergency.cdc.gov/radiation/pdf/Infographic_Decontamination.pdf.

179. CDC, Population Monitoring in Radiation Emergencies: A Guide for State and Local Public Health Planners, https://www.cdc.gov/nceh/radiation/emergencies/pdf/population-monitoring-guide.pdf.

180. RMU Planning and Operational Guidance, https://www.gov.uk/government/uploads/system/uploads/attachment_data/file/340153/HPA-CRCE-017_for_website.pdf.

181. Useful information from a parliamentary select committee. https://publications.parliament.uk/pa/cm201617/cmselect/cmsctech/163/163.pdf.

182. IAEA EPR Medical/T 2014 Medical Preparedness and Response for a Nuclear or Radiological Emergency, Lecture 44.

183. IAEA, Pocket Guide for Medical Responders to a Radiation Emergency Resulting in a Mass Casualty Event, https://www-pub.iaea.org/MTCD/Publications/PDF/EPR-Medical_T_2011/PDF/Pocket_Guides.pdf.

184. NHS England, Emergency Preparedness, Resilience and Response, Concept of Operations for managing Mass Casualties, https://www.england.nhs.uk/wp-content/uploads/2018/03/concept-operations-management-mass-casualties.pdf.

185. NHS England, Emergency Preparedness, Resilience and Response, Guidance for the initial NEPLG management of self-presenters from incidents involving hazardous materials, https://www.england.nhs.uk/wp-content/uploads/2019/03/guidance-for-the-initial-management-of-self-presenters-from-incidents-involving-hazardous-materials.pdf.

186. Oak Ridge Institute for Science and Education, Guidance for Radiation Accident Management, Procedure Demonstrations, https://orise.orau.gov/resources/reacts/guide/procedure-demonstrations.html.

187. IAEA, EPR-First Responders 2006, Manual for First Responders to a Radiological Emergency, https://www-pub.iaea.org/MTCD/Publications/PDF/EPR_FirstResponder_web.pdf.

188. IAEA, E-Learning Tools for First Response to a Radiological Emergency, https://www.iaea.org/publications/8178/e-learning-tools-for-first-response-to-a-radiological-emergency.

189. US NRC, Review of NUREG-0654, Supplement 3, "Criteria for Protective Action Recommendations for Severe Accidents" Technical Basis for Protective Action Strategies (2010) https://www.nrc.gov/docs/ML1023/ML102380087.pdf.

190. IAEA Safety Standards Series No. GS-G-2.1 (2007), Arrangements for Preparedness for a Nuclear or Radiological Emergency, https://www.iaea.org/publications/7503/arrangements-for-preparedness-for-a-nuclear-or-radiological-emergency.

191. NRC, Emergency Planning Zones, (2018). https://www.nrc.gov/about-nrc/emerg-preparedness/about-emerg-preparedness/planning-zones.html.

192. B. Musico, NSIR/DPR, Emergency Planning, Emergency Planning Zone Sizing for Small Modular Reactors – Regulatory History & Policy Considerations https://www.nrc.gov/docs/ML1817/ML18177A386.pdf.

193. Oak Ridge Institute for Science and Education, Guidance for Radiation Accident Management, https://orise.orau.gov/resources/reacts/guide/how-to-detect-radiation.html

194. N P McColl & P Kruse, NAIR Technical Handbook 2002 Edition, NRPB-W7, https://assets.publishing.service.gov.uk/government/uploads/system/uploads/attachment_data/file/347760/2002_NrpbW7.pdf .

195. Nuclear Emergency Planning and Response Guidance, https://www.gov.uk/government/publications/national-nuclear-emergency-planning-and-response-guidance.

196. Nuclear Installation Act 1965, https://www.legislation.gov.uk/ukpga/1965/57.

197. The Health and Safety at Work Act (1974), http://www.legislation.gov.uk/ukpga/1974/37/contents.

198. Energy Act (2013), http://www.legislation.gov.uk/ukpga/2016/20/contents/enacted.

199. Defence Nuclear Safety Regulator (DNSR), https://www.gov.uk/government/groups/defence-nuclear-safety-regulator-dnsr.

200. IAEA, Convention on Nuclear Safety, https://www.iaea.org/publications/documents/infcircs/convention-nuclear-safety.

201. HMG, How we regulate radiological and civil nuclear safety in the UK, https://assets.publishing.service.gov.uk/government/uploads/system/uploads/attachment_data/file/978867/how-we-regulate-radiological-and-civil-nuclear-safety-in-the-UK.pdf.

202. Compliance with the Convention on Nuclear Safety Obligations: 7th national report, https://www.gov.uk/government/publications/compliance-with-the-convention-on-nuclear-safety-obligations-7th-national-report.

203. ONR, Licence condition handbook, February 2017, http://www.onr.org.uk/documents/licence-condition-handbook.pdf.

204. MOD, JSP 518 Regulation of the Naval Nuclear Propulsion Programme Part 1: Directive, https://assets.publishing.service.gov.uk/government/uploads/system/uploads/attachment_data/file/341330/20140801_JSP_518_V4_1_Pt_1.pdf.

205. OECD, Multilateral agreements in nuclear energy, https://www.oecd-nea.org/law/multilateral-agreements/liability-compensation.html

206. Convention on third party liability in the field of nuclear energy of 29 July 1960, as amended by the additional protocol of 28 January 1964, by the protocol of 16 November 1982, and by the protocol of 12 February 2004 , https://assets.publishing.service.gov.uk/government/uploads/system/uploads/attachment_data/file/267036/356-paris-brussels-conventions-1.pdf.

207. The Nuclear Installations (Liability for Damage) Order 2016, https://www.legislation.gov.uk/uksi/2016/562/pdfs/uksi_20160562_en.pdf

208. Explanatory Memorandum to The Nuclear Installations (Liability for Damage) Order 2016 2016 No. 562, https://www.legislation.gov.uk/uksi/2016/562/pdfs/uksiem_20160562_en.pdf.

209. Clifford Chance, Changes to the UK Nuclear Liability Regime: Implications For The Industry https://www.cliffordchance.com/content/dam/cliffordchance/PDFDocuments/Client Briefing - Nuclear liability briefing - September 2017 68083-5-7672 v0.1.pdf

210. Taxpayer liability for safety at the Wylfa Nuclear power project, https://commonslibrary.parliament.uk/research-briefings/cdp-2018-0187/

211. IAEA, The 1997 Vienna Convention on Civil Liability for Nuclear Damage and the 1997 Convention on Supplementary Compensation for Nuclear Damage — Explanatory Texts IAEA International Law Series No. 3 (Rev. 2) (2020), https://www-pub.iaea.org/MTCD/Publications/PDF/P1906_web.pdf

212. Health and Safety at Work Act 1974. http://www.legislation.gov.uk/ukpga/1974/37/contents

213. The Management of Health and Safety at Work Regulations 1999. http://www.legislation.gov.uk/uksi/1999/3242/contents/made

214. Work with ionising radiation, Ionising Radiations Regulations 2017, Approved Code of Practice and guidance (2018) https://www.hse.gov.uk/pubns/priced/l121.pdf

215. Management of Health and Safety at Work Regulations, http://www.hse.gov.uk/pubns\books\l21.htm.

216. Civil Contingencies Act 2004, https://www.legislation.gov.uk/ukpga/2004/36/contents

217. Civil Contingencies Secretariat, Civil Contingencies Act 2004: a short guide (revised) https://www.the-eps.org/wp-content/uploads/civil-contingencies-act-2004-a-short-guide-revised.pdf

218. The Civil Contingencies Act 2004 (Contingency Planning) Regulations 2005, https://www.legislation.gov.uk/uksi/2005/2042/contents/made.

219. Control of Major Accident Hazards Regulations 2015 (COMAH), http://www.legislation.gov.uk/uksi/2015/483/contents/made.

220. Control Of Major Accident Hazards Regulations 2015 (COMAH), https://www.hse.gov.uk/comah/background/comah15.htm

221. US National Threat and Hazard Identification and Risk Assessment (THIRA) (2019) https://www.fema.gov/sites/default/files/2020-06/fema_national-thira-overview-methodology_2019_0.pdf.

222. ONR, The Purpose, Scope, and Content of Safety Cases, 2019. http://www.onr.org.uk/operational/tech_asst_guides/ns-tast-gd-051.pdf

223. WASH-1400, The Reactor Safety Study. The Introduction of Risk Assessment to the Regulation of Nuclear Reactors (2016), https://www.nrc.gov/docs/ML1622/ML16225A002.pdf.

224. IAEA, Basic Safety Principles for Nuclear Power Plants 75-INSAG-3 Rev. 1 INSAG-12 (1999), https://www-pub.iaea.org/MTCD/Publications/PDF/P082_scr.pdf.

225. UK-EPR, PCSR – Sub-chapter 15.0 – Safety requirements and PSA Objectives, http://epr-reactor.co.uk/ssmod/liblocal/docs/PCSR/Chapter%2015%20-%20Probabilistic%20Safety%20Analysis/Sub-Chapter%2015.0%20-%20Safety%20requirements%20and%20PSA%20objectives.pdf.

226. ONR, Determination of the Off-Site Emergency Planning and Public Information Areas for the AWE Plc Nuclear Licensed site of Burghfield, Project Assessment Report ONR-TD-PAR-17-06 Revision 0, February 2018 http://www.onr.org.uk/pars/2018/awe-burghfield-17-055.pdf.

227. BBC News, Leicester lockdown: Boundary line branded 'stupid' by residents, https://www.bbc.co.uk/news/uk-england-leicestershire-53247817.

228. ONR, Off-Site Emergency Arrangements, NS-INSP-GD-067 Revision 1, http://www.onr.org.uk/operational/tech_insp_guides/ns-insp-gd-067.pdf.

229. Directors of Public Health in Local Government, Roles, Responsibilities and Context, https://assets.publishing.service.gov.uk/government/uploads/system/uploads/attachment_data/file/860515/directors-of-public-health-in-local-government-roles-responsibilities-and-context.pdf.

230. Association of Publlic Health Directors. What is a Director of Public Health (DPH)? https://www.adph.org.uk/wp-content/uploads/2016/10/Role-of-a-Director-of-Public-Health.pdf

231. Policy paper, PHE CRCE radiation emergency response arrangements (2014). https://www.gov.uk/government/publications/radiation-emergency-response-arrangements-public-health-readiness/phe-crce-radiation-emergency-response-arrangements.

232. HMG, Government creates new National Institute for Health Protection, https://www.gov.uk/government/news/government-creates-new-national-institute-for-health-protection.

233. Cabinet Office, Communicating with the public: the ten step cycle, https://www.gov.uk/government/publications/communicating-with-the-public-the-ten-step-cycle.

234. Carriage of Dangerous Goods (Amendment) Regulations 2019, https://www.legislation.gov.uk/uksi/2019/598/pdfs/uksi_20190598_en.pdf.

235. ONR, Prior Information to the Public, Radiation Emergencies during the Transport of Radioactive Material, http://www.onr.org.uk/transport/prior-information-cdg09.pdf.

236. Nuclear Industries Security Regulations 2003, http://www.legislation.gov.uk/uksi/2003/403/contents/made.

237. IAEA, Integrated Management Systems for Safety and Security, https://www.iaea.org/topics/integrated-management-systems-for-safety-and-security.

238. HM Government, Food and Environment Protection Act 1985, https://www.legislation.gov.uk/ukpga/1985/48.

239. ONR, Security Assessment Principles for the Civil Nuclear Industry, http://www.onr.org.uk/syaps/security-assessment-principles-2017.pdf.

240. ONR, Emergency preparedness and response, http://www.onr.org.uk/emergency-arrangements.htm.

241. Defence Safety Authority, About Us, https://www.gov.uk/government/organisations/defence-safety-authority/about.

242. MOD, Regulation of the naval nuclear propulsion programme (JSP 518) https://www.gov.uk/government/publications/jsp-518-regulation-of-the-naval-nuclear-propulsion-programme

243. MOD, Regulation of the nuclear weapons programme (JSP 538) https://www.gov.uk/government/publications/jsp-538-regulation-of-the-nuclear-weapons-programme-nwp.

244. Regulatory Guidance Series, No EPR 7, Environmental Permitting Regulations (England and Wales) 2010, Regulatory Guidance Series, No RSR 2, The regulation of radioactive substances activities on nuclear licensed sites, https://assets.publishing.service.gov.uk/government/uploads/system/uploads/attachment_data/file/296397/geho0310bsgf-e-e.pdf.

245. Cabinet Office, Revision to Emergency Preparedness, https://assets.publishing.service.gov.uk/government/uploads/system/uploads/attachment_data/file/61027/Chapter-4-Local_20Responder-Risk-assessment-duty-revised-March.pdf

246. IAEA, TECDOC Series, https://www.iaea.org/publications/search/type/tecdoc-series.

247. IAEA Safety Standards Series No. TS-G-1.2 (ST-3), Planning and Preparing for Emergency Response to Transport Accidents Involving Radioactive Material (2002), https://www.iaea.org/publications/6348/planning-and-preparing-for-emergency-response-to-transport-accidents-involving-radioactive-material.

248. IAEA, Emergency Preparedness and Response Publications search, https://www.iaea.org/publications/search/type/emergency-preparedness-and-response.

249. Actions to protect the public in an emergency due to severe conditions at a light water reactor, EPR-NPP Public Protective Actions, https://www-pub.iaea.org/books/iaeabooks/10362/Actions-to-Protect-the-Public-in-an-Emergency-due-to-Severe-Conditions-at-a-Light-Water-Reactor.

250. Communication with the public in a nuclear or radiological emergency, EPR-Public Communications 2012, https://www-pub.iaea.org/books/iaeabooks/8889/Communication-with-the-Public-in-a-Nuclear-or-Radiological-Emergency.

251. IAEA, Welcome to the Nuclear Safety and Security Online User Interface, https://nucleus-apps.iaea.org/nss-oui/home.

252. IAEA, Specific Safety Guide, Safety Standards Series No. SSG-16 (Rev. 1) Establishing the Safety Infrastructure for a Nuclear Power Programme (2020). https://www.iaea.org/publications/8636/establishing-the-safety-infrastructure-for-a-nuclear-power-programme.

253. IAEA, Fundamental Safety Principles, IAEA Safety Standards Series No. SF-1 (2006), https://www.iaea.org/publications/7592/fundamental-safety-principles.

254. IAEA Safety Standards Series No. GSR Part 1 (Rev. 1), Governmental, Legal and Regulatory Framework for Safety, https://www.iaea.org/publications/10883/governmental-legal-and-regulatory-framework-for-safety.

255. IAEA Safety Standards Series No. GSR Part 3, Radiation Protection and Safety of Radiation Sources: International Basic Safety Standard, https://www.iaea.org/publications/8930/radiation-protection-and-safety-of-radiation-sources-international-basic-safety-standards.

256. IAEA, Transport (Website), https://www.iaea.org/topics/transport.

257. Generic procedures for assessment and response during a radiological emergency, IAEA-TECDOC-1162, https://www-pub.iaea.org/MTCD/Publications/PDF/te_1162_prn.pdf.

258. Local authorities' Preparedness for Civil Emergencies: A Good Practice Guide for Chief Executives, November 2018, https://www.gov.uk/government/publications/local-authorities-preparedness-for-civil-emergencies.

259. LGA, A Councillor's Guide to Civil Emergencies, 2018, https://local.gov.uk/councillors-guide-civil-emergencies.

260. Cabinet Office, Expectations and Indicators of Good Practice Set for Category 1 and 2 Responders (2013), https://assets.publishing.service.gov.uk/government/uploads/system/uploads/attachment_data/file/252341/Expectation_and_Indicators_of_Good_Practice_Set_for_category_1_2_Responders.pdf#page=14&zoom=100,78,625.

261. Torness Off--Site Emergency plan, https://www.eastlothian.gov.uk/downloads/file/23702/torness_off_site_plan_2017-18.

262. National Steering Committee on Warning & Informing the Public, https://www.gov.uk/government/groups/national-steering-committee-on-warning-informing-the-public.

263. ONR principles for determination of offsite emergency planning areas (formerly known as DEPZs), http://www.onr.org.uk/depz-onr-principles.htm.

264. Suffolk Joint Emergency Planning Unit, Sizewell Off Site Emergency Plan, Issue 3.6 dated Oct 18, https://www.suffolkresilience.com/uploads/NPM_-_Sizewell_Off_Site_Plan_-_Issue_3.6_dated_Oct_18.pdf.

265. HMG, Nuclear Emergency Planning and Response Guidance, Concept of Operations (2015), https://assets.publishing.service.gov.uk/government/uploads/system/uploads/attachment_data/file/472419/NEPRG00_-_Concept_of_Operations.pdf.

266. Devonport Off-Site Emergency Plan, Version 7.0 May 2020, https://www.plymouth.gov.uk/sites/default/files/Devonportoffsiteemergencyplanv7.0.pdf.

267. Joint Emergency Services Interoperability Principles, https://jesip.org.uk/home.

268. NARU, National Ambulance Service Command and Control Guidance, https://naru.org.uk/wp-content/uploads/2021/03/NARU-COMMAND-AND-CONTROL-GUIDE-V3.1-07.2019.pdf.

269. Motorola Solutions, Airwave, https://www.airwavesolutions.co.uk/home/.

270. Home Office, Guidance Emergency Services Network: overviewhttps://www.gov.uk/government/publications/the-emergency-services-mobile-communications-programme/emergency-services-network.

271. Cabinet Office, Resilience Direct, https://www.gov.uk/guidance/resilient-communications#resiliencedirect.

272. National Response Plan Major Nuclear or Radiological Accidents Number 200/SGDSN/PSE/PSN – February 2014, http://www.sgdsn.gouv.fr/uploads/2018/02/plan-national-nucleaire-fevrier2014-anglais.pdf.

273. State Council of the People's Republic of China, Full text of China's Nuclear Emergency Preparedness, http://english.www.gov.cn/archive/white_paper/2016/01/27/content_281475279484672.htm.

274. Wang Zhe, Emily Yy Chan, Kevin Sida Liu, May Ps Yeung, Collaborating Centre for Oxford University and CUHK for Disaster and Medical Humanitarian Response (CCOUC), The Chinese University of Hong KongThe Disaster and Emergency Management System in China, http://ccouc.org/_asset/file/policybrief-disaster-andemergmxsysinchina-final20161020.pdf.

275. Government of Canada, How Canada prepares for radiological and nuclear emergencies, https://www.canada.ca/en/health-canada/services/health-risks-safety/radiation/radiological-nuclear-emergencies/how-canada-prepares.html.

276. Kansai Electric Power, Nuclear Power Information, Measures against Potential Natural Disasters, https://www.kepco.co.jp/english/energy/nuclear_power/severeaccident_bousai.html.

277. HMG, Preparation and planning for emergencies, https://www.gov.uk/guidance/preparation-and-planning-for-emergencies-the-capabilities-programme.

278. Communicating with the public: the ten step cycle, https://www.gov.uk/government/publications/communicating-with-the-public-the-ten-step-cycle.

279. HMG, National Statistics, Energy Trends: UK weather, https://www.gov.uk/government/statistics/energy-trends-section-7-weather.

280. NAIR, https://www.gov.uk/guidance/national-arrangements-for-incidents-involving-radioactivity-nair.

281. National Arrangements for Incidents involving Radioactivity, NAIR User's Handbook - 2009, Edition. https://assets.publishing.service.gov.uk/government/uploads/system/uploads/attachment_data/file/347744/NAIR_April__S_.pdf.

282. Portable Digital Assistant for First Responders to a Radiological Emergency, https://www-pub.iaea.org/MTCD/publications/PDF/EPR-First_Res-PDA/html/ti1-transport.htm.

283. Association of University Radiation Protection Officers, Transport of Radioactive Materials by Road, https://www.gla.ac.uk/media/Media_309373_smxx.pdf.

284. Rear Admiral E S J Larken DSO, Practical Emergency Management, https://www.icheme.org/communities/subject_groups/safety and loss prevention/resources/hazards archive/~/media/Documents/Subject Groups/Safety_Loss_Prevention/Hazards Archive/S139 - Major Hazards II/S139-24.pdf.

285. Preparing Scotland Scientific and Technical Advice Cell (STAC): guidance, 2013, https://www.gov.scot/publications/preparing-scotland-scientific-technical-advice-cell-stac-guidance/.

286. Operations Manual for IAEA Assessment and Prognosis during a Nuclear or Radiological Emergency, EPR-A&P (2019), https://www.iaea.org/publications/12362/operations-manual-for-iaea-assessment-and-prognosis-during-a-nuclear-or-radiological-emergency.

287. Cabinet Office, Expectations and indicators of good practice set for category 1 and 2 responders, https://www.gov.uk/government/publications/expectations-and-indicators-of-good-practice-set-for-category-1-and-2-responders.

288. ONR, Local Authorities for sites, Off-Site Emergency Plans under the Radiation (Emergency Preparedness and Public Information) Regulation, http://www.onr.org.uk/local-authorities-sites.htm.

289. Cabinet Office, The role of Local Resilience Forums: A Reference Document (2013), https://assets.publishing.service.gov.uk/government/uploads/system/uploads/attachment_data/file/62277/The_role_of_Local_Resilience_Forums-_A_reference_document_v2_July_2013.pdf.

290. National Resilience Standards for Local Resilience Forums (LRFs) (2020) https://assets.publishing.service.gov.uk/government/uploads/system/uploads/attachment_data/file/912761/NRS_for_LRFs_V3.0__Aug2020.pdf.

291. Nuclear Emergency Planning and Response Guidance, Part 2 - Response (2015), https://assets.publishing.service.gov.uk/government/uploads/system/uploads/attachment_data/file/472422/NEPRG02_-_Response.pdf

292. NHS, Emergency Preparedness, Resilience and Response (EPRR), https://www.england.nhs.uk/ourwork/eprr/.

293. NHS England, Emergency Preparedness, Resilience and Response, Incident Response Plan (National), Updated 2017, https://www.england.nhs.uk/wp-content/uploads/2017/07/NHS-england-incident-response-plan-v3-0.pdf.

294. NHS Resilience Scotland, Preparing For Emergencies: Guidance for Health Boards in Scotland, (2013). https://www.gov.scot/binaries/content/documents/govscot/publications/advice-and-guidance/2013/09/preparing-emergencies-guidance-health-boards-scotland/documents/nhsscotland-resilience-preparing-emergencies-guidance-health-boards-scotland/nhsscotland-resilience-preparing-emergencies-guidance-health-boards-scotland/govscot%3Adocument/00434687.pdf.

295. NHSScotland - Emergencies, Responding to Emergencies: Annex J, https://www.sehd.scot.nhs.uk/emergencyplanning/Documents/annex_j.htm.

296. Centre for Science and Policy, https://www.csap.cam.ac.uk/organisations/civil-contingencies-secretariat/.

297. The Ontario (Canada) Radiation Health Response Plan dated 2014, https://www.health.gov.on.ca/en/pro/programs/emb/rhrp/docs/radiation_health_response_plan.pdf.

298. NI, The Purpose of the Safety Directors' Forum (SDF), https://www.nuclearinst.com/tof.

299. ONR, Fukushima and the UK nuclear industry, http://www.onr.org.uk/fukushima/.

300. ONR, Japanese Earthquake and Tsunami: Update on UK 'National Action Plan' UK response to ENSREG National Action Plan A Statement on ONR's Actions Extracted From the UK Post Japanese Earthquake and Tsunami Implementation Plan: 2nd Updated Progress Reporthttp://www.onr.org.uk/fukushima/ensreg-report-2017.pdf.

301. NARU, Guidance for Dealing with Radiological Incidents (2020), NARU 045.

302. OneResponder Network, https://manager.oneresponder.net/home/landing/#/about.

303. National Incident Management System Guideline for the National Qualification System (2017), https://www.fema.gov/sites/default/files/2020-05/fema_nims_nqs_guideline_0.pdf.

304. IAEA, Training in Emergency Preparedness and Response, https://www-ns.iaea.org/downloads/iec/er-training-booklet.pdf.

305. Joint Agency Modelling: A process to deliver emergency response national guidance for a radiological atmospheric release, http://www.harmo.org/Conferences/Proceedings/_Bruges/publishedSections/PPT/SCM_HARMO19_JointAgencyModelling.pdf.

306. The application of atmospheric dispersion modelling for the provision of health protection advice in the event of a radiological incident ADMLC Seminar, 12/03/2020 Peter Bedwell & Sarah Millington

307. AWE Off-Site Emergency Response Plan, Version 1.8 January 2019, https://info.westberks.gov.uk/CHttpHandler.ashx?id=42490&p=0.

308. PHE Radiation Protection Advisor Services, https://www.phe-protectionservices.org.uk/rpa.

309. US, CDC, Radiological Terrorism: Just in Time Training for Hospital Clinicians, https://emergency.cdc.gov/radiation/justintime.asp.

310. National Operational Guidance, Contamination and decontamination, https://www.ukfrs.com/pdf/print/promos%3A210072.

311. Effect of Radiological Countermeasures on Subjective Well-Being and Radiation Anxiety after the 2011 Disaster: The Fukushima Health Management Surveyhttps://www.ncbi.nlm.nih.gov/pmc/articles/PMC5800223/.

312. Fukushima Daiichi: ANS Committee Report http://fukushima.ans.org/report/Fukushima_report.pdf.

313. IAEA, GSG-14, Arrangements for Public Communication in Preparedness and Response for a Nuclear or Radiological Emergency, https://nucleus-apps.iaea.org/nss-oui/Content/Index?CollectionId=m_6b73046e-346a-456d-aaa2-9122561f3b62&type=PublishedCollection

314. IAEA, International Nuclear and Radiological Event Scale (INES) https://www.iaea.org/topics/emergency-preparedness-and-response-epr/international-nuclear-radiological-event-scale-ines.

315. Kyshtym disaster, https://ipfs.io/ipfs/QmXoypizjW3WknFiJnKLwHCnL72vedxjQkDDP1mXWo6uco/wiki/Kyshtym_disaster.html.

316. Tom McKenna, IAEA, Strengthening Defence in Depth in Emergency Preparedness and Response by Pre-establishing Tools and Criteria for the Effective Protection of the Public During a Severe Reactor Emergency, https://www-pub.iaea.org/iaeameetings/cn205p/Tuesday/McKenna.pdf.

317. NRPB, Board Statement on Emergency Reference Levels, Doc NRPB Vol. 1. No 4. 1990, https://www.gov.uk/government/uploads/system/uploads/attachment_data/file/423879/NRPB_Vol_1_No_4_1990_-_for_website.pdf.

318. BEIS, Radioactive incident monitoring, https://www.gov.uk/government/collections/radioactive-incident-monitoring.

319. Met. Office, Numerical Atmospheric-dispersion Modelling Environment (NAME), http://www.metoffice.gov.uk/research/modelling-systems/dispersion-model.

320. Met. Office, CBRN incident management, https://www.metoffice.gov.uk/services/public-sector/cbrn.

321. Cabinet Office, Resilience Direct, https://www.gov.uk/guidance/resilient-communications#resiliencedirect.

322. IAEA, EPR-NPP Public Protective Actions 2013 https://www-pub.iaea.org/MTCD/Publications/PDF/EPR-NPP_PPA_web.pdf.

323. IAEA, Placing the Radiological Health Hazard in Perspective in an Emergency due to Severe Conditions at a Light Water Reactor https://www-ns.iaea.org/downloads/iec/health-hazard-perspec-charts2013.pdf.

324. IAEA, Monitoring/decontamination of vehicles and equipment, https://www-pub.iaea.org/MTCD/publications/PDF/EPR-First_Res-PDA/html/ti8.htm.

325. IAEA Technical Volume 5, 'Post-Accident Recovery', https://www.iaea.org/sites/default/files/technical-volume-5-williams.pdf.

326. IAEA, GSG-11, Arrangements for the Termination of a Nuclear or Radiological Emergency, https://www.iaea.org/publications/12269/arrangements-for-the-termination-of-a-nuclear-or-radiological-emergency.

327. IAEA, Management of Large Volumes of Waste Arising in a Nuclear or Radiological Emergency, IAEA-TECDOC-1826, https://www-pub.iaea.org/MTCD/Publications/PDF/TE-1826_web.pdf.

328. IAEA Webpage, Radiological crime scene management, https://www.iaea.org/topics/radiological-crime-scene-management.

329. IAEA, Nuclear Security Series No. 22–G, Radiological Crime Scene Management Implementing Guide (2014) https://www-pub.iaea.org/MTCD/Publications/PDF/Pu1672web-85447671.pdf.

330. IAEA, Radiological Crime Scene Management, e-learning. https://elearning.iaea.org/m2/course/view.php?id=330.

331. Safecast: successful citizen-science for radiation measurement and communication after Fukushima: Azby Brown et al 2016 J. Radiol. Prot. 36 S82, https://iopscience.iop.org/article/10.1088/0952-4746/36/2/S82.

332. UK Recovery Handbooks for Radiation Incidents 2015, Inhabited Areas Handbook, https://www.gov.uk/government/publications/uk-recovery-handbooks-for-radiation-incidents-2015

333. Cabinet Office, Human Aspects in Emergency Management Guidance on supporting individuals affected by emergencies. (2016) https://assets.publishing.service.gov.uk/government/uploads/system/uploads/attachment_data/file/564306/human_aspects_guidance_2016_final.pdf

334. HMG, Human aspects in emergency management, https://www.gov.uk/government/publications/human-aspects-in-emergency-management

335. Radsafe, https://www.radsafe.org.uk/index.

336. Transportation Emergency Preparedness Program (TEPP), Model Procedure for First Responder Initial Response to Radiological Transportation Accidents, http://energy.gov/sites/prod/files/em/TEPP/2-b-1FirstResponderInitialResponse.pdf.

337. 2016 Emergency Response Guidebook, http://www.dgtraining.com/Resources/ERG2016.pdf.

338. ICRP, 20011, Fukushima Nuclear Power Plant Accident, http://www.icrp.org/docs/Fukushima%20Nuclear%20Power%20Plant%20Accident.pdf.

339. K.I.Pearce, Mobile Measurements, Nuclear Engineering International, https://www.neimagazine.com/features/featuremobile-measurements-7539961/.

Appendix A - Past accidents

This appendix looks at some aspects of the history of nuclear power and, in particular, a disappointing history of accidents and mistakes.

A.1 Introduction

There are at least two distinct forms of lessons that could be learned from real events. The first is that they provide an opportunity to challenge the underlying assumptions, values and beliefs upon which emergency plans are made. An example of this is Chernobyl which bought into focus the possibility, no matter how remote, of a more severe accident than is considered to be reasonably foreseeable. This lesson was backed up by both Buncefield and Fukushima but still struggles to be taken seriously.

It is widely recognised that this learning does not prevent future accidents. "*Despite efforts to the contrary, emergencies continue to outflank attempts at prediction, anticipation and prevention. Accidents, emergencies, and crises present by their nature an apparently endless ability to surprise*" (E S J Larken).

The second form of learning is confirmation that accepted procedures do, or do not, work in the real event or that responders do, or do not, manage their roles in the ways expected. This is often the easier form of learning to extract and, in fact, the only form of learning that is offered by exercises based on event sequences gleaned from safety cases.

It should also be noted that exercises can only give an impression of how the public may react to the situation and how that might affect the political situation. Real events give an opportunity to examine the reactions of the public, of the traditional media and of social media.

A number for radiation emergencies have been logged by the IAEA and lessons extracted (Ref. 2). Table 22 gives a summary of the accidents discussed in Ref. 2. It is worth spending some time considering how you as an individual and your emergency response organisation as a whole might the responded to these events.

Ref. 1 Rear Admiral E S J Larken DSO, Practical Emergency Management, https://www.icheme.org/communities/subject_groups/safety and loss prevention/resources/hazards archive/~/media/Documents/Subject Groups/Safety_Loss_Prevention/Hazards Archive/S139 - Major Hazards II/S139-24.pdf

Ref. 2 International Atomic Energy Agency, Lessons Learned from the Response to Radiation Emergencies (1945–2010), IAEA, Vienna (2012), http://www-pub.iaea.org\books/IAEABooks/8920/Lessons-Learned-from-the-Response-to-Radiation-Emergencies-1945-2010

Type of accident	Number worldwide	Number in UK
Criticality	27	
Nuclear operations	14	1 (Windscale)
Transport (Sea)	13	
Transport (air)	5	
Transport (space)	6	
Industrial sources	68	7
Industrial accelerators and X-ray generators	24	1 (injury to fingers of one hand due to improper procedures and failure in warning systems).
Research with sources	7	
Research with accelerators and X-ray generators	13	1 (injury to two fingers. Shutter removed prior to and during servicing).
Orphan sources	37 (about 30 deaths)	
Medical applications	42	7

Table 22 Summary of radiation accidents reported in IAEA EPR-Lessons Learned

A.2 Being unprepared - The Windscale Fire

In 1957 the Windscale works were supporting the British nuclear weapons programme. They were operating at the edges of knowledge and technology. The site operated two natural uranium reactors that were air cooled and graphite moderated. These were large structures of graphite bricks with a number of channels that admitted both fuel and the materials used to generate the isotopes the weapons programme needs. They each had a number of instruments that measured parameters such as temperature within the core. But these were few compared to modern reactors and left a lot of gaps. The reactors were controlled by moving fuel and control rods around and by controlling the air flow through the channels. They were slow to respond. This slow response could be seen as a desirable design feature as it prevents fast transients causing problems but it also makes fine control difficult. Lack of computing power meant the positioning of fuel and control rods was based on experience, relatively simple estimates of effect, and on trial and error.

For a number of days a process had been underway to heat one of the two nuclear reactors (or piles, as the two devices were known) to anneal the reactor and release Wigner energy. This is a process that was needed periodically and this is the ninth time it has been done. In the early afternoon of the 10th October a down-wind air sampler was found to have about 10 times the normal background level of radioactive dust on it. After this more sampling at different locations confirmed that a release was taking place. At 16.30 a visual inspection revealed that the core was on fire and measurements confirmed that radioactivity was being released into the building and into the environment.

> **Wigner Energy.** Potential energy stored in a crystal structure, such as graphite, resulting from atoms being displaced by collision with neutrons. Release, which can be caused by a temperature exceeding a critical value, can cause a rapid rise in temperature.

There was no emergency plan. The reactor core was on fire and radioactivity was being leaked into the environment. The local authority received a phone call saying that a cloud of radioactive gas was moving across their county.

They improvised. On-site, they tried to remove the burning elements from the reactor, but with limited success, and then to remove elements around the fire to produce a fire break. They tried to cool the core and extinguish the fire by injecting carbon dioxide and finally by turning off the air supply and flooding the core with water. By the afternoon of Saturday 12th October, two days after it started, the on-site event was over.

Meanwhile a hastily organised survey of radiation in the area was conducted to determine if there was (1) an external radiation hazard, (2) an inhalation hazard and (3) an ingestion hazard. At a mile downwind they registered external levels of radiation up to 400 times the normal background. (The highest radiation level measured was 4 milli-roentgens per hr (40 μSv/hr). The normal natural background radiation level in the area is about 0.1 μSv/hr). With the wind direction variable and the survey team moving they were able to measure cloud dose and ground dose throughout the event, but it was "*apparent that the accident had not caused any significant external radiation exposure to individuals living in the district*" (Ref. 1).

Air contamination levels were measured (12,000 on-site and 1,000 off-site measurements according to (Ref. 1)) these were used to conclude that "*the air contamination resulting from the incident rose on occasion to worrying, but not dangerous levels on the site, while the dilution resulting from wind variations considerably reduced the hazard in the district*".

The transcript of the Penney enquiry following the Windscale Fire is now available (Ref. 2) and makes an interesting read.

Based on the environmental monitoring concern was expressed about the potential contamination of milk and a number of samples taken from different milkings from farms in the area showed increasing levels of radioiodine. As a result the consumption of locally produced milk was, with agreement of the Ministry of Agriculture Fisheries and Food, banned over an area. This ban was based on estimates of the impact of radioiodine on children, the critical group, but applied to all age groups. The ban was enforced by the police.

For the protection of the public, it would have been helpful to have had an agreed off-site monitoring programme detailing what should be measured, where it should be measured, who should measure it and why. It would have been helpful to know what levels of radiation and/or contamination should trigger protective actions to protect the public and how to warn and inform the public. It would also help to know who had authority to intervene in food sales.

They did not have any of these. The accident resulted in significant changes to UK law and preparedness.

Ref. 1. H J Dunster, H Howells and W L Templeton, District Surveys following the Windscale Incident, October 1957, J. Radiol. Prot. 27 (2007) 217–230. http://iopscience.iop.org/article/10.1088/0952-4746/27/3/001/pdf. (Accessed 29/8/18).

Ref. 2. A Revised Transcript of the Board of Enquiry into the Fire at Windscale Pile No. 1 at Windscale http://news.bbc.co.uk/1/shared\bsp/hi/pdfs/05_10_07_ukaea.pdf (Accessed 29/8/18).

A.3 The public response and needs - Three Mile Island

Three Mile Island is a site near Middletown, Pennsylvania that, in 1979, had two quite new Pressurised Water Reactors (PWR). These were designed with three barriers to atmospheric release of radioactivity; the fuel pin, the cooling circuit and the containment. The design imperative was to keep the core covered with water and adequately cooled to prevent core damage. Unlike the PWRs currently being built at Hinkley Point C, which has a "core catcher" the containment was not designed to survive a serious core melt[16].

Several accounts of the incident are available on the internet. These include a background report from the US NRC (Ref. 1), a World Nuclear Association (WNA) document (Ref. 2), and an IAEA analysis (Ref. 3). Many more reports are available.

In the early morning (about 04.00) of March 28, 1979 the pumps pushing water into the heat exchangers at Three Mile Island Reactor 2 (TMI-2) failed causing the turbine-generator and then the reactor to shut down. When the nuclear reactor shutdown the chain reaction, which generates most of the heat, ceased but the decay of fission products continued to produce decay heat. With the reactor cooling switched off temperatures and pressures in the primary cooling circuit increased rapidly and a pressure relief valve at the top of the pressuriser opened as intended to release the pressure. This valve then failed to reseat (close) once the pressure had reduced as it should have done. This produced a leak path. A failure in the instrument panel hid this fact from the operators. The loss of water was detected and compensated for by an automatic system starting an injection pump. Misled by the complex, and not entirely correct, instrumentation and under the impression that there was too much water in the system the operators switched

[16] The melting of containment following a core melt was named the China syndrome in the media because the reactor core could "melt through to China".

off this supply and the core was uncovered a few hours later and started to melt. Eventually the operators believed the instruments that were reporting core damage and restored cooling. The containment held and very little radioactivity was released.

A general emergency, defined as having the "*potential for serious radiological consequences to the general public*" was declared at about 06.30. Over the next few days there was confusion about the extent of reactor damage and the possible consequences. Schools were closed, residents were advised to stay indoors, and farmers told to keep animals under cover and fed with stored feed. The local Governor advised evacuation of pregnant women and pre-school children within 5 km of the site (this was now two days after the onset of core damage) and later this distance was increased to 20 km. Technical debates about the size of hydrogen bubbles in the containment and reactor vessel and the risks posed by them were aired within the media which was very unhelpful in terms of reassuring the public. Within days 140,000 people had left the area (voluntary, self-evacuation). Approximately two-thirds of the households within 5 miles (8 km) of TMI-2 had at least one person evacuate. Almost all of whom returned to their homes within three weeks. The NRC report on the incident (Ref. 4) concluded that "*absence of a clear chain of command in the emergency organization and the lack of a disciplined approach in the communication of information to State and Federal agencies led to the evacuation scare on March 30.*"

Further the NRC report states that "*We believe that there need not have been a recommendation to evacuate at 9.15 am Friday, and we feel that there should have been such a decision on Wednesday at 9:00 am*". This is based on core damage and hydrogen and radioactivity build up in containment on Wednesday which could have led to containment failure and a major release of radioactivity - but fortunately didn't. By Friday the core was adequately cooled and temperatures and pressures had fallen.

A US NRC training course (Ref. 5) states that "*Uncertainty about the causes of the accident, confusion about how to deal with it, and contradictory information and appraisals of the level of danger in the days following the accident often made utility and government authorities appear inept, deceptive, or both*".

A US NRC history of nuclear regulation, 1946-1999 (Ref. 6) concluded that "*the Three Mile Island accident had vividly demonstrated the deficiencies in existing procedures for coping with an off-site nuclear emergency. The lack of effective preparation had produced confusion, uncertainty, and panic among members of the public faced with the prospect of exposure to radiation releases from the plant*".

Important learning from this experience covered a range of topics including the value of continuously reading perimeter monitors, the need to have exclusive communications links between key responding organisations, the need to agree the technical situation across responding organisations and the need to ensure that communications with the public are consistent.

One thing that TMI did show very clearly was that if a certain group, in this case pregnant women, were advised to evacuate an area then many more might be expected to voluntarily evacuate and from a wider area. This could, indeed did in the case of TMI, cause traffic problems. This has remained a fear in the nuclear planning community ever since. No one wants to see a population that would be relatively safe and comfortable in their homes stuck in traffic jams and with relatively poor protection from the radioactive plume over and around their cars.

The issue about the message received by the public is important and made more immediate by the emergence of social media. We can recognise the same issues in the handling of the Covid-19 pandemic around the world. Even if the situation was clear to the authorities and the media briefers were at the top of their game, pop up experts on mainstream media and on social media will present alternative views of the situation creating doubt. Fold in a complex and changing situation with delays in the internal communications and a perception of immediate danger and the public's confidence could be lost quickly.

Many changes were made to nuclear emergency planning in the USA and in the UK. These included an aspiration to provide the public with clear, timely and unambiguous information on the event. In the UK this resulted in the formation of the Media Briefing Centre concept and the machinery behind it to collate, agree and promulgate information. This is a mechanism that continues to evolve, particularly in the light of the increasing importance of social media and the increasing tempo of news media in general.

Ref. 1. US NRC, Backgrounder on the Three Mile Island Accident, https://www.nrc.gov/reading-rm/doc-collections/fact-sheets/3mile-isle.html

Ref. 2. World Nuclear Association, Three Mile Island Accident, http://www.world-nuclear.org/information-library/safety-and-security/safety-of-plants/three-mile-island-accident.aspx

Ref. 3. IAEA NS Tutorial, 3.2.3.3. Safety case study: the Three Mile Island accident https://www.iaea.org/ns/tutorials/regcontrol/assess/assess3233.htm

Ref. 4. Three Mile Island, A Report to the Commissioners and the Public, Volume II Part 3. https://catalog.hathitrust.org/Record/003177367

Ref. 5. ERI Consulting, Perspectives on Reactor Safety, , NUREG/CR-6042, Rev. 2 SAND 93-0971. https://www.nrc.gov/docs/ML0210/ML021080026.pdf

Ref. 6. US NRC, J. Samuel Walker and Thomas R. Wellock, A Short History of Nuclear Regulation, 1946–2009, https://www.nrc.gov/docs/ML1029/ML102980443.pdf

A.4 Clear Command and Control - Piper Alpha

The Piper Alpha accident occurred on an oil rig 200 kilometres south-east of Aberdeen on 6th July 1988. The rig had two compressors to pump gas along a pipeline to shore. One of these was being maintained and a safety relief valve had been replaced with a temporary blanking plate. When the other compressor failed in service the operators incorrectly started the one under repair, losing sight of the paperwork forbidding this. Gas leaking from the plate exploded at about 22:00, destroying the control room and causing fires to break out. Mayday signals were sent between 22:00 and 22:08.

According to a survivor "*from the cellar deck downwards, the steel members were red-hot as though they were in a furnace. Beneath, the water was steaming. The sky and the surrounding sea were bathed in a brilliant orange and the raging fire's constant scream of anger obliterated all other sounds*[17]." (Ed Punchard quoted in (Ref. 1))

Gas continued to be pumped onto Piper Alpha and this led to an even bigger explosion at 22:20. Brave attempts were made to rescue the rig's crew but 165 people died out of 226 on board.

The enquiry after the event thought that the accident could have been avoided, that the rig could have been better prepared for a fire and explosion and that the crew could have been better trained in emergency response and personal survival.

It was reported that there was confusion in the accommodation module with no one in charge, no one giving instructions or advice (Ref. 2). In an emergency that affects a wide area it is important that control is taken in those areas where non-involved personnel muster. In the nuclear industry it is now considered good practice to have high-visibility vests stored in muster areas and the expectation is that the most appropriate person present (someone with experience, rank and training) will take command of the area and don the vest for visibility. They will then communicate with the Site Emergency Controller for instruction or advice.

An excellent review of Command and Control based on Piper Alpha can be found in an article by Rear Admiral E S J Larken DSO, (Ref. 3). In this he identifies several failure modes for command systems under pressure:

- *"We may see reality skewed, as through a prism - deny the problems we face, and see instead reflections that we find more malleable, and therefore preferable.*
- *We may see a confused mass of data, deducing nothing of value. We settle for the answer we last used or the one we are offered, or we just guess.*
- *We deny there is a problem, and we deal with trivia. We act with no clear objective, no focus, no priority.*
- *We do not understand the information presented. We misinterpret, draw wrong conclusions, followed by wrong judgements.*
- *We panic. In clear sight of reality and its awful implications, we freeze, do nothing or become emotional.*
- *We interpret the information correctly and make good decisions but organise resources ineffectively and fail to control the situation".*

---

[17] These sort of conditions are very hard to mimic in exercises!

Importantly, for nuclear emergency planning, the enquiry was critical of the delay in stopping the pumps on the neighbouring rigs that were pumping gas into the fire while the responders identified and contacted someone with authority to order them to be stopped. This led to one of the principles of command and control as exercised in the industry that there is always someone available and unambiguously identified by the emergency response process who has full authority to make any decision on behalf of the company. The command-and-control process focusses on giving this person the best situational awareness possible and the best advice available to aid appropriate and timely decision making.

Following the incident, the HSE assumed responsibility for regulating the off shore industry and introduced new regulations and inspections (Ref. 4).

Ref. 1. Piper Alpha: condolences are not enough, https://www.healthandsafetyatwork.com/piper-alpha-lessons

Ref. 2. Jim Thomson, 2013, Safety in Engineering Ltd, http://www.safetyinengineering.com/FileUploads/Piper Alpha_1369665793_2.pdf

Ref. 3. Rear Admiral E S J Larken DSO, Practical Emergency Management, https://www.icheme.org/communities/subject_groups/safety and loss prevention/resources/hazards archive/~/media/Documents/Subject Groups/Safety_Loss_Prevention/Hazards Archive/S139 - Major Hazards II/S139-24.pdf

Ref. 4 HSE, A guide to the Offshore Installations (Safety Case) Regulations 2005, http://www.hse.gov.uk/pubns\books/l30.htm

A.5 Severe accident - Chernobyl

The Chernobyl accident happened on 26 April 1986. This was a soviet designed RBMK-1000 which had several design features that made it dangerous to operate in certain ways. However, the operators were not entirely aware of this. On the day of the accident the reactor was the subject of an experiment. In accordance with the instructions for the experiment several safety systems were switched off. Because of delays in starting the experiment the reactor was being operated outside its normal operating regime.

The design errors and the impact of working outside the normal operation envelope came together disastrously when the experiment was initiated. Two large explosions wrecked the reactor building and then the fuel over heated and caught fire in several places. Two workers died in the explosions.

Water was poured into the core to cool it but this was stopped after a while as there was a risk of flooding the other reactors on the site. From the second to tenth day after the accident, some 5,000 tonnes of boron, dolomite, sand, clay and lead were dropped on to the burning core by helicopter in a heroic effort to extinguish the blaze and limit the release of radioactive particles.

The accident deposited radioactive dust and debris over Europe, but particularly over Belarus, Russia and Ukraine. Iodine-131 was the isotope of greatest concern in the early days after the accident followed by Cs-137 when the short-lived iodine had decayed.

The cloud of radioactive material reached as far as the UK and, while it was over Cumbria and north Wales it rained heavily washing radioactive caesium (Cs-137 and Cs-134) out of the plume and onto the ground. On most soils the caesium would have soon be "fixed" into the soil and not available for uptake by crops but not on the acid soils found in the Cumbrian, Scottish and Welsh uplands where the activity continued to be found in grass and the sheep that grazed that grass for many years following the deposition.

The Food Standards Agency (Ref. 1) monitored the levels of radioactivity in grass and sheep and imposed a limit on the acceptable concentration of radio-caesium in sheep (1000 Bq/kg). Meat above this level could not be sold. Some controls lasted until 2012 (Ref. 2).

According to a WHO report (Ref. 3) about 530,000 people were used in the clean-up of the area around Chernobyl. These received an average additional dose of about 120 mSv. Those evacuated from the area (115,000 people) received an average additional dose of 30 mSv. This, however, is a measure of whole body dose. Many people in the area received high doses to the thyroid gland due to the release of radio-iodine, which was breathed in as the plume passed but, more importantly, entered into the food chain particularly milk over the following weeks. A few people continue to live in areas where the radiation dose is double the background levels experienced before the accident.

Twenty-eight responders died of acute radiation sickness (ARS) within three months of the accident. There was a significant increase in the levels of thyroid cancers, mainly in children, in the areas around Chernobyl. An increase in the incidence of leukaemia in the most highly exposed responders has also been seen.

Day	Date of release	Daily release (10^{15} Bq)
1	26 April	704
2	27 April	204
3	28 April	150
4	29 April	102
5	30 April	69
6	1 May	62
7	2 May	102
8	3 May	107
9	4 May	130
10	5 May	130
	Total	1760

Table 23 - Estimated release of Iodine-131 from Chernobyl

A large number of people were moved from their homes never to return in the aftermath of the accident (although they did not know this at the time. They had been asked to take clothes for a few days). The social and economic disruption caused had a long-term impact on mental health, not helped by the survivors being shunned by the residents of areas they were moved into. It is widely believed that the psychological harm caused by the response to the accident are far greater than the harm caused by the radiation. An important lesson for those preparing against any possible future accidents.

An idea of the duration of the event comes from Table 23 (from Ref. 4) which shows the release of I-131 as a function of time. The report states that "*continuing low-level releases occurred in the following week and for up to 40 days after the accident, particularly on 15 and 16 May, attributable to continuing outbreaks of fires or to hot areas in the reactor. These later releases can be correlated with increased concentrations of radionuclides in air measured at Kiev and Vilnius*". Could your emergency response organisation work flat out for 40 days?

NRPB estimated that the average total UK dose from Chernobyl fallout was 0.05 mSv, although the dose to people living in areas of high deposition was higher (e.g., the average additional dose per person in Cumbria was 6 times the UK average). A parliamentary briefing on the impact of Chernobyl on the UK can be found on the internet (Ref. 5).

The Chernobyl accident refocussed emergency planning thinking on the possibility of a large accident although it was, rather complacently as it turned out, thought very unlikely that such an accident could occur in the western world because of better reactor design and construction, better operator training and better compliance.

One clear outcome of the Chernobyl accident was the introduction of the RIMNET system of fixed radiation monitors around the UK (Ref. 6).

Reference 7 has been written for those who understand a bit of physics, like a challenging read and want to know more about the physics of the Chernobyl accident.

Ref. 1. Food Standards Agency, https://www.food.gov.uk/about-us

Ref. 2. FSA, Post-Chernobyl monitoring and controls survey reports, https://www.food.gov.uk/research/research-projects/post-chernobyl-monitoring-and-controls-survey-reports

Ref. 3. WHO, CHERNOBYL at 25th anniversary, Frequently Asked Questions, April 2011. http://www.who.int/ionizing_radiation/chernobyl/20110423_FAQs_Chernobyl.pdf?ua=1

Ref. 4 NEA, Chernobyl, The release, dispersion and deposition of radionuclides. https://www.oecd-nea.org/rp/chernobyl/c02.html

Ref. 5 Parliamentary Office of Science and Technology, Briefing Note 45, Chernobyl fallout. http://researchbriefings.files.parliament.uk/documents/POST-PN-45/POST-PN-45.pdf
Ref. 6 Radioactive incident monitoring, https://www.gov.uk/government/collections/radioactive-incident-monitoring.
Ref. 7 Keith Pearce, The Physics of the Chernobyl Accident, https://amzn.to/37F8tmQ

A.6 Severe accident II - Buncefield

At about 06.00 on Sunday 11 December 2005 employees at the Buncefield oil storage depot in Hemel Hempstead raised the alarm on being made aware of a developing vapour cloud. Almost immediately there was an explosion which has since been claimed to be the most powerful post-war explosion in Europe. A petrol storage tank was being filled from a pipeline. Although the tank had safety systems fitted to prevent overfilling they failed to operate and petrol was spilt from the vents on the tank roof and formed a cloud of flammable vapour that was ignited by a spark causing a series of large explosions and subsequent fires.

It took about thirty-two hours to extinguish the main blaze and much longer to extinguish all other fires at the Buncefield site. Forty-three people received minor injuries. Extensive damage was caused to surrounding commercial and domestic properties, and local environmental pollution occurred from the use of fire-fighting foam and the spill of petrol and diesel from the storage tanks. The widespread damage to the site also resulted in temporary disruption of fuel supplies in the South East of England (Ref. 1 & 2).

This led to new emergency planning guidelines for Control of Major Accident Hazards (COMAH) sites (Ref. 3).

Buncefield led to another set of reviews of the bases for nuclear emergency plans and, in particular, raised some concerns about the limited planning then in existence for accidents of greater impact than those upon which the detailed emergency plans were based.

> *"Operators of Buncefield-type sites should review their emergency arrangements to ensure they provide for all reasonably foreseeable emergency scenarios arising out of credible major hazard incidents, including vapour cloud explosions and severe multi-tank fires that, before Buncefield, were not considered realistically credible".*
>
> Buncefield Major Incident Investigation Board

Another criticism was about the quality of scientific advice available to Gold Command which led to delays in fighting the fire while differences in opinion about the best environmental technique were discussed. This led directly to the formation of the Science and Technology Advice Cell (Ref. 4).

> *"To ensure timely co-ordinated scientific and technical advice during the response to an emergency, local responders are invited to consider collectively through the relevant Local Resilience Forum (LRF), arrangements for establishing a STAC to provide advice to the Strategic Co-ordination Group (SCG) when required".*
>
> Ref. 4 below

References

Ref. 1 Buncefield: Why did it happen? The underlying causes of the explosion and fire at the Buncefield oil storage depot, Hemel Hempstead, Hertfordshire on 11 December 2005
Ref. 2 The Buncefield Major Incident, 11 December 2005, Ten years on, A report by the COMAH Strategic Forum
Ref. 3 The Competent Authority Control of Major Accident Hazards, Competent authority guidance for inspectors on emergency arrangements for COMAH establishments
Ref. 4 Provision of scientific and technical advice in the strategic co-ordination centre: guidance to local responders, https://www.gov.uk/government/publications/provision-of-scientific-and-technical-advice-in-the-strategic-co-ordination-centre-guidance-to-local-responders

A.7 Severe accident III - Fukushima

The Fukushima accident again led to a great deal of reconsideration of the safety of nuclear power stations, the appropriate scope for off-site emergency plans, support for effected members of the public and recovery.

The accident was brought about by a tsunami which caused significant damage on the nuclear site including the total loss of off-site power and all on-site backup generation capacity. This took so long to fix, because of the radiation levels, site damage, damage to local infrastructure and the immediate needs of the tsunami victims, that reactors and used fuel facilities over heated causing a series of leaks and explosions. A detailed report about what happened at the Japanese nuclear power stations can be found in (Ref. 1)

A detailed review of the circumstances of the Fukushima accident was carried out at all UK nuclear facilities (the Weightman Report) (Ref. 2). Further information was sought by the European Council leading to the "Stress Tests" (Ref. 3).

The Torness Stress Test report, for example, can be found on the internet (Ref. 4).

This includes the conclusion: "*Conclusion TOR 6.1: EDF Energy has detailed robust arrangements for Emergency Response which are subject to a programme of continuous improvement and exercised as required by standard procedures and regulatory demand.*

Based on the learning from the Japanese Earthquake, the subsequent EDF Energy Emergency Nuclear Generation review of Station Safety Cases and examination of the associated risks on plant, it is not believed the fundamental risk profile on the nuclear power station has changed. Therefore, current arrangements remain fit for purpose. However, as part of EDF Energy standards we ensure any lessons identified by real events and exercises (internal/external) are reviewed and built upon within our arrangements".

And the consideration: "*Consideration TOR 6.1: EDF Energy will consider how lessons identified from Japan and credible beyond design basis events can be reflected in our facilities, procedures, training and exercise programmes. Utilising experience from other emergency response organisations and the military, EDF Energy will consider enhancement of its staff welfare, human factors and emotional aspects associated with emergency response.*"

Additional thought was given to how the Company could prepare for an on-site response with significant disruption to off-site infrastructure. This involved consideration of the stocks, equipment, communications and machinery that a damaged reactor site might require and how best it could be stored to ensure its availability against a number of scenarios.

There was also significant focus on the steps that had been taken to protect the public and how successful the protective actions and public reassurance had been.

At Fukushima the public around the site were advised to evacuate with the evacuation distance increasing with time extending to 20 km by the evening of 12th March. Beyond that those who might have been exposed to radiation doses in excess of 20 mSv were also advised to evacuate. Determining who to evacuate proved difficult for a number of reasons including the damage that had been done to the radiation monitoring equipment by the earthquake and tsunami.

The World Nuclear Association presents the following timetable:

11 March
- 14:46 The earthquake occurred.
- 15:42 TEPCO made the first emergency report to the government.
- 19:03 The government announced a nuclear emergency.
- 20:50 The Fukushima Prefecture Office ordered 2 km radius evacuation.
- 21:23 The government ordered 3 km evacuation and to stay inside buildings (shelter) in the area of 3-10 km radius.

12 March
- 05:44 The government ordered 10 km radius evacuation.

18:25 The government ordered 20 km evacuation.
15 March
11:01 The government ordered people to stay inside buildings (shelter) in the area of 20-30 km from the plant
25 March The government requested voluntary evacuation in the area of 20-30 km
21 April The government set the 20km radius no-go area.

There was significant criticism of the authorities for the way the emergency was handled. This included the way in which protective actions were applied and communicated.

The public were concerned about the dose constraints being used to support decisions about short term and continuing protective actions, leading to a response from the ICRP as discussed in Section 4.9.4.

It is claimed that the Japanese government decided that accuracy of information given to the public was much more important than timeliness and that this had serious impacts. For example, "*It is likely that owing to the government's accuracy-first policy, the information was not disseminated in a timely fashion concerning the amount of environmental radioactivity estimated from the measurements and/or computations and the extent of the damage of the reactor core fuels, scientifically inferred from the time elapse since the loss of coolant, because it took a great deal of time to confirm the facts with enough accuracy for the government. As a result, the delayed provision of related information meant that the public lost trust in its reliability*" (Ref. 5).

The UK government (Ref. 6) state that the important learning from Fukushima for the recovery process includes:

- Build public confidence through effective engagement on impacts of the event and options for recovery.
- Involve communities in decision-making.
- Strengthen co-ordination and liaison between central and local government.
- Balance the benefits of dose reduction with the negative impact of on-going remediation.
- Establish realistic criteria for stopping remediation which are linked to radiation exposure.
- Optimise dose reduction and minimise waste in areas where people live and work.
- Contribute to enabling more people to return to their homes outside restricted areas.

Ref. 1 ANS, A Report by the American Nuclear Society Special Committee on Fukushima (March and June 2012) http://fukushima.ans.org/report/Fukushima_report.pdf
Ref. 2 ONR, Fukushima and the UK nuclear industry, http://www.onr.org.uk/fukushima/
Ref. 3 ONR, Stress Tests, http://www.onr.org.uk/fukushima/european-council-stress-tests.htm
Ref. 4 EDF, EU Stress Test Report, Torness https://www.EDFenergy.com/sites/default/files/jer-srt-stt-pub-fin-008_tor_stress_test_v1.1.pdf
Ref. 5 Managing the Fukushima Challenge, Risk Anal. 2014 Jul; 34(7): 1240–1256. https://www.ncbi.nlm.nih.gov/pmc/articles/PMC4145710/

Ref. 6 Nuclear Emergency Planning and Response Guidance - Recovery, https://www.gov.uk/government/publications/national-nuclear-emergency-planning-and-response-guidance

A.8 Criticality Accident - Tokaimura

Plant designed for handling or processing enriched fissile material is carefully designed to avoid inadvertently creating a super-critical assembly. Containers and pipes tend to be long and thin rather than spherical to increase the surface area and enhance neutron leakage, water and other moderators/reflectors are carefully excluded and the quantities being processed in any one batch are carefully limited.

In 1999 at Tokaimura in Japan unauthorised procedures resulted in 16.6 kg of enriched uranium being fed into a precipitation tank. It went critical. One worker received a dose in the range 10 - 20 Gy and died on day 83. A second received a dose between 6 and 10 Gy and died on day 211. The third worker received 1.2 - 5.5 Gy and survived. A larger number of workers and emergency services workers received additional radiation doses.

Approximately 200 residents were evacuation from within 350m of the event. Of these 90% received doses of < 5mSv and the others were below 20 mSv (Ref. 1).

Several criticality accidents have occurred. They release a large amount of radiation in a short time and are often deadly to those nearby. They do not tend to release significant amounts of contamination but clean-up workers and emergency services may be exposed to additional doses of radiation in the course of their duties.

Ref. 1 International Atomic Energy Agency, Lessons Learned from the Response to Radiation Emergencies (1945–2010), IAEA, Vienna (2012), http://www-pub.iaea.org\books/IAEABooks/8920/Lessons-Learned-from-the-Response-to-Radiation-Emergencies-1945-2010

A.9 Industrial Accident - Chilca

A serious radiological accident occurred in January 2012, in Chilca, Lima. A radiography camera was being used to determine the quality of the welds. This used an Iridium-192 source which was housed in a shielded box from which it could be wound out to expose photographic film the other side of the weld from the source.

Three workers undertook 97 such tests during a shift. During the next day they all fell ill with various symptoms including sickness, dizziness and fatigue. It transpired that the source had not been correctly wound back into its shield between each test and the radiation monitors that would have alerted the team had not been deployed (poor training and poor safety culture).

Once the source was wound in correctly there was no further danger and no contamination at the scene. Medical care was protracted, and one worker had parts of his hand amputated on day 101. If this sort of accident were to happen in the UK, the local authority role would be one of communication and reassurance.

Ref. I AEA, The Radiological Accident in Chilca, https://www-pub.iaea.org/books/IAEABooks/11095/The-Radiological-Accident-in-Chilca

A.10 The struggling local authority - Grenfell Tower fire

On 14 June 2017 a fire broke out at Grenfell Tower, a 24-storey residential housing block in North Kensington, London. The tower, which provided social housing, was managed by Kensington and Chelsea TMO (Tenant Management Organisation) on behalf of Kensington and Chelsea Borough Council. It was built in 1974 and contained 129 flats.

The fire was reported shortly after midnight and spread rapidly. It affected the majority of the floors. The Metropolitan Police have stated that 71 died as a result of the fire (Ref. 1).

There was severe criticism of the performance of the local authority Royal Borough of Kensington and Chelsea (RBKC) in the media and in the House of Commons. The Times ran the headline "*Grenfell Tower: fire victims left in lurch by chaotic relief effort*" on the 19th of June (Ref. 2).

On July 5th, an Independent Grenfell Recovery Taskforce was announced. This was to "*provide advice and practical and strategic support to RBKC*" and to "*work alongside the council to make sure that they have the manpower, expertise and resources necessary to take on the longer-term recovery work beyond the immediate Civil Emergency Response*". More details of its composition were given on 26th July (Ref. 3).

While its initial report published on 31st October (Ref. 4) admits that "*no local authority could develop a response to a disaster of this magnitude without help from other local authorities, public services and government. The support required by survivors and their re-housing needs, for multiple hundreds of households would alone challenge any local authority's resilience*" it also states that "RBKC failed its community on the night of 14 June and in the weeks immediately following. While many staff did their best to help, a leadership vacuum aligned with siloed service delivery and a distant council that did not know its residents meant there was little effective and structured support from RBKC".

This report also states that "*RBKC needs greater focus on operational delivery and putting its effort into addressing the immediate needs of those affected by the fire. To help do this RBKC needs to understand the individual needs of survivors, and currently it does not have a co-ordinated central means of understanding the full range of survivor needs. The local authority lacked a clear presence at the scene and lacked leadership in the periods during and after the fire. This led to a lack of co-ordination in attempts to support those affected. This was extensively reported in the media and very quickly resulted in highly voluble anger*".

The report concluded that the Council's failings on the day of the fire and in the days following contributed to a lasting "*trust deficit*" between RBKC and the residents.

A Guardian article (8/12/17) reported that "*Lack of interaction between councils, convoluted system to request aid and poor information sharing were among failings*" (Ref. 5). The Grenfell rest centre was described as "*quite chaotic.... Staff felt there needed to be more information about who was in charge*". There was a lack of co-ordination between rest centres. There were five open at one point but only one formal rest centre in place.

The second independent report (16/3/18) (Ref. 6) mentions empathy and emotional intelligence stating that "*a variety of training has been provided to front-line officers and Councillors, but we still hear of unsympathetic or inappropriate responses from council staff and Councillors. Further training is on offer and this is now mandatory for front-line staff*". It also acknowledges that "*The Taskforce is aware that recovery to a new kind of normality for the bereaved, survivors and the wider community will take time. It is also recognised that recovery is essentially an individual journey. The Council's role is to provide many of the services and support that will help individuals and groups on that journey. This will also take time and careful forward planning*".

The Grenfell Tower was a terrible accident which directly affected 129 households and many more indirectly. The cause was fire. This lacks some of the mystery and fear associated with a nuclear accident, but even so it was felt that "*on-going health concerns, including understanding the health implications from what local people had breathed in as a result of the fire need to be addressed*" (Ref. 4).

We are left to wonder what the impact might be if a local authority ever produced such a poor response to a nuclear accident.

Interestingly the neighbouring Hammersmith and Fulham Borough reports starting to provide information leaflets to their tenants on building cladding and fire risk assessment on 17th June - 3 days after the event. These were followed by "*advice and reassurance sessions*". H&F report that they could not satisfy the number of requests for meetings and discussions with different Tenants' and Residents' Associations (TRA) (Ref. 7). This thirst for knowledge and face to face meetings is likely to be replicated in the aftermath of a nuclear accident. The public will want to understand more about radiation and contamination and about what the Council is doing to protect them.

An important conclusion from this report was that "*The main communications difficulty arising from the Grenfell fire was the Council's capacity to answer large volumes of wide-ranging and detailed questions from the media, government and the public about all aspects of fire safety in council properties. At a time when resources were stretched in dealing directly with the incident on the ground, capacity to meet information demand was stretched. This is an area that all services should consider in their emergency planning*". (Remember that this is a neighbouring council not the directly affected one.)

References

Ref. 1 House of Commons Library, 15/5/17, Grenfell Fire background. https://researchbriefings.parliament.uk/ResearchBriefing/Summary/CBP-8305.

Ref. 2 The Times, https://www.thetimes.co.uk/article/grenfell-tower-fire-victims-left-in-lurch-by-chaotic-relief-effort-plnbxsf8s

Ref. 3 New taskforce to support recovery from Grenfell fire disaster, 26/7/17, https://www.gov.uk/government/news/new-taskforce-to-support-recovery-from-grenfell-fire-disaster

Ref. 4 Independent Grenfell Recovery Taskforce The Taskforce Initial Report – 31 October 2017 https://www.gov.uk/government/publications/initial-report-from-the-independent-recovery-taskforce-31-october-2017

Ref. 5 The Guardian, Grenfell fire: response failings highlighted in report on tragedy https://www.theguardian.com/uk-news/2017/dec/08/grenfell-tower-fire-response-failings-highlighted-in-report-into-tragedy

Ref. 6 Second report from the Independent Grenfell Recovery Taskforce: 16 March 2018 https://www.gov.uk/government/publications/second-report-from-the-independent-grenfell-recovery-taskforce-16-march-2018

Ref. 7 http://democracy.lbhf.gov.uk/documents/g5624/Public%20reports%20pack %2006th-Dec-2017%2019.00%20Audit%20Pensions%20and%20Standards%20Committee.pdf?T=10

A.11 Learning from history

In its analysis of the lessons learned from the responses to radiation emergencies from 1945 to 2010 (Ref. 1) the IAEA conclude that the lessons demonstrate the importance of:

- all States establishing and maintaining arrangements (including exercising them) for dealing with radiation emergencies in the public domain according to the national circumstances;
- those responsible for facilities in which radiation sources are kept or used establishing their own arrangements for emergency response graded according to the level of risk;
- those responsible for facilities/locations where orphan sources might be encountered (e.g. scrap metal yards) establishing arrangements for emergency response;
- appropriate resources being available and responsibilities in an emergency being clearly defined in order for there to be an adequate response to radiation emergencies;
- medical personnel being trained to recognize radiation-induced injuries, since they are frequently the first to encounter such injuries in patients, and encouraged to inform the regulatory body in the event that they suspect the appearance of such an injury; and
- giving clear information promptly to the media and the public in the event of a radiation emergency because of the considerable interest that such events attract, and in order to avoid undue disruption of the response.

Experience shows that radiological emergencies are not just about nuclear power plants. County level and national level emergency preparedness teams need to be aware of the other sources of radiological risk such as orphan sources, dirty bombs, industrial accidents and radioactive satellite re-entry.

Appendix B - Glossary

The nuclear industry and radiological protection use a lot of technical terms. Some are listed and explained below. Many of these definitions can be traced to the national lexicon of UK civil protection terminology.

Term	Definition
Absorbed dose (Gy)	The radiation energy imparted per unit mass of an irradiated body. It is measured in joules per kilogram, a unit which is also called the gray (Gy).
Activity	Decay rate of radionuclides.
AGR	Advanced Gas Cooled Reactor - type of nuclear reactor used to generate electricity in the UK
Alpha radiation	A heavy particulate radiation (a helium nucleus) emitted by a number of heavy elements it has very short range but can cause considerable damage if emitted in or very near living matter due to localized energy deposition.
ANS	American Nuclear Society
ARS	Acute Radiation Sickness (or Syndrome) (sometimes known as radiation toxicity) is an acute illness caused by irradiation of the entire body (or most of the body) by a high dose of penetrating radiation in a very short period of time
Averted dose	The dose saved by implementing a protective action.
Beta radiation	A light particulate radiation (an electron) emitted by neutron heavy nuclei. It has a short range but can penetrate the outer layers of skin and damage the underlying living tissue.
Casualty Bureau	Initial point of contact for receiving and assessing information about casualties.
Cat 1 responder	Category 1 responder - Those organisations at the core of the response to most incidents (emergency services, local authorities, Environment Agency, local health organisation). See CCA
Cat 2 responder	Category 2 responder - Co-operating organisations involved in the response to an incident (Office for Nuclear Regulation, transport and utility companies). See CCA
CBRN	Chemical, biological, radiological and nuclear defence. Protective measures taken in situations in which chemical, biological, radiological or nuclear warfare (including terrorism) hazards may be present.
CCA	Civil Contingencies Act 2004 - Statutory framework for the delivery of civil protection
CESC	Central Emergency Support Centre. A crisis centre in EDF's Gloucestershire headquarters which provides support to sites during emergencies.
CFIL	European council food intervention level. Limits on levels of contamination in food that can be sold.
Containment vessel	A reinforced structure built around a nuclear reactor with the design intent of retaining the majority of radioactivity released from the reactor in an accident.
COBR	Cabinet Office Briefing Rooms. UK Government crisis management centre.
COMAH	Control of Major Accident Hazards. Regulations (2015) that require that businesses "*take all necessary measures to prevent major accidents involving dangerous substances*".
Confinement	Prevention or control of releases of radioactive material to the environment in operation or in accidents. Note: Confinement is closely related in meaning to containment, but confinement is typically used to refer to the safety function of preventing the 'escape' of radioactive materials, whereas containment refers to the means for achieving that function.
Contamination	In this sense, the presence of radioactive material, usually in the form of dust, on surfaces or people and/or their clothes.
Containment	Methods or physical structures designed to prevent or control the release and the dispersion of radioactive substances.
COP	Common Operating Picture - The multi-agency document used to share situational awareness and actions
CPM	Counts per minute. Rate of detection of radioactive event as shown on many detectors.

CRCE	NIPH's Centre for Radiation, Chemical and Environmental Hazards which provides radiological protection advice, services and training.
Critical	A term used in reactor physics to denote the situation where the chain reaction is self-sustaining i.e. each fission produces one more on average.
Criticality accident	The release of energy as a result of accidentally producing a self-sustaining or divergent fission chain reaction.
CRIP	Common Recognised Information Picture - this is a single, authoritative strategic overview of an emergency or crisis that is developed within COBR according to a standard template and is intended for briefing and decision-support purposes. Multi-agency Sitrep information from SCGs may be used to inform the CRIP, and SCGs may receive the CRIP from COBR where it is relevant.
CTA	Company technical advisor. EDF person in SCC, there to provide information and interpretation about event.
Chain reaction (fission)	A sequence of nuclear fission reactions in which fissions are induced by neutrons emerging from preceding fissions. Depending on whether the number of fissions directly induced by neutrons from one fission is on the average less than, equal to, or greater than unity, the chain reaction is convergent (subcritical), self-sustaining (critical), or divergent (supercritical).
CHP	Company health physicist. EDF person in SCC, there to provide information and interpretation about radiological situation.
Decay heat	The name given to the heat generated by the decay of fission products in the fuel assembly. It is because of this that reactor cores must continue to be cooled after shutdown.
Decontamination	Removal or reduction of hazardous materials to lower the risk of further harm to victims and/or cross contamination.
Detailed Emergency Planning Zone	An area around a nuclear licensed site for which a detailed off-site (local authority) plan should exist. A term not currently favoured by the regulators.
Deterministic Effect	An effect of radiation which has a threshold dose below which the effect will not be seen but which then gets more severe as the dose increases. Sun burn is an example of a deterministic effect. (See also Stochastic Effect)
DNSR	Defence Nuclear Safety Regulator - responsible for regulating the nuclear hazards of the Defence Nuclear Enterprise consisting of the Naval Nuclear Propulsion Programme and the Nuclear Weapon Programme, where the Ministry of Defence has specific exemptions from statute.
Dose constraint	A constraint set as a prospective upper bound of individual doses, used to define the range of options considered in the process of optimisation for a given radiation source in a planned exposure situation (EU BSS).
Dose limit	The value of the effective dose (where applicable, committed effective dose) or the equivalent dose in a specified period which shall not be exceeded for an individual.
DPH	Director of Public Health
DRL	Dose Reference Level. A level of radiation dose that should not be exceeded in the relevant situation.
e	Electron charge. Fundamental physical constant expressing the naturally occurring unit of electric charge, equal to $1.6021765 \times 10^{-19}$ coulomb.
EA	Environment Agency
ECC	Emergency Control Centre. Crisis centre from which the site response to an emergency is managed.
EDAS	Emergency Dose Assessment Service, a component of EDF's emergency response responsible for estimating off-site doses.
Effective dose	The effective dose is defined by the ICRP as the sum of the equivalent doses in the principal tissues and organs in the body, each weighted by a tissue weighting factor, w_T. Measured in sieverts (Sv). It is a measure of the harm the radiation does to the person.
Emergency exposure situation (ICRP 103)	Unexpected situations that may require urgent protective actions and perhaps also longer-term protective actions to be considered.
Emergency management system	A legal or administrative framework establishing responsibilities for emergency preparedness and response, and arrangements for decision making in the event of an emergency situation.

ENSREG	The European Nuclear Safety Regulators Group (ENSREG) is an independent, expert advisory group created in 2007 following a decision of the European Commission. It is composed of senior officials from the national nuclear safety, radioactive waste safety or radiation protection regulatory authorities and senior civil servants with competence in these fields from all 28 Member States in the European Union and representatives of the European Commission.
EOP	Emergency Operating Procedure. Advice on how to respond to an alarm.
EPGMS	Emergency perimeter gamma monitoring system. A ring of detectors around the perimeter of AGR sites.
Equivalent dose	Multiplying the absorbed dose by appropriate weighting factors depending on the type of radiation, creates the equivalent dose in the relevant organ or tissue. Measured in sieverts (Sv). It is a measure of the harm the radiation does to the organ or tissue.
ERL	Emergency Reference Level. The quantitative criteria recommended by the NRPB for the introduction of protective actions to protect the public in the event of an accident.
EU BSS	Euratom Community (European Union) basic safety standards to protect workers, members of the public, and patients against the dangers arising from ionising radiation.
Evacuation	Removal, from a place of actual or potential danger to a place of relative safety, of people and (where appropriate) other living creatures.
Exposure	A measure of the ionisation produced in air by x-rays or gamma radiation; the sum of electric charges on all ions of one sign in a small volume of air when all electrons liberated by photons are completely stopped, per unit mass of the air.
Existing exposure situation	An exposure situation that already exists when a decision on its control has to be taken and which does not call or no longer calls for urgent measures to be taken.
Fallout	A term applied to radioactive dust settling on the ground and other surfaces following a release of radioactivity - usually applied to the aftermath of a nuclear bomb.
Family and Friends Reception Centre	A facility to help reunite family and friends with survivors.
FCP	Forward Control Point
FEMA	Federal Emergency Management Agency (USA). Mission: To lead America to prepare for, prevent, respond to and recover from disasters with a vision of "A Nation Prepared".
FEPA	Food and Environmental Protection Act
Food Controls	The combination of legal controls (under FEPA) and advice not to eat food that might have been contaminated by a release of radioactivity.
FSA	Food Standards Agency. An independent Government department working across England, Wales and Northern Ireland to protect public health and consumers' wider interests in food
FSS	Food Standards Scotland. A non-ministerial government department of the Scottish Government. It is responsible for food safety, food standards, nutrition, food labelling and meat inspection in Scotland.
Gamma radiation	A photon (light energy) emitted by many radioactive materials when they decay. Very long range and deposit energy in a spread-out manner. Short-wavelength electromagnetic radiation emitted in the process of nuclear transition or particle annihilation.
Go in, Stay in, Tune in	Generic advice to members of the public on how to respond to an emergency. The intention is to get them into buildings that protect them from the hazard and provide them with information on broadcast media.
(Government Decontamination Service)	The UK Government Decontamination Service (GDS), Guidance and access to capability to recover from a CBRN and major hazmat incident is now provided from Defra's CBRN Emergencies team and the GDS branding ended.
Gold	Strategic level of management of an incident
Gray (Gy)	Unit of absorbed dose. 1 Gy = 1 J/kg
HALES	Highly Active Liquor Evaporation and Storage (Sellafield plant)
Half-life	The time taken for half of the radioactive atoms present to decay. This varies for different isotopes from a very small fraction of a second to many millions of years. The fraction left after each successive half-life is ½, ¼, 1/8, 1/16, 1/32.

HART (Hazardous Area Response Team).	A unit providing standard paramedic care to any persons within a hazardous environment that would otherwise be beyond the reach of NHS care
Hazard	A potential danger. "Potentially hazardous" is redundant. Note that a hazardous facility is not necessarily a high-risk facility.
HSAWA	Health and Safety at Work Act (1974) provides framework for health and safety regulations.
HAZMAT	Hazardous material
HIRE	Hazard identification and Risk Evaluation as required by REPPIR.
IAEA	International Atomic Energy Agency
ICPD	Ingestion and Commodities Planning Distance
ICRP	International Commission on Radiological Protection. Since 1928, ICRP has developed, maintained, and elaborated the International System of Radiological Protection used world-wide as the common basis for radiological protection standards, legislation, guidelines, programmes, and practice.
Information	Pieces of raw, unanalysed data or reports from various sources about an event, criminal activity or subject of interest.
Intelligence	The product of the collation, evaluation and analysis of raw information.
Ionising radiation	Types of radiation (including alpha, beta, gamma and neutron) that can ionise the material they pass through.
IRR-17	Ionising Radiations Regulations (IRR), provides a three-tier system of regulatory control, notification, registration and consent for work with ionising radiations. Requires risk assessments, contingency plans and the minimisation, control and limitation of radiation dose.
INES	International Nuclear Event Scale. International Atomic Energy Agency seven point scale used in order to enable prompt communication of safety-significant information in case of nuclear accidents.
Justification 1	The process of determining for a planned exposure situation whether a practice is, overall, beneficial; that is, whether the expected benefits to individuals and to society from introducing or continuing the practice outweigh the harm (including radiation detriment) resulting from the practice.
Justification 2	The process of determining for an emergency exposure situation or an existing exposure situation whether a proposed protective action or remedial action is likely, overall, to be beneficial; that is, whether the expected benefits to individuals and to society (including the reduction in radiation detriment) from introducing or continuing the protective action or remedial action outweigh the cost of such action and any harm or damage caused by the action.
LBR	Lifetime baseline risk
LET	Linear energy transfer. The amount of energy that an ionising particle transfers to the material traversed per unit distance.
LNT	The linear no-threshold model. A model which assumes that the long-term harm caused by ionising radiation is directly proportional to the dose.
LRF	Local Resilience Forum. Multi-agency partnerships made up of representatives from local public services, including the emergency services, local authorities, the NHS, the Environment Agency and others.
MACA	Military Aid to the Civil Authorities: consisting of following categories: MACC Military Aid to the Civil Communities - Assistance in an emergency; MACM Military Aid to the Civil Ministries - Assistance in the event of industrial action; MAPC Military Assistance to the Civil Powers Assistance to the police.
MBC	Media Briefing Centre, a component of off-site plans.
MCA	Maritime and Coastguard Agency
MCA	Military Co-ordinating Authority
Medical exposure	Exposure to ionising radiations of persons as part of their diagnosis or treatment. (ICRP)
MOD	Ministry of Defence

Moderator	A material in a nuclear reactor that is there to reduce the speed of neutrons produced in fission to make it more likely that they will induce further fissions. Hydrogen, in the form of water is a common moderator. Graphite (carbon) is used in the British Gas Cooled Reactors.
NAIR	National arrangements for incidents involving radioactivity. Provides assistance to the police and other emergency services where no radiation expert is otherwise available
NIA	Nuclear Installations Act 1965 (as amended), requires nuclear power sites to be licensed, controls nuclear materials and governs some aspects of liability.
NIHP	National Institute for Health Protection
NILO	National Interagency Liaison Officer (Ambulance Service)
Noble gases	Very unreactive gases including helium (He), neon (Ne), argon (Ar), krypton (Kr), xenon (Xe), and radon (Rn). Radioactive forms of these can be included in atmospheric releases. They give an external dose but no inhalation or ingestion dose.
NPP	Nuclear Power Plant
NRPB	National Radiological Protection Board. Its remit was to conduct research on radiological protection, and to provide advice and information on the subject to government departments. Now part of PHE.
NRTE	The Vulcan Naval Reactor Test Establishment
Nuclide	A species of atom characterized by its mass number, atomic number, and a possible elevated nuclear energy state if prolonged.
Occupational exposure	Exposure to radiation incurred at work, and principally as a result of work. (ICRP)
Off-site Nuclear Emergency (OSNE)	Declaration state: A hazardous condition which results, or is likely to result, in the need to consider urgent protective actions to protect the public outside the site security fence from a radiological hazard.
OIL	Operational Intervention Level. A system to aid decision with regard to activities and doses.
ONR	Office for Nuclear Regulation - whose mission is: "to provide efficient and effective regulation of the nuclear industry, holding it to account on behalf of the public".
OR	Operating Rule. Limits set on the operation of a reactor and its support equipment.
Orphan source	A radioactive source which is neither exempted nor under regulatory control, e.g. because it has never been under regulatory control or because it has been abandoned, lost, misplaced, stolen or otherwise transferred without proper authorisation.
PACRAM	Procedures and Communications following Release of Radioactive Material. Met. Office system for providing weather and dispersion information in an emergency.
PAG (Protective Action Guides)	Dose guidelines that would trigger public safety measures, such as evacuation or staying indoors, to minimize or prevent radiation exposure during an emergency. Similar to ERLs. US EPA
PHE	Public Health England. An executive agency, sponsored by the Department of Health and Social Care to protect and improve the nation's health and wellbeing, and reduce health inequalities. Possibly to become part of NIHP in 2021.
PHE CRCE	Specialist unit of PHE providing expert advice for radiation, chemical and environmental hazards
PITs	Potassium Iodate Tablets (see SITs)
Planned exposure situation	An exposure situation that arises from the planned operation of a radiation source or from a human activity which alters exposure pathways, so as to cause the exposure or potential exposure of people or the environment. Planned exposure situations may include both normal exposures and potential exposures.
Potential exposure	An Exposure that is not expected with certainty but may result from an event or sequence of events of a probabilistic nature, including equipment failures and operating errors.
PPE	Personal Protective Equipment i.e. coveralls, eye protection, hard hat.

Preparedness	Preparedness is the capability to take actions. Arrangements must be in place to ensure a timely, managed, controlled, co-ordinated and effective response to a radiation emergency at the operator, local, regional and national levels, as well as at the international level.
Prior Passive Measures	Measures which are put in place before the event to provide passive protection should the event occur
Prior Support Measures	Measures put in place in the planning stages to support an active response to the event
Projected dose	The overall exposure to a member of the public in an accident should no protective actions be taken. See residual dose and averted dose
Protective action	Action taken in an exposure situation to reduce or prevent exposure. The action can be taken at the source, at points in the exposure pathway, or occasionally by modifying the location, habits or working conditions of the exposed individuals. (ICRP) "protective action" means an action or actions taken in order to prevent or reduce the exposure of emergency workers, members of the public, the environment or the contamination of property from ionising radiation in the event of a radiation emergency, and includes the provision of appropriate information to the public in accordance with regulations 21 and 22; (REPPIR 19, regulation 2(1))
Public Exposure	Exposure to ionising radiation to members of the public in their everyday life (not at work or undergoing medical exposure). (ICRP)
PWR	Pressurised Water Reactor
Rad (radiation absorbed dose)	A unit of absorbed dose. The absorption of 100 ergs of energy per gram of absorbing tissue. 1 rad = 10^{-2} J/kg of the medium. Note: the gray (1 Gy = 1 J/kg) is the preferred unit of absorbed dose,
RBKC	Royal Borough of Kensington and Chelsea
Radioactive Dispersal Device (RDD)	Dirty Bomb - device intended to disperse a radioactive source in public area.
Radioactive decay	The process whereby an unstable atomic nucleus changes its form and loses energy in the form of radiation. A spontaneous nuclear transformation in which particles or gamma radiation is emitted, in which x-radiation is emitted following orbital electron capture, or in which the nucleus undergoes spontaneous fission.
Radioactivity	The particles or photons (light) which are emitted when unstable atoms decay.
Radiation emergency	"radiation emergency" means a non-routine situation or event arising from work with ionising radiation that necessitates prompt action to mitigate the serious consequences— (a) of a hazard resulting from that situation or event; (b) of a perceived risk arising from such a hazard; or (c) to any one or more of— (i) human life; (ii) health and safety; (iii) quality of life; (iv) property; (v) the environment; REPPIR-19 Reg. 2
Radiation exposure	A measure of the ionisation produced in air by x-rays or gamma rays. It is the sum of the electric charges on all ions of one sign that are produced when all electrons liberated by the radiation in a volume of air are completely stopped, divided by the mass of air in that volume.
Radionuclide	Unstable nuclide. Can decay giving off radiation.
RADSAFE	RADSAFE is a private company limited by guarantee that offers mutual assistance in the event of a transport accident involving radioactive materials belonging to a RADSAFE member

RCIS	The Redgrave Court Incident Suite. A crisis centre in the Office for Nuclear Regulation (ONR) Headquarters in Bootle. This facility will co-ordinate the ONR response to an event.
RCG	Recovery Co-ordinating Group - a sub-group of the SCG
RDD	Radiological dispersal device. A dirty bomb. Designed to spread radioactive material in a public area. Not a nuclear bomb.
Reference level	In an emergency exposure situation or in an existing exposure situation, the level of effective dose or equivalent dose or activity concentration above which it is judged inappropriate to allow exposures to occur as a result of that exposure situation, even though it is not a limit that may not be exceeded;
Reasonably foreseeable	According to REPPIR-01 guidance (paragraph 50) "In the context of a radiation emergency, a reasonably foreseeable event would be one which was less than likely but realistically possible".
Relocation	Relocation is the displacement of a population from a contaminated area for a period of several weeks or longer.
REM (Roentgen Equivalent Man)	A measurement that correlates the dose of any radiation to the biological effect of that radiation (equivalent dose). Note: the preferred unit of equivalent dose is the sievert (Sv).
Remedial measures	The removal of a radiation source or the reduction of its magnitude (in terms of activity or amount) or the interruption of exposure pathways or the reduction of their impact for the purposes of avoiding or reducing doses that might otherwise be received in an existing exposure situation.
REPPIR or REPPIR-19	The Radiation (Emergency Preparedness and Public Information) Regulations, 2019. Applies to any site with more than certain quantities of radioactive materials. Requires risk assessment and report. Requires emergency planning subject to a threshold of risk.
Rest Centre	A facility managed by the local authority for temporary accommodation of evacuees and homeless survivors. With overnight facilities.
Residual dose	The dose to which a member of the public would be exposed to in an accident if protective actions were taken to reduce the dose. See projected dose and averted dose
Response	Response is the performance of actions. These actions are aimed at mitigating the consequences of a radiation emergency for human life and health, property and the environment to include providing a basis to resume normal social and economic activity.
RIFE	Radioactivity in Food and the Environment (RIFE) reports. Annual assessment reports of radioactivity in food and the environment and the public's exposure to radiation.
RIMNET	Nuclear radiation monitoring and nuclear emergency response system. Radiation dose rate readings (gamma plus cosmic) from 96 sites around the UK are collected every hour and checked for any indication of abnormal increase. Any readings of radiological significance for the UK would result in an alert being raised and investigated.
Risk	The cost of a class of accidents over a given period, usually expressed as pounds or fatalities, per year or during plant lifetime. Unless established by experience, risk is estimated as the product of the probability of occurrence and the consequence of the accident type.
RMU	Radiation Monitoring Unit. A RMU is used to determine levels of radioactive contamination in or on people and any subsequent requirement for decontamination
RSA	Reactor Safety Alert (Naval)
Roentgen I:	A unit of exposure. It is the measurement of energy produced by Gamma or X-Ray radiation in a cubic centimetre of air. $1\ R = 2.58 \times 10^{-4}$ coulombs/kg in air.
RPA	Radiation Protection Advisor. IRR-17 and REPPIR-19 have regulations where it is necessary to consult a certified and appointed RPA.
RPE	Respiratory Protective Equipment. i.e. filter masks, breathing apparatus.
RTG	Radioisotope thermoelectric generators. Devices which generate electricity using decay heat.
SAGs	Severe Accident Guidance. Documents providing advice about how to manage a damaged reactor.

SBERG	Symptom Based Emergency Response Guidance. Documents providing advice about how to manage a very damaged reactor.
SAGE	Scientific Advisory Group for Emergencies (SAGE) provides scientific and technical advice to support government decision making during emergencies.
SCC	Strategic Co-ordination Centre. A crisis centre usually located in police headquarters. The function of the strategic co-ordination centre (SCC) is to provide a location and infrastructure to enable the SCG to co-ordinate their response to an event.
SCG	Strategic Co-ordinating Group. The SCG takes overall responsibility for the multi-agency management of the emergency and establishes the policy and strategic framework within which lower levels of command operate.
SDF	Safety Directors' Forum -
SFRS	Scottish Fire and Rescue Service
SGoRR	Scottish Government Resilience Room.
Shelter	Protective action advice for the public to 'Go in, stay in and tune in'. Includes closing windows and doors to minimise the ingress of hazardous materials.
Site Incident	Declaration state: A hazardous condition which is confined in its effect within the boundary of the site security fence.
SORT	Special Operations Response Team. A Specialist team within SFRS.
Special Nuclear Material (SNM)	Radioactive material that can sustain a chain reaction.
Stable Iodine (Potassium Iodate) tablets (SITs)	A stable form of iodine, which can bind to and protect the thyroid gland, preventing uptake of radioactive iodine by the gland and thus reduce the likelihood of thyroid cancer in the future.
STAC	Scientific and Technical Advice Cell. Formed at the SCC to provide scientific and technical advice.
Stay time	The period during which personnel may remain in an area subject to additional radiation dose before accumulating some permissible occupational dose.
Stochastic Effect	A radiation effect for which the probability of the effect being seen is proportional to the dose but for which the dose has no impact on the severity of the effect. Cancer is a stochastic effect. (See also Deterministic Effect).
Survivor Reception Centre	A secure area to which survivors not requiring acute hospital treatment can be taken for short-term shelter and first aid.
THORP	Thermal Oxide Reprocessing Plant (Sellafield plant)
TIAC	Time Integrated Air Concentration (the product of airborne concentration (Bq.m^{-3}) and time (s) hence has units of Bq.s.m^{-3})
TMO	Tenant Management Organisation
UKHSA	United Kingdom Health Security Agency
WENRA	The Western European Nuclear Regulators Association (WENRA) is an association of the nuclear safety regulatory bodies of various countries. It provides a forum for the exchange of information and expertise in nuclear safety and the implementation of European Union (EU) initiatives and programmes oriented toward the harmonisation of the safety of EU nuclear facilities.
WHO	World Health Organisation - whose "primary role is to direct and co-ordinate international health within the United Nations' system".
WNA	World Nuclear Association
W_R	A dimensionless factor by which the organ or tissue absorbed dose is multiplied to reflect the higher biological effectiveness of high-LET radiations compared with low-LET radiations. It is used to derive the equivalent dose from the absorbed dose averaged over a tissue or organ.
W_T	The factor by which the equivalent dose in a tissue or organ T is weighted to represent the relative contribution of that tissue or organ to the total health detriment resulting from uniform irradiation of the body. It is weighted such that it sums to one across all tissues.

A

B

C

D

E

F

G

H

I

J

K

L

M

N

O

P

R

S

Published by Katwab Limited,
ISBN 978-1-9164658-2-4 10 August 2021

Also by Keith Pearce,

This book provides advice to members of the public about how to prepare for and respond to a nuclear emergency that sends a plume of radioactive materials over their homes or places or work (or just threatens to).

This book should be read by anyone who lives near a nuclear site or anyone interested in the potential impact of a nuclear emergency.

It shows how to prepare for, and how to respond to, a nuclear emergency (or many other types of emergency that may require you to leave your home at short notice).

With this information you should be able to better understand any advice being given by the authorities and be more confident about keeping your family and friends safe throughout the ordeal.

It is available as a paperback or Kindle book on Amazon.

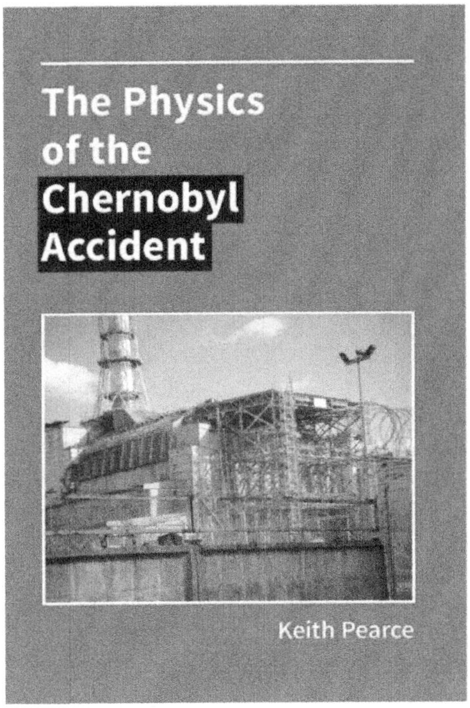

At last, clarity for the general public on what happened at Chernobyl.

This highly readable explanation of what went wrong at Chernobyl neatly slips into the gap between the densely worded, and highly technical, official reports of the accident and the human story presented in many popular books and so intriguingly retold in the recent TV mini-series, Chernobyl.

Did it rouse your curiosity to really understand what really went wrong that April day in 1986, and how it was allowed to happen? Written by industry insider Keith Pearce, "The Physics of Chernobyl" makes fascinating reading, clearly setting out the background to the accident in a way which is accessible to the student of physics as well as those members of the general public who have a basic grasp of physics and are keen to understand the details of the accident.

In just over 100 pages, this fascinating volume sets out the background to the accident, explaining how nuclear reactors work and in particular the RBMK type used in the USSR. It looks at the experiment the operators of the plant were conducting at the time and provides an overview of the physics necessary to be able to understand the issues around the accident and then goes on to examine the detailed timeline of the catastrophe on the fateful day.

The book also reviews other possible explanations for the accident, what has happened since and how the industry, national governments and authorities and the International Atomic Energy Authority learn from such accidents and strive to keep the public and their staff safe.

Made in the USA
Monee, IL
05 November 2026

67383912R00122